SUPERMINDS

STUDIES IN COGNITIVE SYSTEMS

VOLUME 29

SUPERMINDS

People Harness Hypercomputation, and More

by

SELMER BRINGSJORD

Rensselaer Polytechnic Institute (RPI), Troy U.S.A.

and

MICHAEL ZENZEN

Rensselaer Polytechnic Institute (RPI), Troy U.S.A.

KLUWER ACADEMIC PUBLISHERS

DORDRECHT / BOSTON / LONDON

A C.I.P. Catalogue record for this book is available from the Library of Congress.

ISBN 1-4020-1094-X

Published by Kluwer Academic Publishers,
P.O. Box 17, 3300 AA Dordrecht, The Netherlands.

Sold and distributed in North, Central and South America
by Kluwer Academic Publishers,
101 Philip Drive, Norwell, MA 02061, U.S.A.

In all other countries, sold and distributed
by Kluwer Academic Publishers,
P.O. Box 322, 3300 AH Dordrecht, The Netherlands.

Coverdesign paperback edition by Marc Destefano.

Printed on acid-free paper

Printed in the Netherlands.

To Elizabeth (S.B.)

and

To those I love (M.Z.)

Contents

List of Tables

List of Figures

Series Preface

This series will include monographs and collections of studies devoted to the investigation and exploration of knowledge, information, and data-processing systems of all kinds, no matter whether human, (other) animal, or machine. Its scope is intended to span the full range of interests from classical problems in the philosophy of mind and philosophical psychology through issues in cognitive psychology and sociobiology (concerning the mental abilities of other species) to ideas related to artificial intelligence and computer science. While primary emphasis will be placed upon theoretical, conceptual, and epistemological aspects of these problems and domains, empirical, experimental, and methodological studies will also appear from time to time.

According to "the supermind manifesto", human beings possess mental powers that far transcend those possessed by ordinary computing machines. In this sophisticated and fascinating series of studies, Selmer Bringsjord and Michael Zenzen challenge and redefine the limits of some of the most celebrated results in logic, mathematics, and computer science, including but not restricted to the fundamental work of Turing and Church. Anyone who thinks that computationalism has carried the day should be heartened and refreshed by this brilliant and thorough critique of the predominant conceptions in cognitive science today. Displaying mental agility and intellectual brilliance, this book should stimulate rethinking the very foundations of the fields it addresses. It's a superbook!

J.H.F.

Acknowledgments

This book, like most, is born of sustained effort. Yes, sometimes that great luminous idea does suddenly and unaccountably blossom before one's mind, but even then, invariably, there is the painstaking translation of it into language that allows it to be assimilated by others. In the case of the present translation (of perhaps one or two luminous ideas) effort came not only from us, but from many, many others. Accordingly, we have countless debts. They are indeed too many and too great to pay, but we'll try at least to mention most.

Thanks are due once again to Harriet Borton for her matchless LaTeX and TeX expertise, without which there would be no book (since Selmer, if truth be told, knows just enough LaTeX to *think* he knows what he's doing). LaTeX and TeX are amazing things, and Selmer now wonders how he ever typed a syllable without them.

We thank the team at Kluwer for supernatural patience (in the face of a volume they doubtless thought Selmer would *never* get around to delivering) and professionalism: Annie Kuipers, Evelien Bakker, Rudolf Rijgersberg, Charles Erkelens, and especially Jolanda Voogd. When the light at the end of the tunnel finally appeared (in the form of the antepenultimate draft), Jolanda provided indispensable guidance that brought us home. At Rensselaer, Heather Hewitt gave the penultimate version a remarkably meticulous read that resulted in many improvements. Of course, we're massively indebted to Jim Fetzer for his fantastic series. Jim is one of a very select few who contribute in seminal fashion to their fields both by creating historic content *and* by enabling the publishing of content created by others.

We are both grateful for the support Rensselaer has given us through the years. The Rensselaer AI and Reasoning Lab (affectionately known as the RAIR), which Selmer directs at Rensselaer, has in particular been and continues to be an invaluable resource; the lab is piloted, daily, by the incomprehensibly resourceful Marc Destefano, and is an exciting productive place because bright people hang out therein — people like Bram van Heuveln, Jim Fahey, Bettina Schimanski, Paul Bello, Devin Croak, Jason Wodicka, Brad Smith, and many others. Indirectly, support for much of the Bringsjordian thinking that underlies this book came from sponsors of research undertaken by Bringsjord at RPI: the Henry R. Luce Foundation, Apple Computer, IBM, AT&T, the Educational Testing Service, and the National Science Foundation. ETS deserves special credit: were it not for its support of test-related AI, Psychometric AI (explained in the final chapter), proposed as a generalization of the marriage between testing and AI originally established by Turing (1964), would never have been born.

An anonymous referee read a previous draft of the entire book, and provided a brilliant, detailed critique that proved absolutely invaluable.

Some of what appears in this book has been previously published as parts of some papers. In all cases, however, the work has been expanded and refined, sometimes thanks to reaction specifically aimed at our prior work, sometimes thanks to our dealing herein with new work from others that appeared in print after publication of this prior work. As an example of the latter sort, consider Devlin's (2000) *The Math Gene*, a defense of the Pinkerian (1997) view $\mathcal{C}^{\mathcal{E}}$ that human persons are evolved (standard) computers, despite the fact that some persons can do impressive mathematics. Part of our earlier attack on Pinker's view (Bringsjord 2001*a*) is augmented in Chapter 7 so as to include disposing of Devlin. In some cases, we've recently realized that arguments we have refuted did have some redeeming value. For example, it was only after publishing a refutation of Penrose's Gödelian attack on computationalism (Bringsjord & Xiao 2000) that we realized the validity of his intuitions — as we explain in Chapter 2.

Here now is a chapter-by-chapter list of acknowledgments, taking account of our relevant prior work:

In connection with Chapter 1, we're grateful to Michael Costa for inviting Jim Fetzer to organize a symposium on whether minds are computational systems for the annual meeting of the Southern Society for Philosophy and Psychology, in Nashville, Tennessee, April 4–7, 1996. Though Chapter 1 now goes well beyond the paper presented by one of us (Bringsjord) at the Nashville symposium, this meeting marked the genesis of the core idea underlying this chapter. Bringsjord left Nashville in (as his wife would say) a veritable trance, pondering an attempt to formulate a systematic account of machines, supermachines, supermechmachines (= super-mechanical-minds; we explain the neologism in Chapter 1), and superminds. Thanks are of course also due to Jim Fetzer and Bill Rapaport for participating with Bringsjord in the Nashville meeting. [Rapaport's (1998) objection to Bringsjord's Argument from Introspection (in Chapter IX of Bringsjord 1992) was particularly ingenious — but see note 7 in Rapaport's paper.] We are also indebted to to Dave Ferrucci, Jim Fahey, and Pat Hayes for helpful discussions prior to the SSPP conference. In addition, we owe a great debt to Carol Cleland for discussing hypercomputation with us. Out of that discussion has emerged our overall scheme — presented in Chapter 1, and exploited in Chapter 8 — for a (nascent, at this point) formal philosophy of hypercomputation.

Thanks are due to the *Journal of Experimental and Theoretical Artificial Intelligence* for permission to use, in Chapter 2, parts of the paper "A Refutation of Penrose's Gödelian Case Against Strong AI" (Bringsjord & Xiao 2000), co-written with one of Selmer's former computer science graduate students: Hong Xiao. For helping us refine this chapter, we are indebted to Martin Davis, Kelsey Rinella, Marvin Minsky, David Chalmers, Jim Fahey, Michael Zenzen, Ken Ford, Pat Hayes, Bob McNaughton, and Kostas Arkoudas. Selmer would like to express special thanks to Roger Penrose for debate and conversation concerning many of the issues treated in Chapter 2. Bringsjord and Penrose both believe that the mind is beyond computation; both *also* believe that Gödelian results can be used to demonstrate this. However, as this chapter testifies, they differ over how to carry out the demonstration.

Chapter 3, "The Argument From Infinitary Reasoning," would have been impossible without intelligent (and sometimes ferocious: Brngsjord 1982) reaction to the core argument from Paul Feltovich, Bob Hoffman, and Ken Ford, in whose *Expertise in Context* (1997) a much more primitive version of this argument appeared. Selmer has been working off and on for five years to improve this published kernel, but without the *Expertise in Context* volume, the kernel would never have been formulated in the first place. We're indebted as well to others who reacted to written and nuncupative ancestors of this chapter, including: David Israel, Jim Fahey, Drew McDermott, Marvin Minsky, Saul Traiger, Jim Moor, Stevan Harnad, Pat Hayes, Larry Taylor, Jim Fetzer, and a number of anonymous referees. We are also indebted to Clark Glymour, whose elementary but esemplastic *Thinking Things Through* (Glymour 1992) presents some of the background material presupposed in this chapter in a most inspiring way.

Chapter 4, "Supermentalism and the Fall of Church's Thesis," has its roots in "The Narrational Case Against Church's Thesis," an earlier, much more primitive version of the argument that appears as a chapter in (Bringsjord & Ferrucci 2000). We are indebted to Elliot Mendelson not only for his stimulating paper on Church's Thesis (Mendelson 1986), but also for his response to Bringsjord's analysis of it in personal communication. This exchange, at the 1993 Eastern APA Meeting in Atlanta, marks the point at which Bringsjord became convinced that Church's Thesis is false. An anonymous referee is owed thanks for the clever objection that appeals to equivalence classes, the rebuttal to which is one of the new parts of the case for Church's Thesis appearing in the present volume. Another anonymous referee sparked more careful consideration of the confessedly monumental respect we have for Umberto Eco's reader response theory. Janet Folina provided insightful commentary on the ancestor of this chapter presented by Bringsjord at the 1993 APA meeting — comments incorporated in our response to Mendelson's response. We are also grateful to Michael McMenamin for providing, in his unpublished "Deciding Uncountable Sets and Church's Thesis," an excellent objection to our attack on Church's Thesis (which we rebut in the chapter in question). Kevin Ware provided valuable feedback on a previous draft of this chapter as well. And long ago Dave Porush provided innumerable insights about "rule-breaking" belletristic literature. These insights, more than anything else, helped move Selmer toward an outright rejection of Church's Thesis.

We are grateful to *Philosophy and Phenomenological Research* for permission to use, in Chapter 5, parts of the paper "The Zombie Attack on the Computational Mind" (Bringsjord 1999). Insightful comments on this paper and its ancestors were provided by John Searle, Daniel Dennett, Stevan Harnad, Robert Van Gulick (who offered particularly clever comments on the remote ancestor presented at the 1994 Eastern APA Meeting), Peter Smith, Kieron O'Hara, Jim Fahey, Marvin Minsky, David Chalmers, Larry Hauser, Ned Block, and Pat Hayes.

"The Argument from Irreversibility," Chapter 3, part of which appeared in *Synthese* (which graciously gave permission for us to include an expanded version herein), couldn't have been refined without the insights of Bill Rapaport and Stuart Shapiro. Pat Hayes, Ken Ford, Marvin Minsky, Jim Fahey, and many Rensselaer students provided trenchant objections and tolerated the maddeningly slow evolution of this chapter from a rather inauspicious primogenitor.

Part of Chapter 7, "What Are We? Where'd We Come From?", includes parts of (Bringsjord 2001*a*); these parts are used herein with permission from *Philosophical Psychology*.

We've been inspired by countless debates and discussions with many people on issues concerning minds and machines in general. The majority of these thinkers advanced positions at odds with our own, and thereby helped sharpen our thoughts. They include: Jim Moor, Jim Fetzer, David Chalmers, Jack Copeland, Carol Cleland, Stevan Harnad. Also included in this group are Pat Hayes, Stu Shapiro, Marvin Minsky, and Ken Ford, who have staunchly articulated the view that, despite Selmer's arguments to the contrary, cognition *is* ordinary computation. Jim Moor (in a different role), Robert Cavalier, Marvin Croy, Tony Beavers, Jon Dorbolo, Bill Uzgalis, Eric Steinhart, and many others in and involved with the national "Computing and Philosophy" group pretty much every year supply stimulating but (of course!) ultimately unsuccessful objections to supermentalism.

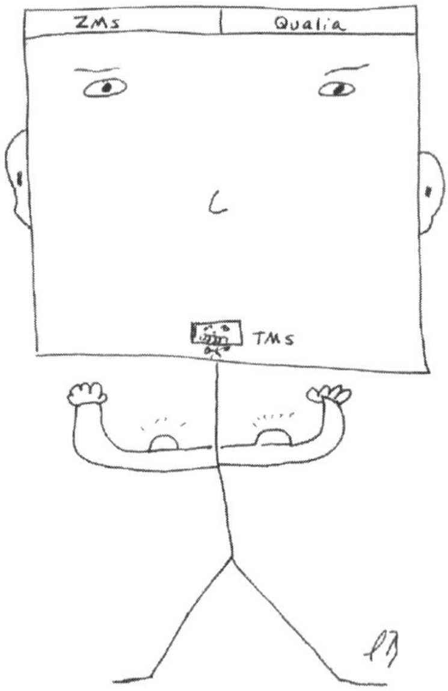

Figure 1: Impressionistic View of Superminds. © SCB

Preface

0.1 The Supermind Manifesto

Super*minds*, yes; supermen or superwomen, no, at least not in the colloquial sense: We don't credit you with the power to turn a chunk of coal into a diamond by squeezing it in your hand, or to fly faster than a speeding bullet, or look — courtesy of X-ray vision — through walls. What we *do* ascribe to you, in general, is power well beyond that possessed by ordinary computing machines, that is, by computing machines operating at or below the level of Turing machines and their equivalents (or, as we say, at or below the **Turing Limit**). To speak of us not merely as minds, but as *super*minds, may nonetheless strike you as hyperbolic, but if so, your impression is in no small part a function of the current intellectual landscape. The fact is, most intellectuals working in the cognitive sciences and in relevant parts of engineering, in this day and age, are rabid computationalists: they go around preaching that minds are ordinary computing machines (embodied in flesh rather than silicon), and that computational artifacts will soon enough match and then soon thereafter exceed the human mind.

For example: In the June 19, 2000 issue of TIME magazine, devoted to "The Future of Technology," we hear from author and inventor Ray Kurzweil that nanobots (microscopic robots) will by 2030 be able to map out a synthetic duplicate of your brain after you swallow (yes, swallow) a few of them. This duplicate will be instantiated in computer hardware 10 million times faster than the sluggish, old-fashioned grey stuff inside your cranium; the result will be an artificial intelligence immeasurably more clever than you. Vernor Vinge, associate professor of mathematics and computer science at San Diego State University, is another example. Prophesying for the *Chronicle of Higher Education* (July 12, 2000; online edition), he gives us a more compressed timeline: By his lights, on the strength of the trend that the speed of computer hardware doubles every 18 months, computers will be more intelligent than all humans at some point within 20 years. This point he calls "The Singularity," which ushers in post-humanity, an age in which humans are left in the dust by machines that get exponentially smarter by the day (if not the nanosecond). For a third example, consider Hans Moravec, who in his recent book, *Robot: Mere Machine to Transcendent Mind* (Moravec 1999), informs us that because hardware is getting faster at the rate Vinge cites, by 2040 "fourth generation" robots will exceed humans in all respects, from running companies to writing novels. Such robots, so the story goes, will evolve to such lofty cognitive heights that we will stand to them as single-cell organisms stand to us today. (Bringsjord & Ferrucci (2000) point out that Moravec's optimism flies in the face of the fact that there isn't a shred of a reason, today, for holding that robots, tomorrow, will be creative.) Many others in the field of Artificial Intelligence (AI) predict the same sensational future unfolding on about the same rapid schedule.

This situation, as of 2002, sets our context. Perhaps you'll agree that if the doctrine that human persons have cognitive powers not exceeding those of rats came to dominate the intellectual scene, then contrarians drawing attention to even *reductio* proofs from *Logic 101* might be seen as suggesting that, at least when it comes to mental powers, we reside atop Olympus. This is the boat we find ourselves in. Supermentalism is jarring only because the silly view that persons are equivalent to (or perfectly simulable by) Turing machines (or, again, their equivalents) is so indefensibly entrenched.

When laying down the foundational concept of a Turing machine, Turing did *not* start with some mathematical description of a machine of some sort; he started with the concept of a sub-*person* — or, to use his term, a 'computist' — carrying out primitive operations. In his inaugural writings on isomorphic points (independent, by the way, of Turing's), Post (1944) spoke of mindless "workers," humans whose sole job was to slavishly follow explicit, excruciatingly simple instructions. Likewise, Charles Babbage modeled the calculating cogs in his Difference Engine (and, for that matter, in the never-built Analytical Engine) on an army of hairdressers hired by Baron Gaspard Riche de Prony, director of the École des Pont et Chaussées, to transform data tables into decimal form when France went decimal in 1799 (Holt 2001). These perfectly clear and seminal schemes, which make plain the wide gulf between person *qua* ordinary computing machine and person in full cognitive glory, have been supplanted by all sorts of obfuscating exotica — Turing machines, register machines, neural nets, and so on —, things which facilitate formal analysis,[1] but hide from the unsuspecting the fact that though we can indeed operate as if we're ordinary computers through and through, doing so is for us to call upon only the dim side of our psyches, and leave the more robust stuff temporarily aside by an act of will. We know what it is to move the beads in an abacus, to carry out functions like addition, to simulate low-level programming languages with pencil and paper, and so on; we know that such activity is essentially what Turing and Post and Babbage (and others) generalized in order to mathematize ordinary computation; we know that to use *only* an abacus is to be intellectually trammeled — we know all this and yet, as Fetzer (2001*a*) eloquently notes, we still tolerate the computationalist telling us that we are always and only abaci in action. (Of course, 'we' here doesn't refer to the authors of the book you're now reading. E.g., Bringsjord, in talks addressing the question "Are We Computing Machines?", usually kicks the presentation off by showing the slide pictured in Figure 2 — which sums up the silly view that human persons are nothing more than birds moving beads around in an abacus.) This must end. Relative to such humble information-processing, we are indeed nothing less than super — as we intend to show you herein.

0.2 This Book in The Supermind Quartet

The present volume is part of a much larger project comprising a quartet of books and a number of stand-alone articles. This overall project has three objectives:

Objective 1 Overthrow computationalism, the view ($\mathcal{C}$) that minds are ordinary computing machines.

Objective 2 Establish the position ($\mathcal{S}_c^\infty$) that minds are *super*minds, i.e., that though minds aren't computing machines, they *include* such machines, *and* two other main things, viz., information-processing machines capable of hypercomputation (as reflected in the superscript $^\infty$), and a capacity to enjoy and exploit phenomenal consciousness,

[1]E.g., one of us (Bringsjord) has considerable success teaching computability theory via abaci. He gets help from the elegant proof of Turing Machine/abaci equivalence in (Boolos & Jeffrey 1989), which draws on (Lambek 1961), which is in turn a simplification of the machines investigated in (Melzak 1961).

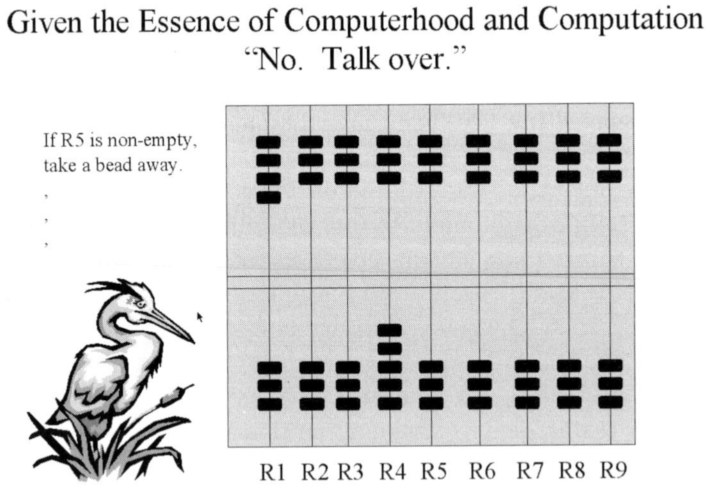

Figure 2: Are Persons Captured by a Bird Moving Beads Around?

which can't be expressed in any third-person scheme, computational or otherwise (as reflected in the subscript $_c$).

Objective 3 Specify and defend a practical, concrete, applied approach to AI and CogSci anchored by the theoretical content produced to reach Objectives 1 and 2. This specification and defense is to revolve around implemented systems, not mere speculation; engineering is to flow from philosophizing. The systems will establish (among other things) both that animal behavior can be simulated by ordinary computation ($\mathcal{B}_a$) and that the behavior of persons, at least to a substantial degree, can also be simulated by such computation ($\mathcal{B}_p$).

With the publication of the book you're holding, the first two objectives have (by our lights) been reached, and Objective 3 has been partially reached as well. Objective 1 was reached via Bringsjord's earlier *What Robots Can and Can't Be* (*WRCCB*; Bringsjord 1992), for therein he completed the bulk of the case against computationalism. There are two “transitional” chapters in *WRCCB*: one builds a bridge to the present book, the other builds a bridge to the book that has partially secured Objective 3. The “transition” chapter in *WRCCB* that paves the way to the present volume is “Chapter VIII: Free Will,” which features an argument for the view that people enjoy — by virtue of having genuine power over a certain range of mental events — the status of “Zeus machines,” which are capable of hypercomputation (e.g., they can solve the halting problem, as explained in *WRCCB*, and in the primer on hypercomputation given in the present volume, section 1.8). Objective 3 was partially reached courtesy of *Articial Intelligence and Literary Creativity: Inside the Mind of* BRUTUS, *A Storytelling Machine* (*AILC*; Bringsjord & Ferrucci 2000). The chapter in *WRCCB* that bridges to *AILC* was "Chapter IV: What Robots Can Be." Objective 2 is met by the present volume (*SM*). Fully securing Objective 3 will require yet another book, one that will complete the quartet. In this final book, AI engineering undertaken on the basis of the view of minds as superminds will be described. Eight methodological prescriptions that capture this coming R&D are presented in the last chapter of the present volume; this last chapter, therefore, builds a bridge to the final book in the supermind quartet.

We end the preface with some more detailed reference information for those interested in understanding and grappling with the supermind project.

So, the book you're holding, then, presents a more systematic version of the view that we are superminds, a view we have labeled $\mathcal{S}_c^\infty$. The first part ($^\infty$) of this view is that we are infinitary information processors; the second part ($_c$) is that we enjoy a form of consciousness which can't be reduced to information processing of *any* sort. Accompanying the supermind doctrine, in our case, is extreme optimism about "Weak" AI, which is to say we affirm the view ($\mathcal{B}_p$) that embodied computing machines with sensors and effectors (i.e., robots), will sooner rather than later display a behavioral repertoire rivaling all, or perhaps nearly all, of our own.

The case for $\mathcal{S}_c^\infty$ and $\mathcal{B}_a$ and $\mathcal{B}_p$ includes arguments given by Bringsjord (along with, sometimes, Zenzen) outside the present book, in various stand-alone papers. Two of these papers are key to the overarching project, and are therefore listed in Table 1. The first is "Animals, Zombanimals, and the Total Turing Test: The Essence of Artificial Intelligence" (Bringsjord 2000*a*);[2] this paper is denoted by '*Zombanimals*' in Table 1. '*Androids*' stands for the paper "Could, How Could We Tell If, and Why Should—Androids Have Inner Lives" (Bringsjord 1995*a*). Some elements of these two papers appear in refined form in the present volume. A listing of the bulk of the arguments

[2]A preprint, replete with figures that didn't make it into the cramped journal version (Bringsjord 2000*a*) is currently available at

`http://www.rpi.edu/~faheyj2/SB/SELPAP/ZOMBANIMALS/zombanimals2.pdf`

Table 1: Finer-Grained Roadmap of Supermind Project

Argument	**Location**	**Goal**
Argument from Mary	*WRCCB*, Ch. 1	$\neg\mathcal{C} \wedge \mathcal{S}_c$
Arbitrary Realization Argument	*WRCCB*, Ch. 6	$\neg\mathcal{C} \wedge \mathcal{S}_c$
"Souped Up" Searlean Argument	*WRCCB*, Ch. 5	$\neg\mathcal{C} \wedge \mathcal{S}_c$
Argument from "Free Will"	*WRCCB*, Ch. 8	$\neg\mathcal{C} \wedge \mathcal{S}^\infty$
Argument from Introspection	*WRCCB*, Ch. 9	$\neg\mathcal{C}$
Victorious Gödelian Argument (promised in *WRCCB*, Ch. VII)	forthcoming	$\neg\mathcal{C} \wedge \mathcal{S}_c^\infty$
Arg. from Infinitary Reasoning	*SM*, Ch. 5	$\neg\mathcal{C} \wedge \mathcal{S}^\infty$
Zombie Attack	*SM*, Ch. 3	$\neg\mathcal{C} \wedge \mathcal{S}_c$
Argument from Irreversibility	*SM*, Ch. 4	$\neg\mathcal{C} \wedge \mathcal{S}_C$
Argument Against Church's Thesis	*SM*, Ch. 6	$\neg\mathcal{C} \wedge \mathcal{S}_c^\infty$
Argument *for* Artificial "Creativity"	*WRCCB*, Ch. 9; *AILT*	$\mathcal{B}_p$
Argument *for* Artificial Detective Work	*WRCCB*, Ch. 4	$\mathcal{B}_p$
Argument Against the Turing Test	*Androids*	$\mathcal{B}_p$
Argument for Artificial Animals	*Zombanimals*	$\mathcal{B}_a$

to date is given in Table 1.[3] Some of the arguments in this table are of course defended in other publications. For example, Bringsjord (2002) gives a "new-century" defense of the Chinese Room Argument in his "Real Robots and the Missing Thought Experiment in the Chinese Room Dialectic," in (Bishop & Preston 2002).

[3]Notice that we say the *bulk* of the arguments. A number of arguments aren't mentioned here. One example is "The Argument Against AI From Imagistic Expertise," by Selmer and Elizabeth Bringsjord (1996).

Chapter 1

What is Supermentalism?

In this chapter we: distinguish the supermind view, which we sometimes refer to by 'supermentalism,' from computationalism; explain the centrality of personhood and related concepts (e.g., cognition) to our project; and provide, as promised, a more detailed preview of the arguments for supermentalism coming later in the book. Along the way we provide a primer on hypercomputation.

1.1 Computationalism is Dead

Computationalism, by Fetzer's formidable lights, isn't only dead, but didn't deserve to be alive in the first place:

> If we want to understand the nature of thinking, then we have to study thought and not just the properties of formal systems. The boundaries of thought far transcend mere computability. (Fetzer 1994, p. 25)

And:

> Sometimes our professional activities become detached from our life experiences to a degree that might astonish empirical scientists. Our enthusiasm for a theoretical position may even appear to be virtually independent of our experiences in life, which, were they only taken seriously, might completely undermine what we take as our best theories. The computational conception that dominates what is known as "cognitive science" provides a remarkable illustration of this point. Even if some of our thought processes are computational, most of them are not, which makes *our best theory* either trivial or false. (Fetzer 2001*b*, p. 1)

We concur. So, the obvious question then is: "Now what?" Well, we have an answer, and we're going to spell it out and defend it for you in the coming pages; in brief for now, the answer is as follows. Computationalism-driven AI (= "Strong" AI) will yield Total Turing Test-passing zombies (in the philosopher's, not *The Night of the Living Dead*'s, sense of 'zombie'), but will fail to reach its goal of replicating persons, for two general reasons. One, persons process information at a "super"-Turing (or hypercomputational) level; two, they enjoy certain properties (e.g., intentionality, self-consciousness, incorrigibilism, qualia, etc.) beyond the reach of *any* mere information-processing machine. Accordingly, "Weak" AI is the way to go; that is, AI should strive to engineer "sub-person" artifacts (which can nonetheless probably be crafted so as to *appear* to be full-fledged persons), and the irreducibly philosophical investigation of personhood can proceed ahead full steam.[1] But let's slow down and ask two questions: (i) Do we seriously hold that computationalism is already dead? (ii) What *is* computationalism?

1.2 Are We Serious?

Many readers will be shocked to hear that we take computationalism to be dead. Are we serious? After all, no official funeral has been held. If anything, many cognitive scientists are working harder than ever on the basis of the presupposition that cognition is computation. Doesn't the explosive productivity of cognitive neuroscience alone show that we must be blind? And what about the fact that increasingly clever robots are appearing every year, as revealed, for example, by the American Association for Artificial Intelligence's annual robot competition?

We make two points in response.

First, it's important to keep in mind that our focus is on *persons*. We are happy to agree that a mouse, at bottom, is a mere computing machine (with sensors and effectors, of course); ditto for rats, cats, armadillos, dogs, horses, lions, chimps, and so on. In fact, one of us is directly involved with robotics aimed at replicating such creatures in the form of embodied computing machines, that is, in the form of robots (we discuss this robotics in the final chapter of the present volume). Consequently, we find it thoroughly unsurprising that robot

---

[1] In our opinion some of the best of this investigation is being carried out by Bill Rapaport and John Pollock (though Pollock, at least, intends to literally build persons: Pollock 1989).

competitions display robots having many of the abilities of mere animals.

Of course, there *are* those who hold that *persons*, not just mere animals, are computing machines. But — and this is our second point — there are two senses of 'dead' used in this discussion. According to one sense, what we might call the "voting sense," if a scientific/engineering approach is dead, then no one (or at best a few isolated individuals) is taking this approach. In this sense, it would be true to say that classical, uncompromising behaviorism is dead: there just isn't anybody around today who believes that our mental lives can be wholly analyzed and explained by way of stimulus-response patterns. This is *not* the sense of death we have in mind. We know full well that there are those who toil every day in hopes of inching the world closer to a time when artificial persons are among us. (John Pollock (1995) is a perfect, perhaps a paradigmatic, example.) We have in mind an intellectual death that is independent of the beliefs of scientists. Behaviorism was dead long before it was buried; that is, it was dead long before pretty much everybody *realized* it was dead.

You may suppose that if we're right and computationalism is dead in our sense, then sooner or later it will die in the voting sense. Not necessarily. Bringsjord has on a number of occasions asked confident proponents of computationalism this question:

> There is no denying that computationalism, *at present*, hasn't produced the kind of fruit it has long declared to be just around the corner. (One has only to recall Herb Simon's sanguine prophecies at the original 1956 Dartmouth conference. See (Russell & Norvig 1994) for a nice presentation of this prognostication in the context of a discussion of the conference.) As we start the new millennium, we still don't have an artificial agent clever enough to carry out a decent conversation with a toddler. So, at what point do you rethink the entire enterprise? How many years of coming up empty (when your artifacts are stacked against persons) will prompt you to question the view that persons are at bottom computing machines?

So far, no one has responded in a way that even remotely suggests that there *could* come a time when proponents of computationalism, faced with decades, or perhaps even centuries, of abject failure, would seriously entertain the possibility that persons are more than computing machines. Given this, it may well be that computationalism will never die in the voting sense, despite the fact that the artifacts it yields perpetually perform at the level not of persons, but of mere animals.

1.3 What is Dead? — Propositional Answer

Now let's turn to setting out computationalism in more detail. Finding published material to draw upon in order to do so isn't hard, for propelled by the writings of innumerable thinkers (this touches but the tip of a mammoth iceberg of relevant writing: Peters 1962, Barr 1983, Fetzer 1994, Simon 1980, Simon 1981, Newell 1980, Haugeland 1981, Hofstadter 1985, Johnson-Laird 1988, Dietrich 1990, Bringsjord 1992, Searle 1980, Harnad 1991), computationalism has reached every corner of, and indeed energizes the bulk of, contemporary AI and cognitive science (CogSci). The view has also touched nearly every major college and university in the world; even the popular media have, on a global scale, preached the computational conception of mind. Despite all this, despite the fact that computationalism has achieved the status of a Kuhnian paradigm, we regard it to be a carcass. But what exactly is the view in question? Unfortunately, computationalism is as protean as it is pandemic. For example, for starters we have *lots* of one-sentence versions of this doctrine floating about:

- Thinking is computing.
- Cognition is computation.
- People are computers (perhaps with sensors and effectors).
- People are Turing machines (perhaps with sensors and effectors).
- People are finite automata (perhaps with sensors and effectors).
- People are neural nets (perhaps with sensors and effectors).
- Cognition is the computation of Turing-computable functions.
- ⋮

There are differences, and in some cases significant differences, between these sorts of locutions. But surely there is a great and undeniable commonality in the works — a commonality captured, for example, by Haugeland:

> What are minds? What is thinking? What sets people apart, in all the known universe? Such questions have tantalized philosophers for millennia, but ... scant progress could be claimed ...

> until recently. For the current generation has seen a sudden and brilliant flowering in the philosophy/science of the mind; by now not only psychology but also a host of related disciplines are in the throes of a great intellectual revolution. And the epitome of the entire drama is *Artificial Intelligence*, the exciting new effort to make computers think. The fundamental goal of this research is not merely to mimic intelligence or produce some clever fake. Not at all. AI wants only the genuine article: *machines with minds*, in the full and literal sense. This is not science fiction, but real science, based on a theoretical conception as deep as it is daring: namely, we are, at root, *computers ourselves*. (Haugeland 1981, p. 2)[2]

This conveys the core spirit of "Strong" AI, which wavers not a bit in the face of questions about whether sensors and effectors must figure in an explicit version of this view, or where in the Chomsky Hierarchy from finite automata to linear bounded automata to full Turing machines people fall. Nonetheless, it will facilitate matters if we have a rather more focussed version of the doctrine on the table. Accordingly, we will say that computationalism consists, essentially, of the following five propositions:

(CTT) A function f is effectively computable if and only if f is Turing-computable. (The so-called Church-Turing Thesis.)

(P=aTM) Persons are Turing machines. (This thesis is provisional, and is designed only to capture the gist of mind-machine equivalence. E.g., we may ultimately need qualifications to the effect that the Turing machines in question are physicalized.)

(TT) The Turing Test is valid. (This too is designed to be provisional. What the validity of the Turing Test amounts to needs to be determined. Some of this investigation is carried out later in this chapter.)

(P-BUILD) Computationalists will succeed in building persons.

(TT-BUILD) Computationalists will succeed in building Turing Test-passing artifacts.

---

[2]A later Haugeland (1998) book is fascinating in the context of this quote, and the supermind doctrines we advance in this volume. In it we read:

> That there is a difference in kind, not merely in degree, between the "cognitive" capacities of people and those of other mammals strikes me as so obvious as to need no argument; and I will offer none. (Haugeland 1998, p. 1)

It's important to note that computationalism consists in the conjunction of these five statements. To qualify as a card-carrying computationalist, it isn't enough to affirm *some* of these propositions. For example, Descartes affirmed (TT) — long before Turing proposed the imitation game that now bears his name:

> If there were machines which bore a resemblance to our body and imitated our actions as far as it was morally possible to do so, we should always have two very certain tests by which to recognise that, for all that, they were not real men. The first is, that they could never use speech or other signs as we do when placing our thoughts on record for the benefit of others. For we can easily understand a machine's being constituted so that it can utter words, and even emit some responses to action on it of a corporeal kind, which brings about a change in its organs; for instance, if it is touched in a particular part it may ask what we wish to say to it; if in another part it may exclaim that it is being hurt, and so on. But it never happens that it arranges its speech in various ways, in order to reply appropriately to everything that may be said in its presence, as even the lowest type of man can do. And the second difference is, that although machines can perform certain things as well as or perhaps better than any of us can do, they infallibly fall short in others, by the which means we may discover that they did not act from knowledge, but only for the disposition of their organs. For while reason is a universal instrument which can serve for all contingencies, these organs have need of some special adaptation for every particular action. From this it follows that it is morally impossible that there should be sufficient diversity in any machine to allow it to act in all the events of life in the same way as our reason causes us to act. (Descartes 1911, p. 116)

We will return to the second of the tests Descartes here proposes in the final chapter of this book, in connection with our discussion of "Psychometric AI," a brand of "Weak" AI that one of us (Bringsjord), with Bettina Schimanski, has devised as a generalization of an approach to building AI systems on the basis of TT, TTT, and so on. As to the first test, perhaps you'll agree that it sounds remarkably close to TT — though Descartes, unlike Turing, places no special emphasis on mendacity as a divider between our minds and mere machines. Of course, it's far from clear in this passage as to what *exactly* Descartes has in mind. Is he simply reporting that, *in his day and age*, no machine can pass (what we might baptize) the **language test**? Probably he is saying something a good deal stronger than this. Is he saying, then, that it will be true *forever* that no machine will be able to pass the language test? It would seem so. It would seem, a bit more precisely, that Descartes maintains that if a machine passes the language test, then that machine has, or perhaps ought to be *regarded* to have, the mental properties of a person. The analysis of TT-relevant conditionals comes later in this chapter.

One more remark about the propositions forming the core of computationalism: We take it that (P-BUILD) implies **agent materialism**, the view that persons are material things. Even if this isn't so, we assume that nearly all readers will agree that nearly all computationalists are in fact agent materialists. Perhaps a few odd computationalists somewhere hold that they will build persons by constructing a physical device that will then — by way of a process outside of their direct control — become associated with an immaterial person. But we know of no such thinkers, and this bizarre position this doesn't correspond to any commonsensical reading of 'building persons.' We will attack an explicitly agent-materialist brand of computationalism in Chapter 6.

1.4 The Centrality and Logic of Personhood and Cognition in the Present Project

You have no doubt noticed by now that personhood is central to the issues with which we're dealing. It's therefore probably not a bad idea if we take a bit of space to set the record straight on what we take personhood to consist in, at least to some degree. In doing so, we can start by simply restating the foundational theses concerning personhood advanced in *What Robots Can and Can't Be* (1992), the first book in the supermind quartet:

(PER_1) Persons are genuine individual things, not logical constructions.

(PER_2) Persons are bearers of psychological, Cartesian, or self-presenting properties (such as *fearing unicorns*, *believing that quantum mechanics is silly*, ...), and when a person has such a property she is said to be in a **mental**, or **conscious**, **state**.

(PER_3) Some persons bring about some states of affairs.

(PER_4) There are persons.

We say this is only a *start*; the reason is that these four propositions were designed to express part of a *minimalist* ontology for *What Robots Can and Can't Be.* The attempt is made in *WRCCB* to cultivate from (PER_1)–(PER_4) a larger list of properties — a list of properties, again, that no computing machine can have. Recapitulating this attempt here is impracticable; we take another, shorter route, which is simply to list the additional concepts and propositions associated with personhood in the present project. Most of these concepts and propositions are listed in (Bringsjord 1997*a*). According to that list, which we can here dub L_p, persons are taken to have the capacity to

- be conscious (in the phenomenal sense; this seems to be implied by PER_1);
- be self-conscious;
- communicate through a language;
- know and believe things (at a third-order level and beyond; e.g., Jones has the capacity to believe that others believe that Jones believes ...)[3];
- desire things (including that one's own character change);
- "will" (this seems to be implied by PER_3); and
- reason in abstract, domain-independent fashion (so called "System 2" cognition; see Stanovich & West 2000) that is beyond the expressive power of ordinary logical systems (such as first-order logic, or even basic quantified modal logic).

To L_p we add

- create things (in the fullest sense of the term),

and there are probably some other attributes that should be added to the list as well. (Haugeland 1998 would put what he calls "existential committment" not only on the list, but at the top.) However, this should do for now (we revisit the list in Chapter 8). The important point at the present juncture is that personhood includes a *number* of properties, and to focus, for example (as we do at times in this book), on phenomenal consciousness, is not to push personhood off center stage. On the contrary, phenomenal consciousness is at the heart of the present essay precisely *because* it's part of personhood. The same would apply to times when we focus on other attributes, for instance

[3]It seems, by the way, to be provably the case that at least some persons have beliefs at the 10th order and above. See pp. 85–86 of (Bringsjord & Ferrucci 2000).

on judging the interestingness of narrative (as we do in Chapter 4 in the present volume).

This is the spot to inform our readers that when we refer to *cognition* in connection with persons we are are referring to those properties in L_p. The slogan "Cognition is computation," then, when maintained in connection with persons, is the view that everything in L_p can be reduced to computation. Some readers may balk at our sense of the term 'cognition.' In particular, those readers familiar with cognitive psychology may declare that things like consciousness are not cognitive, and for evidence they may cite various textbooks in which not the slightest mention of consciousness is made (e.g., Ashcraft 1994). The fact of the matter is that not all texts are in agreement; see, for example, (Schacter 1989), which gives us the picture of cognition shown in Figure 6.4. But at any rate, there is nothing for us to worry about here. The fact is, cognitive psychology and science have exploded with attempts to explain consciousness and qualia in the language of these fields. The study of cognition now includes the study of consciousness; this is a simple fact that can be verified by surfing the web for no more than two minutes — which will, as of now, inevitably result in your finding the annual conference "Toward a Science of Consciousness," replete with attempts to explain subjective awareness in the schemes proffered by the likes of (Ashcraft 1994). Even before the explosion to which we refer, a number of those theorists concerned with the topic of consciousness took it, even when of the phenomenal variety, to include phenomena traditionally classified as cognitive. For example:

> So how should we point to P-consciousness [= phenomenal consciousness]? Well, one way is via rough synonyms. As I said, P-consciousness is experience. P-conscious properties are experiential properties. P-conscious states are experiential states, that is, a state is P-conscious if it has experiential properties. The totality of the experiential properties of a state are "what it is like" to have it. Moving from synonyms to examples, we have P-conscious states when we see, hear, smell, taste and have pains. P-conscious properties include the experiential properties of sensations, feelings and perceptions, *but I would also include thoughts, wants* and emotions. (Block 1995, p. 230; emphasis ours)

On the other hand, we are happy to admit that cognition doesn't *have* to include those things in L_p. When we turn away from persons to mere animals this becomes obvious. The cognition of a mouse no doubt lacks abstract, domain-independent reasoning in first-order logic (let alone in some logical system beyond FOL). But when *people* are in the picture, cognition includes domain-independent reasoning (though not necessarily of a normatively correct form; e.g., see Rips 1994, Braine 1998*b*, Braine 1998*a*, Yang, Braine & O'Brien 1998).

Now a critic might at this point object as follows: "But surely you don't mean to suggest that your list L_p makes a set of necessary and sufficient conditions! What if a creature has all of the attributes in question, save for *second*-order beliefs? Would you then say that that being isn't a person?"

In fact, one of us (Bringsjord) *is* inclined to say that a creature lacking a capacity for second-order beliefs isn't a person, but there is no need to get bogged down in this micro-controversy. It seems clear enough that some of the properties in question *are* necessary. For example, it seems clear that consciousness is essential (though we do evaluate, momentarily, the contrary view) — and many of the arguments we give against computationalism and for supermentalism exploit this particular fact. Furthermore, that we are talking about *capacities* shouldn't be ignored. Probably many of those creatures who would be put forward as problematic cases designed to bolster the present objection do in fact have, say, the *capacity* for n-order beliefs. If Jones has a brain lesion, and therefore can't X, it may nonetheless be true that he retains the capacity to X. It is of course notoriously difficult to pin down the meaning of 'capacity,' as is made clear by familiar thought-experiments. If Joe Jackson is a great baseball player, but gets hit in the eye by a baseball in the Spring, he may need surgery to return to form in the Summer. We probably wouldn't want to say that between Spring and Summer he lacked the capacity to throw a baseball well — because in all likelihood lots of money will have been spent on repairing his vision by those convinced that he retains the capacity for first-rate ballplaying, and when he returns to form the money will have been well-spent. Likewise, though some will disagree, it seems to us that those who seem to be persons, but can't at time t produce the behaviors associated with the list above, nonetheless have the capacity to produce these behaviors at some time after t. We return to this issue a bit later in this chapter.

At this point, following the lead of a recipe explained elsewhere in a brief note (Bringsjord 2000*b*), we can be clearer about how personhood allows for refutations of computationalism. Collect all the attributes P_i that compose the essence of personhood; call the resultant set $\mathcal{P} =$

$\{P_1, P_2, \ldots, P_n\}$. Let $P_i^\star$ denote the technical, precise correlate for P_i. Then here is part of the pairing we deploy:

P_i	$P_i^\star$
autonomy	'iterative agent causation' (Ch. 8 of Bringsjord 1992)
self-consciousness	'hyper-weak incorrigibilism' (Ch. 8 of Bringsjord 1992)
creativity	a capacity for judging whether stories are interesting (Chs. 5 & 4 of the present book)
consciousness	P-consciousness (Ch. 6 of the present book)
⋮	⋮

Now, set Px iff x is a person, and Ax iff x is an automaton (at or below the Turing Limit). Where '$\Box$' indicates 'logically necessarily' (in some broad sense that needn't detain us; an "S5-ish" interpretation is affirmed in *WRCCB*), the central argument schema aimed against computationalism is

$$\begin{array}{rl} & \Box\forall x(Px \rightarrow P_i^\star x) \\ & \Box\forall x(Ax \rightarrow \neg P_i^\star x) \\ \therefore & \Box\forall x(Px \rightarrow \neg Ax) \end{array}$$

Of course, this schema implies that the capacities listed in L_p (or, more precisely, the technical correlates of these capacities) are in fact *essential* to personhood. Perhaps given this, a critic might seek to sustain and broaden the previous objection, as follows:

> "It isn't at all obvious that qualia (e.g.) are essential to personhood. Suppose my qualia are knocked out, but my mathematical career, my marriage, my friendships — *everything else* about me — is unimpaired. Am I no longer a person?"

The answer is obviously an affirmative one. If there is no longer "something it feels like" to be with your spouse on an especially amorous night, or to craft your seminal proofs, or to spend leisure time with your friends, if there is no longer "something it feels like" to be you, then you are dead: you are a zombie. If anything is at the heart of personhood, it is subjective awareness.

Our critic might at this point demand to know what phenomenal consciousness is *for*. "It isn't enough," she might say, "to declare that consciousness is central to personhood. Remember, your book concerns the foundations not only of AI, but of cognitive *science* — and science deals in what can be observed. Consciousness is invisible, so what tangible things does it enable?" This is a question one of us has answered elsewhere, by answering a directly related question, viz., "Why did evolution engineer phenomenal consciousness?" (Bringsjord, Noel & Ferrucci 2002). Our answer, in a nutshell, is that the function (or at least one of the chief functions) of P-consciousness is to allow for genuine creativity, and a deductive argument in support of this answer is provided in our paper. That argument is an instantiation of the argument-schema

$\mathbf{Arg}_1$

(1) If person S loses x over an interval of time during which S loses the ability to ϕ, and there are substantive reasons to think x is centrally employed when people ϕ (in part because (i) attempts to replicate ϕ-ing in AI systems lacking x have failed, and show no appreciable promise of succeeding in the future; and (ii) subjects report that they need x in order to ϕ), then a function of x is to at least facilitate ϕ-ing.

(2) S loses x over an interval of time during which S loses the ability to ϕ.

(3) There are substantive reasons to think x is centrally employed when people ϕ (in part ... that they need in order to ϕ).

$\therefore$ (4) A function of x is to facilitate ϕ-ing. 1,2,3

where x is set to 'P-consciousness,' and ϕ = 'write belletristic fiction.' Details can be found in (Bringsjord et al. 2002).

Perhaps our detractors will shift to other attributes. Perhaps, for example, it will be claimed that a loss of hypercomputational capacity would not disqualify someone as a person. This is a more intelligent objection, but it's a weak one. We are not saying that minus hypercomputation *per se*, personhood is no longer in the cards. Rather, we are saying something a good deal more specific: viz., minus some particular capacities, personhood is impossible, and these capacities require hypercomputation. So, yes, without hypercomputation, a creature is at best a sub-person (or perhaps just a *non*-person), but the point is that such a creature would be so classified because of a loss of some particular capacities. What capacities? Well, we discuss some of these particular capacities later on. For example, in Chapter 3 we discuss the capacity for infinitary logico-mathematical reasoning (which falls under one of the bullets in L_p, as you can verify). Inevitably, our opponents will persist in their complaints; perhaps they will say, "Well, but why ever would you think that infinitary reasoning is necessary for personhood?!" We counter this objection at the start of Chapter 3. It suffices here to point out that if we are superminds, hypercomputation is far from peripheral; it would on the contrary be part of the *essence* of being a person. Now, how hypercomputation *manifests* itself — that is another matter. As we pointed out in the Preface, one of us has already provided a careful deductive argument for the view that free will, another element of L_p, is possible only because we are, at least in part, hypercomputers (see Chapter VIII of Bringsjord 1992). Certainly it would not be too plausible to claim that our capacity to freely make decisions is not at the center of what it means to be a person.

1.5 The Turing Test

In addition to personhood, the Turing Test figures prominently in the quintet offered above as a declarative encapsulation of computationalism. We now proceed to discuss this test, with help, again, from some of our prior work.

The case presented in Bringsjord's *What Robots Can and Can't Be* (1992) for TT-BUILD is, as Bringsjord readily admits in that monograph, inchoate. The book *Artificial Intelligence and Literary Creativity: Inside the Mind of* BRUTUS, *A Storytelling Machine* (Bringsjord & Ferrucci 2000), mentioned in the Preface, is devoted in earnest to bolstering TT-BUILD. This book features a storytelling agent, BRUTUS, able to (among other things) "understand" and manipulate the concept of betrayal, in the automatic production of stories.

Actually, TT-BUILD is short for a thesis that makes crucial reference not only to the Turing Test (T_1 or, to use the acronym introduced above, TT) but to what has been called (Bringsjord 1992) the **Turing Test sequence**. There are obviously tests easier to pass than T_1, but which still capture Turing's empiricism. For example, we might index T_1 to some particular domain — golf, say — and stipulate that questions moving significantly beyond this domain aren't fair game.[4] We

[4]Admissible question: "What is 'Hogan's dream"? (Answer: A round of golf with a birdie on each hole.) Inadmissible: "Is the feeling one gets from scoring an eagle on a par 5 anything like what it feels like to serve an ace in tennis?" Such restricted versions of T_1 are reminiscent of contests in which programs compete against each other to see which among them can fool the most humans into thinking that they (the programs) are humans. Each program is indexed to some domain.

might insist that the syntax of queries, and the syntax of responses, conform to a certain range of patterns. Or we might decide to modify the scoring that is part of Turing's original proposal — so that, for example, if a system can fool the judge 25% of time, rather than the full 50%, it is judged to have passed.

It's also easy enough to devise tests more stringent than T_1. The most famous of these tests is Stevan Harnad's (1991) "Total Turing Test," which we can baptize as T_2 (we referred to it by 'TTT'). In this test not only is the linguistic behavior of a robot (or android) scrutinized, but its sensorimotor conduct is evaluated as well. In order to pass T_2, a robot must not only have the capacity to be an articulate (to use one of Harnad's terms) pen-pal; it must also *look* human. In the same paper, Harnad considers more demanding tests than T_2, for example tests in which the judge is allowed to "look under the hood," that is, to look at the physical stuff out of which the synthetic contestant is made. Other thinkers, for example Peter Kugel (1990), have proposed tests so demanding that they cannot, as a matter of mathematics, be passed by Turing machines. [These tests and others are discussed and rejected (as tests for phenomenal consciousness) in (Bringsjord 1995*a*).] Given that there are *many* tests that can be devised, let's suppose that there is some sequence

$$T_0, T_{.25}, \ldots, T_1, T_2, \ldots$$

which underlies the relevant modification of TT-BUILD, namely,

TT-BUILD$_{\diamond_h T_i}$ It is humanly possible to build robots (or androids, etc.) capable of ascending (perhaps all of) the Turing Test sequence.

We think TT-BUILD$_{\diamond_h T_i}$ is true, and we hope to provide increasingly powerful evidence for this proposition as the years unfold (we return to this hope in the final chapter, where it's discussed in more detail). But we think a claim of the form 'if computer c (or robot r) passes T_i then c (r) is conscious,' that is, TT$_i$, is false.(We discuss below the question of whether a claim of this form is one Turing originally made when introducing his test.) We have two types

This kind of contest was held at Turing 2000, the conference at Dartmouth held to commemorate the birth of AI and Turing's prediction that by the year 2000 a machine would be built that could pass the famous test that now bears his name. The conference was masterfully hosted by Jim Moor, author of a classic treatment of TT: (Moor 1976).

of arguments for this position; each is based on a series of thought-experiments (explained in Bringsjord 1995*a*). The first series begins with a gedanken-experiment in which an excruciatingly simple Lisp-based random sentence generator serendipitously passes T_1. Would you want to say, in such a situation, that as luck would literally have it, a new consciousness has popped into existence?[5] Probably not. After all, such a sentence generator is easy enough for an AI novice to write — an observation that underlies this argument:

Arg$_2$

(5) If TT_1 is true, then students in courses like *Intro to AI* write programs sufficient for realizing AI's grandest dreams.

(6) It's not the case that students in courses like *Intro to AI* write programs sufficient for realizing AI's grandest dreams.

∴ (7) $\neg\mathrm{TT}_1$ 1,2

This argument is specified and defended in (Bringsjord 1995*a*), and is then extended in that same paper so as to refute TT_2 and beyond.[6] The extension allowing for the refutation of TT_2, i.e., a refutation of

- If c (r) passes the *Total* Turing Test, then c (r) is conscious,

is easily produced: the trick is to combine the random sentence generator (from above) with a random "behavior" generator, i.e., a program that randomly outputs not only text, but also commands to effectors. Now imagine that this composite computer program, when hooked to some real effectors, serendipitously passes Harnad's Total Turing Test. Would anyone want to say that the system in question is conscious? It would seem not. But then it follows that the conditional in question has been overthrown, since its antecedent is true while its consequent isn't.

Of course, we're assuming that the conditional in TT_i is stronger than a *material* conditional, for if the conditional is a material one, then since no machine can currently pass the test (= TT $\vee$ TTT $\vee$

[5]In fact, this is exactly what some defenders of computationalism want to say, e.g., Cole & Foelber (1984). For a refutation of this "multiple person" view, see Chap. V of (Bringsjord 1992).

[6]Access to info about the book in which this paper appears can be had through Bringsjord's web site.

any test further up the sequence) in question, the thesis is (vacuously) true. What Turing therefore had in mind was no doubt something a bit stronger — probably something like

- Necessarily, if c (r) passes the Turing Test, then c (r) is conscious.

Now, it may strike you as odd, if not flatly wrong, that we have turned the focus upon consciousness. Someone might object, specifically, as follows: "Turing presents his test as a test for 'intelligence' or 'thought.' Your interpretation of the TT cheats over toward where the two of you want to go: phenomenal consciousness and qualia." But a careful reading of Turing's (1964) paper supports our interpretation. Specifically, we draw your attention to the section therein entitled "(4) The Argument from Consciousness." The argument from consciousness is one Turing takes to be well-expressed in Professor Jefferson's Lister Oration of 1949:

> Not until a machine can write a sonnet or compose a concerto because of thoughts and emotions felt, and not by the chance fall of symbols, could we agree that machine equals brain — that is, not only write it, but know that it had written it. No mechanism could feel (and not merely artifically signal, an easy contrivance) pleasure at its successes, grief when its valves fuse, be warmed by flattery, be made miserable by mistakes, be charmed by sex, be angry or depressed when it cannot get what it wants. (From p. 17 of Turing 1964.)

What is Turing's response? He gives an absolutely dreadful argument:

> If one refuses to agree that passing TT (when, say, sonnets are requested by the judge) implies consciousness, one must accept solipsism. Since solipsism is false, one cannot refuse in the manner indicated by Jefferson.

Though the two of us, along, no doubt, with you, reject solipsism, we can't for the life of us imagine where Turing gets his first premise. To take a guess, perhaps the underlying idea is that if one rejects the passing/consciousness implication, one will have no solution (or candidate solution) to the problem of other minds save for solipsism, but since the literature is filled with proposed solutions to the problem of other minds that make no appeal to TT, the underlying idea is fatally flawed. At any rate, Turing's argument is beside the point we've by now made, which is that Turing clearly holds that if a computer (or robot) passes his test (in part by producing sonnets, etc.), it follows that it's conscious.

Another objection that might be brought against us at this point runs as follows:

> "You have said that computationalism is supposed to be the *conjunction* of the five propositions you gave above, but as you understand 'validity of the Turing Test,' *no* computationalist then subscribes to computationalism! Computationalism says that thought is essentially computation; TT *definitionally understood*, which is how you understand it, targets behavioral output, regardless of (computational) means. So, anyone accepting TT as definitional (as 'valid,' in your terms) is a *behaviorist*, not a *computationalist*."

There are two reasons why this objection fails. First, note that the stronger thesis, that is, our way of expressing the claim that the TT is valid, *isn't* a *definition*. It's a modalized conditional. Definitions, of course, are at heart *bi*conditionals. Hopefully no one is dim enough to hold the converse in this case. After all, lots of conscious beings (e.g., paralyzed but otherwise well-functioning humans) are unable to pass TT. Someone might retort that a modalized conditional has all the drawbacks of a definition, but then there is the second problem: we specifically refer to a computer c and/or robot r in the theses in question. This implies that it's not *only* behavioral output that is important. The output must be produced by the normal operation of a computer and/or robot; that is, the output must be produced by computation.

We move now to the fact that Harnad has produced a more sophisticated thesis to accompany his Total Turing Test, one he specified near the end of a long debate with Bringsjord. Here's the thesis:

TT_2-H If x is TTT-indistinguishable from someone with a mind, one has no non-arbitrary reason for doubting that x has a mind when told that x is a machine.

Unfortunately, this thesis is false — given that Harnad accepts a modalized interpretation of the conditional, and the argument from serendipity (and Searle (1980)'s Chinese Room Argument) against the original Turing Test. In order to see this, begin by supposing that x is TTT-indistinguishable (= TT_2-indistinguishable) from someone with a mind. To make this a bit more vivid, suppose that roboticist Smith escorts a robot around with him which is TTT-indistinguishable from someone with a mind, but suppose that Smith cloaks this robot in some way (with synthetic skin that passes for the real thing, artificial eyes that look human, etc. — use your imagination) and refers to it as 'Willie.' So, you carry out conversations with Willie for years; *and*, since we're dealing here with the *Total* Turing Test, you also play baseball (and the like) with Willie over the same long period of time. And then one day Smith says, "Aha! Watch this!" And he tears off the skin over Willie's forehead, and therein sits a computer busy flashing and humming along.

So far so good. How do we get the falsity of TT_2-H's consequent built into the thought-experiment? No problem. Let's begin by reminding ourselves of what we need. We need the Willie scenario to include the falsity of

> One has no non-arbitrary reason for doubting that Willie has a mind.

Let's instantiate this to one of us (Michael, say); that seems harmless enough. So we need the scenario to include the falsity of

> Michael has no non-arbitrary reason for doubting that Willie has a mind.

Here's how it works. Michael says to Smith:

> "Look, Willie is still just a bunch of symbols swimming around inside a box. There's no person in the picture. It's easy enough to show that there's no principled connection between the symbols and sensors/effectors working wonderfully and there being a person in the picture, for consider a situation (alluded to above) wherein Willie's program is a *composite* program: a lucky random sentence generator *and* a lucky program for processing information received via sensors *and* a lucky program for processing information sent to effectors — all code that's based on technology students master in the first few weeks of *Intro to AI.*"

(It's important to note here that (i) all that is needed to overthrow Harnad's TT_2-H is a *non-arbitrary* reason for denying that Willie is conscious; we *don't* need a *compelling* reason. And (ii) Harnad agrees that TT_1 is shown to be false by the original version of the argument from serendipity.)

In the second type of attack on the Turing Test, we argue that zombies, *contra* Dennett, are logically — and, in fact, physically — possible, and that this fact suffices to refute certain versions of computationalism (the argument appears in Chapter 5 of this volume).

The initial thought-experiment in the series of those that underly this attack is provided by Searle in his *The Rediscovery of the Mind*.[7]

Before moving on to describe the supermind view in broad strokes, we end this section by considering an objection that takes account of what we have said about personhood and hypercomputation, and Turing Testing as well:

> "You insist that the Turing Test and its variants are invalid. But you also insist that hypercomputation is central to personhood, in the sense that without this powerful brand of information processing, some of the capacities that are part and parcel of personhood would be impossible. This puts you in a nasty bind; let me explain. If the capacities that require hypercomputation are significant, as you claim, then surely there is a way to test for them in Turing's scheme. For example, suppose that it's on the strength of analog chaotic information processing (a particular sort of hypercomputation, if I understood the primer you give below, which I've peeked ahead at) that Gödel was able to give us the incompleteness results that Bringsjord is on record as venerating (e.g., Bringsjord 1998*b*). This would mean that we could probe, in a Turing Test, for such behavior! So you can't have it both ways: you can't hold that Turing Testing is invalid, and yet that hypercomputation is scientifically significant."

This objection is easily and swiftly surmounted, for it's plagued with many fatal flaws, all easily revealed. The first such flaw is that it presupposes verificationism (a doctrine that is discussed and rejected later in connection with, arguably, its most able proponent — Daniel Dennett — in Chapter 5). For now, it suffices to point out that it's

---

[7] The thought-experiment runs as follows. You have been diagnosed with inoperable brain cancer. Desperate, you seek out experimental techniques for combating the disease, and find a neurosurgeon willing to try to gradually replace your brainflesh with (healthy) silicon. Searle writes:

> As the silicon is progressively implanted into your dwindling brain, you find that the area of your conscious experience is shrinking, but that this shows no effect on your external behavior. You find, to your total amazement, that you are indeed losing control of your external behavior ... [You have become blind, but] you hear your voice saying in a way that is completely out of your control 'I see a red object in front of me.'... We imagine that your conscious experience slowly shrinks to nothing, while your externally observable behavior remains the same. (Searle 1992, pp. 66–7)

in it. What we find amazing is that though the mathematical side of the entire landscape is as real and uncontroversial as can be, many of those systematically pondering the human mind seem to be unaware of this side. For example, consider Johnson-Laird (1983)'s reasoning about human consciousness in his seminal *Mental Models*:

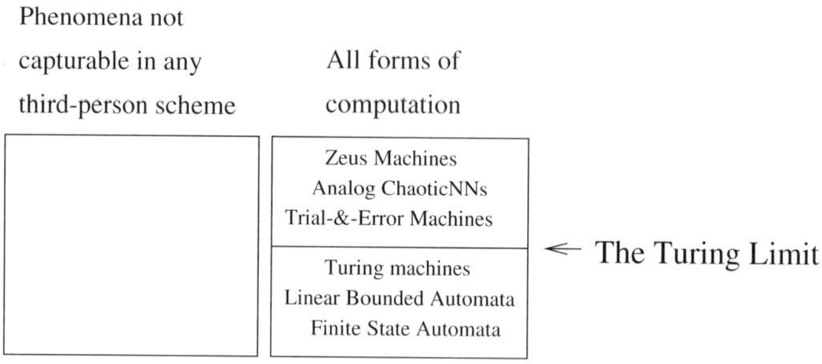

Figure 1.1: The Entire Landscape

> No one really knows what consciousness is, what it does, or what function it serves. There is, however, a clear series of alternatives. Either consciousness is wholly mysterious or else it is open to scientific explanation. Either Turing's thesis that effective procedures are computable is correct or else consciousness may depend on hitherto unknown effective procedures that are not computable. (Johnson-Laird 1983, pp. 448–449)

Johnson-Laird says that in light of these alternatives, consciousness must fall into one of four categories:

> 1. It is a supernatural phenomenon — a real 'ghost' in the machine that is totally beyond scientific explanation.
> 2. It can be explained, but only by recourse to as yet unknown non-computable effective procedures.
> 3. Like the weather, it is explicable by theories that can be simulated by computer programs, but it can no more be embodied within a computer than can an actual anti-cyclone: only organisms with brains can be conscious.
> 4. It is computable: computers can in principle be programmed so as to be conscious. (Johnson-Laird 1983, pp. 449)

The remarkable thing about this analysis is that Johnson-Laird does take account of the side of the entire landscape that is non-mathematical, and yet seems oblivious to the space of functions above the Turing Limit — a space that has nothing mystical about it, and can be traced back to formal work carried out well before Johnson-Laird's commentary (e.g., Turing 1938, Turing 1939). As our readers will now certainly understand, there is another possibility for consciousness (and, of course, for other powers on L_p as well), namely, that it corresponds to non-effective hypercomputational processes. (Of course, as a matter of fact, we place consciousness on the "ineffable" side of the overall landscape — though we are reluctant to say with Johnson-Laird that such a placement implies that consciousness is supernatural.)

The entire landscape (again, shown in Figure 1.1) can be used to catalyze discussion of other foundational points. For example, this picture can be used to show that the scientific credentials of computationalism are suspect: Given this landscape, why would anyone ever have an unshakeable intuition from the beginning, as most computationalists do, that human persons should be located exclusively within the lower right portion of the picture? After all, the right side of the picture has been discovered and charted by none other than persons, and, as we've noted, the upper right side is *much* larger than the lower right. Why does the computationalist assume that cognition is only information processing below the Turing Limit? One of us has asked this question of many computationalists — and a compelling answer has never been provided. This book, along with the three other books composing the supermind quartet described in the Preface (and also the deductive arguments published outside this quartet; some of these arguments are also mentioned in the Preface; see Table 1), furnishes a detailed, rigorous case for supermentalism. But what are the arguments in favor of computationalism, over and above supermentalism? In our experience, only two arguments have been provided. In the first, one argues that the brain can be modeled as a standard artificial neural network, so persons must correspond to such things. (A standard ANN, as discussed in the final chapter, is equivalent to a standard Turing machine in information processing power.) This argument is refuted in (Bringsjord 1992). The second argument goes like this: The simplest view of the nature of the human mind is that that nature is mere computation. One ought to prefer the simplest theory, *ceteris paribus*. Ergo, one ought to affirm computationalism. Unfortunately, this argument is astonishingly weak, for three general reasons. The first reason is that all things *aren't* equal in the present dialectic. This is true, for starters, because there are many arguments against computationalism, and many arguments for supermentalism. The second reason why this pro-computationalist argument is anemic is that, obviously, it will be mired in unresolved debates about Ockam's razor, and like principles. What is simplicity, exactly? This is a question for philosophy of science, and it's a very, very tricky one. Even if we assume that we have a tolerably clear account of theoretical simplicity, it would seem that any number of currently dominant theories about the physical world are anything but simpler than their cast-aside competitors. The third reason why the present argument doesn't provide solid support for computationalism is that it's far from clear that supermentalism is at bottom more complex than computationalism. Is the mathematical space below the Turing Limit more complex than the one above? The answer isn't evident.

Pictorially, supermentalism can be summed up in Figure 1.2, where the circle therein represents persons. We move now to a propositional presentation of supermind view.

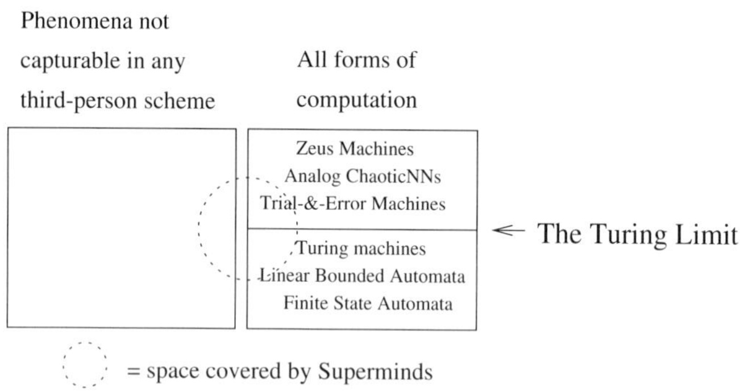

Figure 1.2: Super-Minds (which have TMs, ZMs as parts)

1.7 Propositional Overview of Supermentalism

In the Preface we said that the supermind view is composed of two general propositions, $\mathcal{S}_c^\infty$ and $\mathcal{B}$. These two propositions can in turn be decomposed into the following seven propositions. The first three correspond to $\mathcal{B}$, the fourth to $\mathcal{S}_c$, and the fifth and sixth to $\mathcal{S}^\infty$. Here's the list:

P→TM Persons have Turing machines at their disposal (i.e., persons harness TMs).

$\neg$TT$_i$ $\forall i$

TT$_i$-BUILD Computationalists will succeed in building T$_i$-passing artifacts, $\forall i$.

"P$_i$"-BUILD Computationalists will succeed in building T$_i$-passing zombies, $\forall i$.

P$\neq$axM Part of what personhood entails cannot be reduced to any information-processing scheme.

$\neg$CTT Church's Thesis is false.

P$\rightarrow$ZM Persons have "super"-Turing machines (hypercomputers) at their disposal (i.e., persons harness hypercomputation).

No one should dispute P$\rightarrow$TM. After all, persons regularly build TMs in order to solve problems, and they also engage in cognition intended to perfectly map to TM processing. (As we pointed out previously, Turing and Post started not with the concept of a comput*er*, but rather with the concept of a comput*ist*: a human who carries out — or perhaps just *is* — an effective procedure.[9]) What about the remaining propositions? They are established by arguments given in subsequent chapters, each of which is previewed later in the present chapter.

1.8 A Primer on Hypercomputation

Now it's time to tell you, in broad strokes, what hypercomputation is.

The story (at least the contemporary version) begins with Turing, who in his dissertation (Turing 1938 1939) pondered the possibility of so-called **oracle machines**. These machines are architecturally identical to Turing machines, but are assumed to be augmented with an oracle which, upon being consulted about a Turing machine m and input i, returns a correct verdict as to whether m halts on i. Oracle machines are part of the canon of computer science today. For example, here's a quote from a recently updated classic textbook on computability and uncomputability:

> Once one gets used to the fact that there are explicit problems, such as the halting problem, that have no algorithmic solution, one is led to consider questions such as the following:

[9]Sieg & Byrnes (1996), as we also noted earlier, calls computists comput*ors*.

> Suppose we were given a "black box" or, as one says, an *oracle*, which can tell us whether a given Turing machine with given input eventually halts. Then it is natural to consider a kind of program that is allowed to ask questions of our oracle and to use the answers in its further computation ... (Davis, Sigal & Weyuker 1994, p. 197)

How do Davis et al. transform this figurative scheme into a mathematically respectable one? To answer this question, note that instead of Turing machines, Davis et al. use an equivalent programming language $\mathcal{L}$, the programs of which are composed of lists of statements with optional labels. [There are of course an infinite number of ways to specify computation at and below the Turing Limit. See (Bringsjord 1994) for a discussion of this point. $\mathcal{L}$ is similar to the language for programming register machines given in (Ebbinghaus, Flum & Thomas 1984).] $\mathcal{L}$ allows for three types of statements: adding one to a variable V ($V \leftarrow V + 1$), subtracting one from a variable V ($V \leftarrow V - 1$), and moving by a conditional to a line labeled with L in a program (IF $V \neq 0$ GOTO L). With just these three statements, it's possible to write a program that computes every Turing-computable function. Traditionally, to make it easier to see this, "macros" $V \leftarrow V'$ and GOTO L are allowed. The first macro moves the contents of variable V' to variable V; the second is an unconditional branch that moves the active line to the one with label L; both macros can be easily decomposed into a program written with only the three fundamental statements. (Readers new to this material are encouraged to carry out the decomposition.) As an example of an excruciatingly simple program in $\mathcal{L}$, consider a program that computes the function $f(x_1, x_2) = x_1 + x_2$:[10]

	$Y \leftarrow X_1$
	$Z \leftarrow X_2$
$[B]$	IF $Z \neq 0$ GOTO A
	GOTO E
$[A]$	$Z \leftarrow Z - 1$
	$Y \leftarrow Y + 1$
	GOTO B

At this point we're in position to see how Davis et al. formalize

[10]Note that a conditional or unconditional branch that directs flow to a label not present in the program causes halting. In the program here, then, the label E can be read as "exit."

oracles. The trick is simply to allow a new statement (an **oracle statement**) of the form

$$V \leftarrow O(V)$$

into the syntax of $\mathcal{L}$: "We now let G be some partial function on $\mathbf{N}$ [the natural numbers] with values in $\mathbf{N}$, and we shall think of G as an oracle" (Davis et al. 1994, p. 198). So if the value of variable V is m before an oracle statement is encountered, when the statement is then reached, the value of V changes to $G(m)$ (assuming that G is defined for this argument). In connection, specifically, with the halting problem, where $m_1, m_2, \ldots$ enumerates all Turing machines,the function

$$h(m,n) = \begin{cases} 1 & \text{if } m_m \text{ halts with input } n \\ 0 & \text{otherwise} \end{cases}$$

can be solved by a program in $\mathcal{L}$ in which an encoding u of m and n is given as an argument to G.

Unfortunately, this leaves the *nature* of oracles completely out of the picture. All that matters, at this point, is that oracles return verdicts that can be used by Turing machines (and their equivalents). The idea so far, logico-mathematically speaking, is to set up a scheme for investigating *relative* computability; and in such a scheme, how oracles do their thing is wholly irrelevant. For the two of us, and other philosophers interested in hypercomputation, this situation is unacceptable. What is an oracle? How does it pull off these amazing feats? Fortunately, there are answers; in fact, there are *many* answers. Myriad logico-mathematical devices have been suggested as candidates for how oracles accomplish their amazing work. In fact, just as there are an infinite number of mathematical devices equivalent to Turing machines (machines running programs from the language $\mathcal{L}$ visited above, Register machines, the λ-calculus, abaci, ...; these are all discussed in the context of an attempt to define computation in Bringsjord 1994), there are an infinite number of devices beyond the Turing Limit. As you might also guess, a small proper subset of these devices dominates the literature. In fact, three kinds of hypercomputational devices — analog chaotic neural nets, trial-and-error machines, and Zeus machines — are generally featured in the literature. We discuss only the latter two devices here.[11]

[11]Analog chaotic neural nets are characterized by Siegelmann & Sontag (1994). For cognoscenti, analog chaotic neural nets are allowed to have irrational numbers

Trial-and-error machines have their roots in a paper by Hilary Putnam (1965), and one by Mark Gold (1965); both appeared in the same rather famous volume and issue of the *Journal of Symbolic Logic*. So what *are* trial-and-error machines? Well, they have the architecture of Turing machines (read/write heads, tapes, a fixed and finite number of internal states), but produce output "in the limit" rather than giving one particular output and then halting. Here is a trial-and-error machine $\mathcal{M}$ that solves the halting problem. Take some arbitrary Turing machine m with input u; let $n^{m,u}$ be the Gödel number of the pair m, u; place $n^{m,u}$ on $\mathcal{M}$'s tape. Now have $\mathcal{M}$ print 0 immediately (recall the function h, defined above), and then have it simulate the operation of m on u. If $\mathcal{M}$ halts during the simulation, have it proceed to erase 0 in favor of 1, and then have it stop for good. It's as easy as that.[12]

Zeus machines (or "Weyl machines" from (Weyl 1949); see also Bertrand Russell's (1936) discussion of the possibility of his embodying such devices) are based on the character Zeus, described by Boolos & Jeffrey (1989). Zeus is a superhuman creature who can enumerate **N** *in a finite amount of time* (in one second, in fact). He pulls this off by giving the first entry, 0, in $\frac{1}{2}$ second, the second entry, 1, in $\frac{1}{4}$ second, the third entry in $\frac{1}{8}$ second, the fourth in $\frac{1}{16}$ second, ..., so that, indeed, when a second is done he has completely enumerated the natural numbers. Obviously, it's easy to (formalize and then) adapt this scheme so as to produce a Zeus machine that can solve the halting problem: just imagine a machine which, when simulating an arbitrary Turing machine m operating on input u, does each step faster and

for coefficients. For the uninitiated, analog chaotic neural nets are perhaps best explained by the "analog shift map," explained in Siegelmann (1995), and summarized in (Bringsjord 1998*c*) and Chapter 4 of the present book. There are many supermachines that don't fall (or at least don't *easily* fall) into these three categories. E.g., the movement of a read/write head, if "jazzed up," can yield supermachines; for two examples, see (Giunti 1997) and (Bringsjord 2001*b*). In a very interesting twist, Kugel (1986) explains that a supermachine (a **trial-and-error machine**) can be produced by simply changing how 'output' is understood: To "solve" the halting problem, let a TM — $m^{\star\star}$ — with n^m as input (Gödel number of arbitrary TM m) output **N** (for "No") immediately, and then let it simulate m. If m halts, $m^{\star\star}$ erases **N**, outputs **Y** (for "Yes") and halts itself. If we count $m^{\star\star}$'s *last output* as its output, the full halting problem is solved! However, compare this notion of output with that specified below for the programs of $\mathcal{L}$.

[12]For full exposition, along with arguments that human persons are trial-and-error machines, see (Kugel 1986), a seminal paper that situates trial-and-error machines nicely within both the formal context of the Arithmetic Hierarchy and the philosophical context of whether minds are computing machines.

faster ... (There are, as we reminded readers earlier, countably many Turing machines, and those that don't halt are trapped in an unending sequence of the same cardinality as **N**). If, during this simulation, the Zeus machine finds that m halts on u, then a 1 is returned; otherwise 0 is given.

1.9 An Alternative Characterization of Supermentalism

In this section, using the material presented in our primer on hypercomputation, we set out an array of predicates and operators that we exploit in order to express supermentalism, its competitors, and some additional "interesting" propositions concerning the mind and hypercomputation.

1.9.1 A Scheme for Expressing the Supermind Doctrine and Competitors

Please see Table 1.1 for a set of operators and predicates that can be used to express supermentalism and related theories. (Abbreviations are used, but should be easy enough to decipher, given our coming commentary.) The leftmost column in this table lists three main modal operators that prove useful in discussions of hypercomputation: $\diamond$ (logically possibly), $\diamond_p$ (physically possibly), and $\diamond_h$ (humanly possibly). The column "Computation Predicates" is based on the fact that computational information processing can be divided into the three categories that will by now be familiar to all of our readers: processing that can be carried out by machines having less power than Turing machines (designated by the predicate F, which is intended to suggest finite automata); processing that can be carried out by machines with the power of Turing machines and their equivalents (T); and, finally, processing that can be carried out by machines more powerful than Turing machines (O; selected in honor of oracles), for example, trial-and-error machines.

The rightmost column, "Partitioned Domain," simply presents the core of a sorted calculus allowing us to conveniently refer to mentations, computations, and persons. Alternatively, we could introduce predicates to range over these sub-domains.

The column "Engineeringish Predicates" is a bit trickier. Consider

Operators	Comp. Predicates	Engineeringish Preds	Partitioned Domain
$\Diamond$: log. poss.	F: FSA comp.	$\mathcal{H}^u$: uncon. harnessable	$m_1, m_2, \ldots$: mentations
$\Diamond_p$: phys. poss.	T: TM comp.	$\mathcal{H}^c$: con. harnessable	$c_1, c_2, \ldots$: computations
$\Diamond_h$: human. poss.	O: "Oracle" comp.	$\mathcal{A}$: actualizable	$p_1, p_2, \ldots$: persons

Table 1.1: Relevant Operators and Predicates

a finite automaton[13] F_1 designed to process strings over the alphabet $A = \{a_1, a_2\}$; specifically, F_1 is designed to accept the language L_1 composed of those strings having three consecutive a_1's. Let c denote a computation carried out by F_1. Obviously, we could build a physical device to incarnate F_1; this device could, say, be a toy railroad system whose track is divided into squares upon which either of the characters a_1 or a_2 can be written before F_1 begins to operate. We thus say that F_1 is *actualizable*; abbreviated: $\mathcal{A}F_1$. What about the two other predicate letters in the column in question? Well, suppose that we have actualized F_1, and suppose as well that we would like to investigate whether $a_1a_1a_2a_2a_1a_1a_1 \in L_1$. We say that F_1 can be *harnessed* because we can use the automaton to divine the answer (which is of course that the string is a member of L_1). More specifically, we say in this case that F_1 can be *consciously* harnessed, because a human can formulate a plan that he or she deliberately follows in order to secure the answer. On the other hand, sometimes the solution to a problem just "pops into one's head" through a process that isn't accessible to conscious thought. In this case we say that the process is *unconsciously* harnessable. Overall, what we verified in the case of F_1, where $O_1/O_2/\ldots/O_n$ means that any one O_i of the operators can be selected, is that

(11) $\Diamond_h/\Diamond_p/\Diamond\exists c(Fc \wedge \mathcal{A}c \wedge \mathcal{H}^c c)$.

1.9.2 Interesting Propositions Related to Hypercomputation

There are myriad propositions of great interest that can be expressed on the basis of Table 1.1, all of which will need to be investigated in a truly mature formal philosophy of hypercomputation. (The present book, we are happy to concede, is only an early step on the road toward this mature account.) From our standpoint, what is interesting, and ultimately profitable, is to reflect upon patterns created by the

[13] Readers who are unfamiliar with FAs can peek ahead to section 1.10.1.

various permutations of the machinery encapsulated in Table 1.1, and then to head to full-blown logics to capture these patterns, and their descendants. These new logics would of course inherit patterns from established intensional logics; such logics are part of the purely formal investigation of mentation recommended in the final chapter (see prescription P7 in particular). For example, it's because we know that $\Diamond_h \phi \rightarrow \Diamond_p \phi$ and $\Diamond_p \phi \rightarrow \Diamond \phi$ that we can assert (11). These new logics would also no doubt inherit much from physics. Here's a small fragment of the theses that would presumably be investigated in a mature philosophy of hypercomputation:

(12) $\Diamond / \Diamond_p \exists c Oc$ (These seem to be obviously true.)

(13) $\Diamond / \Diamond_p / \Diamond_h \exists c(\mathcal{A}c \wedge Oc)$

(14) $\Diamond / \Diamond_p / \Diamond_h \exists c(\mathcal{H}^c / \mathcal{H}^u c \wedge Oc)$

(15) $\forall c(\mathcal{A}c \rightarrow \mathcal{H}^c / \mathcal{H}^u c)$

Proposition (14) is, of course, at the heart of the present book. Supermentalism is committed, at its core, to suitable parameterizations of (14). We speak of parameterizing (14) because this proposition, like some of the others we've listed, is actually shorthand for a number of distinct theses. These theses can be specified by, first, making the conscious/unconscious parameter explicit, and then by selecting a modal operator. For example, here are two specific propositions that can be derived from (14):

(14^h_a) $\Diamond_h \exists c(\mathcal{H}^c \wedge Oc)$

(14^h_b) $\Diamond_h \exists c(\mathcal{H}^u c \wedge Oc)$

We believe, on the strength of arguments given in this book (and elsewhere; recall the Preface) that (14^h_b) is without question true. However, (14^h_a) is much less obvious to us. Bringsjord affirms this proposition (because he believes that some human persons can consciously carry out, in thought, supertasks; see Chapter 3); Zenzen is agnostic.

There are many other interesting propositions, most of which we haven't time and space to explore herein. For example, some readers might wonder whether, in the future, people will be able to purchase a "hypercomputation card" to insert into their personal computers, for (to put it mildly!) added power. What these readers are wondering about, in the scheme we've introduced, is whether it's humanly possible to build a consciously harnessable hypercomputer, that is, whether this proposition is true:

(16) $\Diamond_h / \Diamond_p / \Diamond \exists c(Oc \wedge \mathcal{A}c \wedge \mathcal{H}^c c)$.

Note that because the operators and predicates we've allowed ourselves also allow reference to mentations and persons, it's easy to express propositions that serve to encapsulate "Weak" AI, "Strong" AI (= computationalism), and the supermind view to which we subscribe. "Strong" AI can be identified with the proposition that

(SAI) $= \mathcal{C}\ \forall m \exists c(Tc \wedge m = c)$.

"Weak" AI can be be identified with the proposition that

(WAI) $\forall m \exists c(Tc \wedge m \approx c)$,

where '$\approx$' stands for the relation of *simulation*. Roughly put, the idea here is that all mentations can be perfectly simulated by Turing computation.[14]

Finally, as to our supermind view, it can be encapsulated by the following four propositions. The basic idea behind these propositions, as indicated by the previously introduced Figure 1.2 (wherein, again, the circle represents superminds) is that human persons comprise at least three "parts:" one part that engages in information processing at and below the Turing Limit, one part that engages in information processing above the Turing Limit, and one part that cannot be expressed in any third-person scheme whatsoever. This last part includes such things as subjective awareness and qualia.

(SUPER1) $\exists m \exists c(m = c \wedge Tc)$

(SUPER2) $\exists m \exists c(m = c \wedge Fc)$

(SUPER3) $\exists m \exists c(m = c \wedge Oc)$

(SUPER4) $\exists m \neg \exists c m = c$

The kernel of the supermind view, it should be emphasized, is based upon well-known material. Computation at the level of Turing machines and below is in large part set out in every comprehensive

[14] Someone might ask what "perfect simulation" amounts to. Ostensive definitions have been provided elsewhere. For example, Bringsjord & Ferrucci (2000) provide a robust example of a Turing machine-level simulation of the mentation involved in producing belletristic fiction. Simulation, in our sense, piggybacks on the sense of simulation firmly in use in computability theory; see, e.g., (Lewis & Papadimitriou 1981). For a discussion of simulation see section 2.3 in this book, and also (Bringsjord 2000*c*).

textbook on computability theory (e.g., Lewis & Papadimitriou 1981). The supermind view says that human persons can perform feats at this level. Information processing *above* the Turing Limit, as we've indicated, is also well-understood mathematically, and the supermind view includes the proposition that human persons can perform feats at this level. Finally, there are many well-known arguments for the position that human persons do things that can't be described in any symbolic scheme whatsoever (e.g., Bringsjord 1992, Searle 1992, Jacquette 1994).

1.10 Classifying Supermachines/Superminds...

In this section we briefly discuss hierarchies for progressively more powerful machines, supermachines, super-mechanical-minds (or, for short, supermechminds), and superminds. Please note that the next three subsections (sections 1.10.1–1.10.3) are dense, and optional. Readers should feel free to skip to section 1.10.4.

1.10.1 The Hierarchy from the Turing Limit Down

Table 1.2 shows the hierarchy for machines at and below the Turing Limit. As such, this table provides some of the details in one part of the "entire landscape" shown in Figure 1.1. Note that we have included reference to an alternative method of specifying the information processing in question: viz., the method of grammars. We will not discuss this alternative method here; interested readers should consult (Partee, Meulen & Wall 1990). We mention only that from the standpoint of grammars, the hierarchy shown in Table 1.2 is known as the Chomsky Hierarchy.

—	The Turing Limit	—	—
more power	Turing machines	Type 0 Grammars	more power
↑	Linear Bounded Automata	Type 1 Grammars	↑
↑	Push-Down Automata	Type 2 Grammars	↑
less power	Finite Automata	Type 3 Grammars	less power

Table 1.2: Hierarchy of Machines from the Turing Limit Down

We assume readers to be familiar with Turing machines. However, because the details of such machines are needed to articulate the argument of Chapter 3, in that chapter we provide a precise characterization of such machines (and readers unfamiliar with TMs are therefore encouraged to consult the start of that chapter now, and to return to the present point thereafter). Linear bounded automata (LBAs) are simply Turing machines, with the restriction that only a certain portion of the tape is usable; the number of usable squares is a linear function[15] of the length of the input (hence the name *linear* bounded automata). Pushdown automata are basically finite automata with a *stack* that operates in LIFO (last in, first out) fashion. Symbols can be stored in the stack, but the most recently added symbol must be the first to be removed. What about the least powerful class of machine in the hierarchy shown in Table 1.2? Finite automata (FAs) are simple. Like Turing machines, they have a tape at their disposal, but it's only an **input tape**. FAs have a **finite control**, which is at all times in one of a finite number of discrete states $\{q_1, \ldots, q_n\}$, and can somehow sense what symbol is on a square by way of a movable **reading head**. When processing starts, this head is assumed to be positioned on the leftmost square of the input tape, and the FA is in its **initial state** $q^\star = q_i$, $1 \leq i \leq n$. FAs work by simply reading the symbol on the currently scanned square of the tape, and entering a new state based solely on the scanned symbol and the present state. In light of this, FAs can be described by simply presenting their **transition function** τ. As an example, here is a finite automaton with two states q_1 and q_2 that accepts the subset of $\{a, b\}^*$ composed of those strings having an even number of b's:

$$\begin{aligned}\tau(q_1, a) &= q_1\\ \tau(q_1, b) &= q_2\\ \tau(q_2, a) &= q_2\\ \tau(q_2, b) &= q_1\end{aligned}$$

From the standpoint of someone trying to place persons within the entire landscape, FAs and PDAs can seem rather uninteresting. However, most of the machines to which humans ascribe intelligence are simply implemented FAs and PDAs. For example, most of the so-called "non-player characters" in current computer games are FAs and PDAs. No doubt many primitive biological creatures are ultimately nothing more than FAs and PDAs. We will have more to say about this later in this chapter, and again in the final chapter of the book.

---

[15]Where a and b are numbers, $f(x)$ is such a function if it's of the form $ax + b$.

1.10.2 Additional Needed Concepts

In order to move above the Chomsky Hierarchy and present the Arithmetic Hierarchy for classifying problems beyond the reach of Turing machines, some concepts beyond those given in connection with the aforementioned programming language $\mathcal{L}$ from (Davis, Sigal & Weyuker 1994) are needed. In notation, we continue to follow (Davis et al. 1994).

The first thing we need to do is make the concept of a standard **computation** clear. (By *standard* computation we mean computation at or below the Turing Limit.) First, note that we give inputs to a program $\mathcal{P}$ written in $\mathcal{L}$ by making assignments of numbers to the variables manipulated by $\mathcal{P}$. (Examine the program given earlier that computes the function $f(x_1, x_2) = x_1 + x_2$. You will see that the variables in this case are X_1, X_2, Y, and Z.) We say that a **state of a program** $\mathcal{P}$ is simply a list of equations of the form

$$V = m,$$

where V is a variable and m is a number, where an equation is included for every variable occurring in $\mathcal{P}$. For another example, consider a program that computes the function $f(x) = x$ by copying the value of X into Y, for all initial values of X:

$$\begin{array}{ll} [A] & \text{IF } X \neq 0 \text{ GOTO } B \\ & Z \leftarrow Z + 1 \\ & \text{IF } Z \neq 0 \text{ GOTO } E \\ [B] & X \leftarrow X - 1 \\ & Y \leftarrow Y + 1 \\ & Z \leftarrow Z + 1 \\ & \text{IF } Z \neq 0 \text{ GOTO } A \end{array}$$

The list

$$X = 4, Y = 3, Z = 3$$

is a state of this program, but the list

$$X = 3, Z = 3$$

isn't, since no equation for Y is included.

Now, where σ is a state of a program $\mathcal{P}$ and V is a variable occurring in σ, we say that the **value of V at** σ is the (unique) number q where $V = q$ is one of the equations in σ. Given a program $\mathcal{P}$ of length

n and a state σ of this program, we need to say what the next state is. In order to make this precise, Davis et al. (1994) invoke the concept of a **snapshot**, which is simply a pair (i, σ) where i is between 1 and n, and σ is a state of $\mathcal{P}$. The number i simply gives the line number that is about to be executed. To completely formalize things, the **successor** (j, τ) of a given snapshot (i, σ) can be defined by settling each possible case, where the cases are limited to the type of instruction at line i. We leave aside the trivial details here. The important thing is that we can now define a **computation** of a program $\mathcal{P}$ to be a sequence or list $s_1, s_2, \ldots, s_k$ of snapshots of $\mathcal{P}$ such that s_{i+1} is the successor of s_i, and s_k is terminal. (A snapshot (i, σ) is terminal provided that $i = n + 1$.)

Next, recall the introduction of oracles into the language $\mathcal{L}$; we covered this introduction above, in section 1.8. When $G(m)$ is defined (written $G(m) \downarrow$; an up arrow indicates 'undefined'), execution of an oracle statement in $\mathcal{L}$ answers the question "G(m) = ?". When $G(m) \uparrow$ no terminal snapshot is reached. Of course, given that G is a total function, every nonterminal snapshot has a successor.

But how do we render the notion of a program $\mathcal{P}$ **computing a function** f precise? In the case of Turing machines — rather than the programs, following (Davis et al. 1994), we have been talking about — such a machine m computes a function f just in case, when the argument to f is placed upon m's tape at the start of processing, m leaves $f(x)$ on the tape after halting (certain niceties do have to be specified; see Boolos & Jeffrey 1989). How does it work for programs in $\mathcal{L}$? The idea is fundamentally the same. We have only to regard a variable Y in a program $\mathcal{P}$ to be the output variable. We write

$$\psi_{\mathcal{P}}^{(m)}(r_1, r_2, \ldots, r_m)$$

to refer to the value of ("output") variable Y at terminal snapshot s_k. Of course, it's possible that there be an unending sequence $s_1, s_2, \ldots$, and in this situation

$$\psi_{\mathcal{P}}^{(m)}(r_1, r_2, \ldots, r_m)$$

is undefined. Now, we say that g, some m-ary function, is **partially computed** if and only if there exists some program $\mathcal{P}$ such that

$$g(r_1, r_2, \ldots, r_m) = \psi_{\mathcal{P}}^{(m)}(r_1, r_2, \ldots, r_m)$$

for all $r_1, r_2, \ldots, r_m$. A function g is **computable** provided that it's both partially computable and total. Partially computable functions are also often called **partial recursive**, and computable functions are often termed **recursive**.

It's now easy to adapt the standard notion of computation and associated concepts in light of oracles; we still follow the elegant (Davis et al. 1994). A **G computation** is defined as a computation, except that 'successor' is replaced with ***G*-successor**. In keeping with our earlier account of oracles, we say that the number m replaced with $G(m)$ in a G-computation is an **oracle query of the *G*-computation**. The expression

$$\psi_{\mathcal{P},G}^{(m)}(r_1, r_2, \ldots, r_m)$$

is defined as is its standard predecessor, except that we are now talking about ***G*-computation**. If G is a total function, then the partial function

$$\psi_{\mathcal{P},G}^{(m)}(r_1, r_2, \ldots, r_m)$$

is said to be ***G*-computed** by $\mathcal{P}$. A partial function g is **partially *G*-computable** or ***G*-partial recursive** exactly when there is a program that G-computes it. Partially G-computable functions, when total, are deemed to be simply ***G*-computable** or ***G*-recursive**. At this point we have adapted the basic scheme to account for oracles. But in order to present the Arithmetic Hierarchy, we first need to explain how the basic scheme applies to predicates and sets.

A **predicate** on a set S is simply a total function mapping S to {TRUE, FALSE} – but customarily we replace this range with $\{0, 1\}$, where 1 denotes TRUE and 0 FALSE. Given this, the truth-functional predicates can be applied to predicates in the expected ways. For example, where P and P' are predicates on some set S, the predicate $P \wedge P'$ is true iff both P and P' are true. Where R is a subset of a set S, the expression $x \in R$ defines this predicate on S:

$$P(x) = \begin{cases} 1 & \text{if } x \in R \\ 0 & \text{if } x \notin R \end{cases}$$

Here, P is the **characteristic function** of the set R.

Now, we can say that a set B that is a subset of $\mathbf{N}$, the natural numbers, is **recursively enumerable** (or just r.e.) exactly when there exists some partially computable function g such that

$$B = \{x \in \mathbf{N} : g(x) \downarrow\},$$

i.e., B is r.e. just in case, where $\mathcal{P}$ computes g, B is the set of all inputs to $\mathcal{P}$ on which this program eventually halts. You can conceive of $\mathcal{P}$ as a **semi-decision procedure** for B: when $x \in B$, $\mathcal{P}$ returns a **Y** for "Yes." Given this it's easy to fix what it is for a set S to be **recursive**: such a set, *and* its complement, are r.e. In other words, S in this case is such that some program $\mathcal{P}$ produces **Y** when $x \in S$, *and* can produce **N** when $x \notin S$. It's easy enough to say what it is for a set to be recursively enumerable or recursive relative to an oracle: We say that a set $B \subseteq \mathbf{N}$ is ***G*-r.e.** iff there is a partially G-computable function g that yields a definite output (interpreted as "Yes") for each member of B. A ***G*-recursive** set B is defined in the expected manner: there is an oracle-augmented program $\mathcal{P}$ that provides a complete decision procedure for B.

We turn now to the Arithmetic Hierarchy, a classification of Turing-uncomputable problems. This classification scheme allows one to speak of a hierarchy of increasingly powerful supermachines. In addition, the Arithmetic Hierarchy is at the heart of our argument (in Chapter 4) for the view that Church's Thesis is false in light of our "supercognitive" ability to judge narrative.

1.10.3 The Arithmetic Hierarchy

Σ_0 is the set of recursive sets. For every $n \in \mathbf{N}$, Σ_{n+1} is the set of sets that are A-r.e. for some set $A \in \Sigma_n$. Furthermore, for all $n \in \mathbf{N}$, $\Pi = \text{co-}\Sigma_n$, and $\Delta_n = \Sigma_n \cap \Pi_n$ (where W is a set of subsets of $\mathbf{N}$, co-W is the set of all subsets A of $\mathbf{N}$ where the complement of A is in W). By Kleene's Hierarchy Theorem (Davis et al. 1994, pp. 216–217) we have

$$\Delta_n \subset \Sigma_n, \Delta_n \subset \Pi_n;$$

$$\Sigma_n \subset \Sigma_{n+1}, \Pi_n \subset \Pi_{n+1};$$

$$\Sigma_n \cup \Pi_n \subset \Delta_{n+1}.$$

The set-theoretic presentation of the Arithmetic Hierarchy (AH) is not the one we prefer. The predicate-and-quantifier-based presentation is one which, in Bringsjord's experience, makes more intuitive sense for those with training in symbolic logic. We proceed now to this kind of presentation.

To start, remember that each predicate $P(x_1, \ldots, x_p)$ is associated with the set

$$A = \{x \in \mathbf{N} : P((x)_1, \ldots, (x)_p)\}.$$

Now suppose we have some **totally computable** predicate $S(P, u, n)$ iff TM m, running program P on input u, halts in exactly n steps (=

$M_P : u \rightarrow_n$ halt). (Recall that throughout AH our machines, architecturally speaking, are always simply TMs.) Predicate S is totally computable in the sense that, given some triple (P, u, n), there is some program $P^\star$ which, running on some TM $m^\star$, can infallibly give us a verdict, **Y** ("yes") or **N** ("no"), for whether or not S is true of this triple. ($P^\star$ could simply instruct $m^\star$ to simulate m for n steps and see what happens.) This implies that $S \in \Sigma_0$, i.e., that S is a member of the starting point in AH, a point composed of totally computable predicates. But now consider the predicate H, defined by

$$H(P, i) \text{ iff } \exists n S(P, i, n).$$

Since the ability to determine, for a pair (P, i), whether or not H is true of it, is equivalent to solving the full halting problem, we know that H is not totally computable. Hence $H \notin \Sigma_0$. However, there is a program which, when asked whether or not some TM m run by P on u halts, will produce **Y** iff $M_P : u \rightarrow$ halt. For this reason H is declared **partially computable**, and hence in Σ_1. To generalize, informally, the quantifier-based representation of AH is:

Σ_n set of all predicates definable in terms of totally computable predicates using at most n quantifiers, the first of which is *existential*

Π_n set of all predicates definable in terms of totally computable predicates using at most n quantifiers, the first of which is *universal*

Δ_n $\Sigma_n \cap \Pi_n$

This is of course very informal. To put things a bit more precisely, we follow (Davis et al. 1994) and call a predicate $P(x_1, \ldots, x_p)$ **arithmetic** iff there is a recursive predicate $R(x_1, \ldots, x_p, y_1, \ldots, y_n)$ such that

$$\text{(F)} \quad P(x_1, \ldots, x_p) \leftrightarrow Q_1 y_1 Q_2 y_2 \cdots Q_n y_n R(x_1, \ldots, x_p, y_1, \ldots, y_n)$$

Each of the Q_i in (F) is either the universal or existential quantifier. The Q_i are **alternating** if when Q_i is $\exists$, then Q_{i+1} is $\forall$, and vice versa. It is not all that difficult to prove that

- A predicate is Σ_n iff it can be represented in the form (F) with $Q_1 = \exists$ and the Q_i alternating.
- A predicate is Π_n iff it can be represented in the form (F) with $Q_1 = \forall$ and the Q_i alternating.

Readers are encouraged to verify that the sample predicates S and H given above correspond to this scheme.

The Arithmetic Hierarchy allows us to classify unsolvable problems. But such a classification in turn allows us to devise a hierarchy of supermachines: the supermachines able to "solve" problems in higher and higher parts of AH. Full treatment of these matters is beyond the scope of this book, but we provide two examples:

Define

$$\Phi^{(n)}(x_1, \ldots, x_n, y) = \Phi_{\mathcal{P}}^{(n)}(x_1, \ldots, x_n),$$

where y is the Gödel number of program $\mathcal{P}$. Now consider the set

$$\text{TOT} = \{y \in \mathbf{N} : \forall x \Phi^{(1)}(x, y) \downarrow\}.$$

This set is composed of the set of all Gödel numbers of programs that compute total functions. We introduce now the predicate

$$\text{STP}^{(n)}(x_1, x_2, \ldots, x_n, y, t)$$

defined by

> Program with Gödel number y halts after t or fewer steps on inputs $x_1, x_2, \ldots, x_n$.

It should be clear that the predicate STP is computable. But note that the set TOT is

$$\{y \in \mathbf{N} : \forall x \exists t \text{STP}^{(1)}(x, y, t)\},$$

which by form (F) allows us to immediately conclude that TOT $\in \Pi_2$. But while no machine at or below the Turing Limit can solve TOT, a Zeus machine $\mathcal{Z}$ would have no trouble deciding TOT. As is well-known, given a Gödel number y and an input x, even an ordinary TM m can simulate the operation of the TM m^y on x (e.g., Lewis & Papadimitriou 1981). The challenge here is to ascertain whether or not, for a given y, the TM m^y halts after t or fewer steps, with x as input. $\mathcal{Z}$ can simply simulate m^y, but with each step taking less time than that taken up by its predecessor. $\mathcal{Z}$ will also not be troubled by the number of possible inputs, because (universal) quantification here is only over a countably infinite set of possible inputs. Figure 1.3 points the way toward the procedure for $\mathcal{Z}$ to succeed. $\mathcal{Z}$ dovetails two countably infinite sets, the set of inputs (lexicographically ordered, given the underlying finite alphabet) and the set of Turing machines.

At each stop, processing — signaled by a • — takes less time than the previous step, as indicated. But note that the processing at each step may itself be infinite. If the m_i x_i pair simulated at a particular step halts, then $\mathcal{Z}$'s infinitary powers will be superfluous at that step; but if m_i doesn't halt on x_i, then $\mathcal{Z}$ will need to do a "supertask" simulation at this step. The supertasks will need to be suitably timed. For example, if a step at $m_i x_i$ in Figure 1.3 is slated to take $\frac{1}{512}^{\text{th}}$ of a second, then the first step of simulating m_i on x_i can take $\frac{1}{1024}^{\text{th}}$ of a second, the second can take $\frac{1}{2048}^{\text{th}}$ of a second, and so on *ad infinitum*. If at every step $\frac{1}{n}$ in Figure 1.3 $\mathcal{Z}$ embarks on its simulation with a first move taking $\frac{1}{2n}^{\text{th}}$ of a second, $\mathcal{Z}$ will have a lot of "free time" even in cases where the Turing machine in question is a non-halter. That is no matter. $\mathcal{Z}$ can wait idly until the time set aside for each step in Figure 1.3 is consumed. Finally, it's worth noting that $\mathcal{Z}$ can solve many famous unsolvable problems, because a number of such problems are reducible to the halting problem. For example, $\mathcal{Z}$ can determine, given some first-order formula ϕ, whether or not it's a theorem (see the reduction given in Ebbinghaus et al. 1984). For another example, $\mathcal{Z}$ can solve unsolvable tiling problems (see the reduction given in Lewis & Papadimitriou 1981).

	x_1		x_2		x_3	$\cdots$
m_1	$\frac{1}{2}$ •	→	$\frac{1}{4}$ •		$\frac{1}{32}$ •	$\cdots$
			↓		↓	$\cdots$
m_2	$\frac{1}{16}$ •	←	$\frac{1}{8}$ •		$\frac{1}{64}$ •	$\cdots$
					↓	$\cdots$
m_3	$\frac{1}{512}$ •	←	$\frac{1}{256}$ •	←	$\frac{1}{128}$ •	$\cdots$
$\vdots$	$\vdots$	$\vdots$	$\vdots$	$\vdots$	$\vdots$	

Figure 1.3: Dovetailing Procedure for Zeus Machine $\mathcal{Z}$ That Solves TOT

Here is a second example. Define

$$W_n = \{x \in \mathbf{N} : \Phi^{(1)}(x, n) \downarrow\}.$$

The set

$$\text{COF} = \{x \in \mathbf{N} : \bar{W}_x \text{is finite}\}$$

is the same as

$$\{x \in \mathbf{N} : \exists n \forall k (k \leq n \vee k \in W_x)\}.$$

But then by conformity to (F) we can immediately infer that COF is in Σ_3. We leave it to motivated readers to devise a supermachine able to decide COF.

1.10.4 Zombies and Superminds: a Hierarchy

Recall that earlier in the chapter we explained both the centrality of personhood to the present project, and — by our lights, anyway — the essence of personhood. (You may specifically recall that we sought to encapsulate this essence by way of the list L_p.) We are somewhat reluctant to call something a 'mind' if it isn't a person, but for the sake of exposition we can temporarily shed this attitude. So, we will countenance talk of of cat minds and dog minds and horse minds and chimp minds, and so on. However, we do refuse to count something as a mind unless it is conscious. This means that at least the higher mammals are candidates (under our more relaxed reading of 'mind') for being classified as minds, whereas a single-cell organism isn't — so at some point between the two there is breakage. On the other hand, single-cell creatures display determinate behavior, and one can certainly consider how such creatures can be replicated as mechanical artifacts.

To make things a bit more definite, we assume that biological animals fall across a spectrum. (In the final chapter, we will specifically imagine that this spectrum is isomorphic to some initial sequence of the natural numbers.) The first animal, b_1, would perhaps be a simple single-cell organism, then we would move on to a more sophisticated single-cell organism b_2, to $\ldots$, to a multi-cellular organism b_k, to a more sophisticated multi-cellular organism b_{k+1}, to $\ldots$, then perhaps up to an ant b_{k+p}, to $\ldots$, to (say) a parrot, to $\ldots$, and eventually up to the animals that would seem to be just "beneath" human persons: apes and the like. The entire spectrum so far would then be

$$b_1, b_2, \ldots, b_m = \text{ape}.$$

We assume that every creature up to this point can be replicated in the form of mechanical artifacts operating at or below the Turing Limit. Where b_i is a (possibly mindless) biological creature, the mechanical correlate to it produced by AI is denoted by m^{b_i}. Mechanical correlates here are in other words **zombies** — or, since they are artificial versions of *animals* that aren't conscious, we can refer to these correlates as **zombanimals**. (We discuss zombanimals in more detail in the final chapter. A full discussion can be found in Bringsjord 2000*a*.)

It is now possible to provide a picture of the overall hierarchy: see Table 1.3. We say 'overall,' because this hierarchy includes zombanimals, zombies, superzombies (= supermachines), mindless biological creatures, "minded" biological creatures, and superminds — where this last category includes human persons. You will notice that we show therein the breakage to *bona fide* minds at some point, that persons are shown at a point above the Turing Limit and in the Arithmetic Hierarchy, that we assume there to be a Maximal Zombie (a maximally powerful information processing entity; that is, a maximally powerful supermachine), and that we assume there to be a Maximal Mind (which has all the powers of the Maximal Zombie, and then some: recall our argument, given earlier in this chapter, for powers that result from phenomenal consciousness).

We end this section by pointing out that our readers may find it profitable to study the hierarchy shown in Table 1.3 in connection with some work carried out by others. Specifically, readers are advised to consult (Steinhart forthcoming), where information processing is considered in a framework more "physics oriented" than the one underlying the present chapter. Steinhart (forthcoming) concludes that there is a Maximal Mind, but doesn't distinguish between mere information processing and mental powers deriving from what cannot be expressed in any third-person scheme.[16]

1.11 Previewing What's To Come

We now present the kernel of each chapter to come. Note that none of these kernels can be judged in and of themselves: in each case only the fully developed and defended arguments are compelling.

[16]Steinhart attempts the reduction of the physical to the mathematical via the sequence: physical fields are functions from space-time points to n-tuples of numbers, space-time is an n-dimensional manifold of points, regions of space-time are classes of points, points are n-tuples of numbers, and numbers are pure classes. At bottom, this is the well-known Quinean research program. We do not deny the heuristic value of this approach, but a methodology does not comprise a metaphysics. The inherent Pythagorean commitment of Steinhart's reduction needs to be measured against the alternatives of platonism, nominalism, and naturalism as alternatives. Moreover, it is problematic whether physical fields are an adequate representation of current physical theory.

narrative. After the foundation is in place, we press forward in the hope, indeed, of refuting CT. We end by considering some other attacks on CT (some of which, as we shall see, Mendelson himself tried to derail over forty years ago), and the relation between these attacks and our own.

Our attack on Church's Thesis is an instantiation of a certain schema — one which makes reference to the Arithmetic Hierarchy introduced above. It is an easy and well-known theorem that successfully deciding some set A can be adapted to successfully enumerate A; the schema in question, $\mathbf{Arg}_4$, exploits this elementary theorem, a move that is inspired by Peter Kugel (Kugel 1986). In order to refute Church's Thesis one needs to set A (in the schema) to some set that makes all the premises in question true. We think the set of all "interesting" stories, $\mathcal{S}$, fits the bill, as we explain in Chapter 4. The core of this argument is really quite straightforward: People who can *decide* $\mathcal{S}$, that is, people who can decide whether something is an interesting story, can't necessarily *generate* interesting stories. Put baldly, it's easy to see that stories like *King Lear* are interesting, but it's not so easy to *write* drama of this caliber. Here's the schema:

$\mathbf{Arg}_4$

	(17)	If $A \in \Sigma_1$ (or $A \in \Sigma_0$), then there exists a procedure P which adapts programs for deciding members of A so as to yield programs for enumerating members of A.	
	(18)	There's no procedure P which adapts programs for deciding members of A so as to yield programs for enumerating members of A.	
∴	(19)	$A \notin \Sigma_1$ (or $A \notin \Sigma_0$).	MT 10,11
	(20)	$A \in$ AH.	
∴	(21)	$A \in \Pi_1$ (or above in the AH).	DS on AH
	(22)	A is effectively decidable.	
∴	(23)	CT is false.	RAA

1.11.4 Chapter 5: The Zombie Attack on Computationalism

Many proponents of the view that cognition is computation are busy trying to practice what they preach: they are trying to build artificial persons. Two such proponents are the philosophers John Pollock and Daniel Dennett. In two of his books, *How to Build a Person*

(Pollock 1989) and *Cognitive Carpentry: A Blueprint for How to Build a Person* (Pollock 1995), Pollock argues that in the future his OSCAR system will be a full-fledged person. For Dennett, the person-to-be is the robot COG, or a descendant thereof, a being taking shape with Dennett's help at MIT. What sort of artifacts will these creatures be? We offer an unflattering one-word response: Pollock, Dennett, and like-minded researchers are busy building ... *zombies*. We refer to *philosophers'* zombies, not those creature who shuffle about half-dead in the movies.[17] We also hold — for reasons to be specified in Chapter 5 — that the mere *possibility* of zombies is enough to explode the computational conception of mind.

1.11.5 Chapter 6: The Argument from Irreversibility

We refute computationalism in Chapter 6 via specification and defense of the following new argument: Computation is reversible; cognition isn't; ergo, cognition isn't computation. The specification of the argument involves a quartet:

(i) certain elementary theorems from computability theory, according to which computation is reversible;

(ii) the doctrine of agent materialism, according to which, contrary to any sort of dualistic view, human agents (= human persons) are physical things whose psychological histories are physical processes;

(iii) the introspection- and physics-supported fact that human cognition is *not* reversible; and

(iv) the claim — fundamental to AI and Cognitive Science, and put roughly for now — that cognition is computation.

The basic structure of the argument is straightforward: the conjunction of (i), (ii) and (iii) entails the falsity of (iv).

1.11.6 Chapter 7: What, at Bottom, are Superminds? Where'd They Come From?

In this chapter we discuss two questions: "But what, at bottom, are we?" and "How'd we get here?". The first of these questions is short

[17]Actually, the zombies of cinematic fame apparently have real-life correlates created with a mixture of drugs and pre-death burial: see (Davis 1985), (Davis 1988).

for a query regarding which ontology of mind supermentalism is to be alligned with. We follow Jacquette (1994) in setting out these ontologies (which at the highest level fall into three basic categories: **dualism**, **eliminativism**, and **reductivism**), and work by the process of elimination over them to arrive at the view that superminds are either physical things that have non-physical properties that cannot be reduced to behavioral, material, or functional properties (**property dualism**), or they are literally non-physical objects (**substance** or **agent dualism**). One of us (Zenzen) is inclined to accept an explicitly emergentist version of property dualism, while the other is at least very sympathetic toward agent or substance dualism. We defend in Chapter 7 the Chisholmian brand of agent dualism known as the "simple souls" view. We show that Chisholm's position fares very well when matched against the fascinating view, recently propounded by Andy Clark and David Chalmers Chalmers, D. (1998), that persons are not only embodied, but *extensibly* emobodied. Chisholm's simple souls position, for reasons we explain, is in line with the Chisholmian pre-analytic data on persons explicitly set out in Bringsjord's *What Robots Can and Can't Be*, and in the present book (see section 1.4).

The received answer to the "How'd we get here?" question is "Evolution," and the most eloquent proponent of this answer is Steven Pinker (1997) (with Daniel Dennett 1995*b* coming in a worth-considering second), who states and defends it at great length in his landmark *How the Mind Works* (1997). We provide encapsulated versions of two arguments against the Pinkerian view that we are evolved machines, and mention the only other plausible answers to the "How'd we get here?" question.

1.11.7 Chapter 8: Supermentalism and the Practice of AI/CogSci

In the final chapter we very briefly discuss the consequences of supermentalism for the *practice* of AI and CogSci. Accordingly, this chapter bridges from the present book to the final volume in what we've called the Supermind Quartet (recall section 0.2). The chapter ends with eight methodological prescriptions for AI and CogSci carried out in the light of supermentalism.

Chapter 2

A Refutation of Penrose's Gödelian Case

2.1 Introduction

Those who go in for attacking AI can be indefatigable. John Searle tirelessly targets at least one brand of AI ("Strong" AI) with variations on his Chinese Room. (Searle fired the first shot in (Searle 1980). Two decades later, CR is alive and well, as the arrival of a new book (Bishop & Preston 2002) reveals.). Bringsjord has at last count published 16 formal arguments against "Strong" AI. (For a sampling, recall Table 1.) And Roger Penrose appears to have the same endless energy when it comes to producing Gödelian attacks on AI: Having, as it is generally agreed, failed to improve on Lucas' at-best-controversial primogenitor (Lucas 1964) with the argument as formulated in his *The Emperor's New Mind* (*ENM*) (Penrose 1989),[1] Penrose returned, armed with a new Gödelian case, expressed in his *Shadows of the Mind* (*SOTM*) (Penrose 1994). This case, unlike its predecessor, does more than recapitulate Lucas' argument, but it nonetheless fails, as we shall see. The great irony is that this case is based on Penrose's near-deification of logico-mathematical reasoning, but such reasoning, as we show herein, can be used to refute Penrose's case.

The heart of *SOTM*'s Chapter 2 is a diagonal argument designed to show that there is no "knowably sound" algorithm for classifying computations as non-halters. About this diagonal argument Penrose

[1] For an explanation of why standard Gödelian attacks fail, see "Chapter VII: Gödel" in (Bringsjord 1992).

says:

> Admittedly there is an air of the conjuring trick about the argument, but it is perfectly legitimate, and it only gains in strength the more minutely it is examined. (Penrose 1994, p. 75)

Unfortunately, we *have* examined the argument minutely, and Penrose is stone cold wrong: at best, it's enthymematic, and when formalized, a remarkable number of technical glitches come to light. Over and above these defects, the argument, even when viewed charitably, is an instance of either the fallacy of denying the antecedent, the fallacy of *petitio principii*, or the fallacy of equivocation.

In Chapter 3, Penrose (working under the assumption that the argument of Chapter 2 is sound) tries to rule out the remaining possibility: viz., that there *is* an algorithm, but not a knowably sound one, for classifying computations as non-halters. Here again he fails — and once more the problems are formal in nature.

More recently, writing in response to his critics in the electronic journal *Psyche* (Penrose 1996), Penrose has offered a Gödelian case designed to improve on the version presented in *SOTM*. But this version is yet again another failure.

In falling prey to the errors we uncover, Penrose's new Gödelian case is unmasked as the same confused refrain J.R. Lucas initiated 35 years ago. The only redeeming aspect of this refrain is an intuition at its core — one we discuss at the end of the present chapter.

Our plan for this chapter is as follows. In section 2.2 we set out the main foundational positions on AI from Penrose's perspective, chief among which are Penrosean versions of "Strong" and "Weak" AI. In section 2.3 we explain why "Weak" AI is pretty much invulnerable, and therefore Penrose's target should be "Strong" AI only. In section 2.4 we review the mathematical background presupposed by Penrose's Gödelian case. Section 2.5 covers the core diagonal argument for this case. In section 2.6 we review the formal machinery we use to expose the invalidity of this diagonal argument, in section 2.7 we formalize this argument, and in section 2.8 we explain why the argument is fallacious. Section 2.9 is devoted to considering and rebutting replies on behalf of Penrose. In section 2.10 we show that even if the diagonal argument is sound, Penrose's overall case fails. In section 2.11 we give Penrose a last chance: we consider the version of his Gödelian case given in the electronic journal *Psyche* in an attempt to "mop up" defects in the version featured in *Shadows of the Mind.* We sum up the fix Penrose has gotten himself into in section 2.12, and briefly discuss there the future of Gödelian attacks on computationalism — a discussion that includes, specifically, the recent claim (LaForte, Hayes & Ford 1998) that no such attack can *possibly* work. (Bringsjord is nearly done with a Gödelian case that *does* work, and will in due course publish it. See Table 1.) Finally, in section 2.13, we charitably distill a promising core intuition from Penrose's failed argumentation. The next chapter can be viewed as a victorious specification of this intuition in the form of a sound deductive argument against computationalism from mathematical reasoning — but as we show, the reasoning isn't Gödelian *per se*; rather, in a word, it's simply *infinitary*.

2.2 The Main Positions on AI

Penrose begins by setting out in propositional form what he sees as the four fundamental positions on AI (p. 12, Penrose 1994):

$\mathcal{A}$ All thinking is computation; in particular, feelings of conscious awareness are evoked merely by the carrying out of appropriate computations.

$\mathcal{B}$ Awareness is a feature of the brain's physical action; and whereas any physical action can be simulated computationally, computational simulation cannot by itself evoke awareness.

$\mathcal{C}$ Appropriate physical action of the brain evokes awareness, but this physical action cannot even be properly simulated computationally.

$\mathcal{D}$ Awareness cannot be explained by physical, computational, or any other scientific terms.

$\mathcal{A}$ is intended to encapsulate "Strong" AI; put in terms of future robots and the *Total* Turing Test,[2] the thesis boils down to the claim that robots able to pass TTT will arrive, and will moreover have full-blown conscious mental states. $\mathcal{B}$ is supposed to encapsulate "Weak" AI; again, put in terms of future robots, the idea is that TTT-passing robots are headed our way, but despite impressive *behavior*, they will lack consciousness: they will be zombies. $\mathcal{C}$ is Penrose's position; and

[2]Recall our discussion of TT and TTT in Chapter 1. The straight Turing Test, again, tests only for linguistic performance. Harnad (1991)'sTTT, on the other hand, requires that the human and robot (or android) players compete across the full range of behavior. For example, the judge in TTT can ask questions designed to provoke an emotional response, and can then observe the facial expressions of the two players. We discussed all of this, of course, in Chapter 1.

$\mathcal{D}$ is what Penrose calls the "mystical" stance, one apparently affirmed by, among others, Kurt Gödel. (Bringsjord doesn't think of himself as anything even *close* to a mystic, but does affirm $\mathcal{D}$ — as our presentation of supermentalism in Chapter 1 made plain.)

Penrose's four-fold breakdown, upon reflection, is disturbingly imprecise. (As you will recall, in Chapter 1 we gave our own versions of Strong and Weak AI, to which readers are encouraged to compare Penrose's versions.) The main problem is that $\mathcal{B}$ needlessly unites *three* seemingly separate claims, viz.,

$\mathcal{B}_1$ Awareness is a feature of the brain's physical action.

$\mathcal{B}_2$ Any physical action can be simulated computationally.

$\mathcal{B}_3$ Computational simulation cannot by itself evoke awareness.

Some thinkers deny $\mathcal{B}_1$ but affirm $\mathcal{B}_2$. Indeed, Bringsjord is such a specimen. At the moment, as Chapter 7 in the present book reveals, he tends to believe that substance or property dualism is true (and hence is agnostic on $\mathcal{B}_1$), but he wholeheartedly affirms $\mathcal{B}_2$. It seems possible that someone could coherently hold to

$$\mathcal{B}_2 \wedge \neg\mathcal{B}_3 \wedge \mathcal{A}$$

as well, but in the interests of economy we leave this possibility aside. As to $\mathcal{B}_2$ itself, this thesis entails

$\mathcal{B}_2^M$ Physical action relevant to mentation can be simulated computationally.

But there might well be thinkers who affirm $\mathcal{B}_2^M$ but reject the stronger $\mathcal{B}_2$.

For the remainder of this chapter, in a nod in Penrose's direction, let's identify "Weak" AI with $\mathcal{B}_2^M$. Part one of *SOTM* is a sustained argument for

$$\neg(\mathcal{A} \vee \mathcal{B}_2^M).$$

Given this, it follows by propositional logic that if $\mathcal{A} - \mathcal{D}$ exhaust foundational takes on AI, and if the niceties we've noted in connection to $\mathcal{B}$ are ignored,

$$\mathcal{C} \vee \mathcal{D}.$$

If we assume with Penrose that the mystical $\mathcal{D}$ is unacceptable to nearly all scientists and engineers, we're left with $\mathcal{C}$ by disjunctive syllogism; and Penrose spends the second part of *SOTM* exploring and explaining the "uncomputable" physics needed (given $\mathcal{C}$) in order to explain consciousness. Obviously, if the argument of part one fails, part two is little more than a curiosity.

2.3 Why "Weak" AI is Invulnerable

Though Penrose focuses on human mathematical reasoning of a most abstract and esoteric sort (witness the example pertaining to hexagonal numbers discussed in the next section), and though the $\mathcal{A}$–$\mathcal{D}$ quartet is itself rather abstract, there is a firm connection between this reasoning and "Strong" AI: If people are computing machines and cognition is computation, then mathematical reasoning, however far removed it may or may not be from "everyday" cognition, must in the end *be* computation. As Penrose points out:

> It might well be argued that the building of a robot mathematician is very far from the immediate aims of AI; accordingly, the finding of such an F [= a theorem-proving machine on par with human mathematicians] would be regarded as premature or unnecessary. However, this would be to miss the point of the present discussion. Those viewpoints which take human intelligence to be explicable in terms of algorithmic processes implicitly demand the potential of such an F (Penrose 1994, p. 137).

Of course, by *modus tollens* it follows that if no such F exists, AI, at least of the "Strong" variety, cannot be right, that is $\neg\mathcal{A}$. Unfortunately for Penrose, the connection between human mathematical reasoning and $\mathcal{B}_2^M$ is nothing like what he thinks it is; here's why.

The problem for Penrose is that a machine might *appear* to be doing all sorts of mathematical proofs of the type that Penrose venerates, and yet might be doing so on the strength of "mindless" simulation. Bringsjord has such a simulation available to him on the machine he is currently typing this sentence into: this is a simulation, by the theorem prover known as OTTER, of Gödel's first incompleteness theorem (Gödel I).[3] Bringsjord can run OTTER and after a bit of time, bingo, (an encoded version of) this theorem is proved and printed. The important point to realize is that this simulation has none of the types of mental states Gödel instantiated when he carried out his famous proof. For that matter, the simulation has none of the mental states logic instructors like Bringsjord instantiate when they prove Gödel I for their students. (For details on machine proofs of Gödel I see Bringsjord 1998*b*.)

---

[3] OTTER can be obtained at

`http://www-unix.mcs.anl.gov/AR/otter/`

We can bring the point here directly against Penrose, as follows. (This objection is one Searle has ingeniously articulated as well, in slightly different form: Searle 1997.) The key phenomenon for Penrose, the one he believes to be beyond computation, is that of a mathematician "ascertaining mathematical truth." As an example consider this proposition:

(SUM) The sum of two even numbers is always an even number.

When a mathematician attempts to ascertain whether (SUM) is true, he or she is attempting to decide whether or not a certain computation will ever halt. What computation? This one:

$$0 + 2 \text{ odd?}, 2 + 2 \text{ odd?}, 0 + 4 \text{ odd?}, 4 + 2 \text{ odd?}, 2 + 4 \text{ odd?}, 0 + 6 \text{ odd?}, \ldots$$

Obviously, this computation will never halt; knowing just a little bit about arithmetic is enough to grasp this fact. Of course, professional mathematicians would be engaged with propositions rather more challenging than (SUM). Let P be such a proposition, and let P_C denote the corresponding computation. Now suppose that Penrose carries out mathematical reasoning over a stretch of time from t_1 to t_{10} which eventuates in his correct declaration that P_C doesn't halt.[4] Assume that we take a snapshot B_{t_i} of Penrose's brain at each t_i, and that each snapshot has a corelate B'_{t_i} in an artificial neural network N that we build to process an encoding of P from t_1 to t_{10}. Suppose as well that N yields the answer "Doesn't halt" at t_{10}. The problem for Penrose is that N, for every proposition like P, can *in fact* be built. This is evident once one realizes that N needn't have any of the actual mental states Penrose has from t_1 to t_{10}. N, after all, is just a *simulation.* From the perspective of $\mathcal{B}_2^M$, *even if* human mathematical reasoning produces a verdict on whether computation C halts via information processing beyond that which is computable (i.e., hypercomputational processing), it is a trivial matter to build a system that yields this verdict through standard computation. (Analogously: Kasparov may for all we know routinely surpass the Turing Limit when playing chess as he does, but Deep Blue can still beat him by running standard search algorithms.)

[4]Here and hereafter we leave aside categorization of such propositions as Π_1^0, etc.

The predictable rebuttal on behalf of Penrose is that N here isn't *really* a simulation, because there isn't a sufficiently close correspondence between Penrose's rationcination from t_1 to t_{10} and the states of N through this interval. There is a fatal problem afflicting this rebuttal: the B'_{t_i} can approximate the B_{t_i} to a degree of fidelity that far exceeds what we normally demand in cases of "real world" simulation. To see this, consider the grandest and greatest computational architecture used to simulate human cognition: ACT-R (Anderson & Lebiere 1998). ACT-R is intended by John Anderson to mark the fulfillment of Alan Newell's dream of "a unified theory" of all human cognition.[5] ACT-R is composed of two elementary formalisms and one overarching algorithm, a trio used routinely in AI (fully covered, e.g., in Russell & Norvig 1994). The first formalism is a frame-based representation system, which is merely another way of expressing facts in first-order logic. (Alternatively, this system can be viewed as a primitive semantic network.) The second formalism is a production system, which is merely, again, a system that allows for conditional reasoning in first-order logic. The most recent version of ACT-R, version 4.0, is set out in a book that explains, in painstaking detail, how this architecture simulates humans carrying out elementary arithmetic (see Chapter 9 of Anderson and Lebiere 1998.) Chapter 11 of this book is devoted to providing experimental evidence for the view that ACT-R 4.0 can be used to simulate the cognition involved in human scientific discovery. Both simulations involve an exceedingly weak correspondence between real, human cognition and inferencing in first-order logic. Indeed, the correspondence is a good deal weaker than that between Penrose's decision with respect to P_C and the behavior of N.

Interestingly enough, in *SOTM* Penrose does consider a (weaker — because 'simulation' is used in a sense not relevant to AI) version of the objection from simulation that we have given immediately above. Here's how Penrose expressed the objection in *SOTM*:

> **Q7. The total output of all the mathematicians who have ever lived, together with the output of all the human mathematicians of the next (say) thousand years is finite and could be contained in the memory banks of an appropriate computer. Surely this particular computer *could*, therefore, simulate this output and thus behave (externally) in the same way as a human mathematician**

[5]Newell expressed his dream for a unified (production system-based) theory for all of human cognition in (Newell 1973).

> — **whatever the Gödel argument might appear to the tell us to the contrary.** (Penrose 1994, p. 82–83)

Penrose responds to this objection as follows.

> While this is presumably true, it ignores the essential issue, which is how we (or computers) know which mathematical statements are true and which are false. ... The way that the computer is being employed in **Q7** totally ignores the critical issue of *truth judgement.* (Penrose 1994, p. 83)

As a rebuttal against a proponent of "Weak" AI, what Penrose says here is, alas, worthless. "Weak" AIniks, proponents of $\mathcal{B}$ (or, more precisely, $\mathcal{B}_2^M$), explicitly ignore such genuinely mental phenomena as truth judgment. They only care about *simulating* such phenomena. On the other hand, no proponent of $\mathcal{A}$ would articulate **Q7** in the first place.

The upshot of all this is that from this point on we view Penrose's Gödelian case as an argument exclusively against $\mathcal{A}$, for it's just a brute fact that he will do no damage to $\mathcal{B}_2^M$.

Ironically, Penrose's work is likely to catalyze "Weak" AI projects; specifically, formidable attempts to implement systems capable of establishing results currently within the reach of only human mathematicians. (Hubert Dreyfus' widely publicized claim that chess grandmasters would forever be restricted to *homo sapiens* seemed to motivate many to prove him wrong.) This is not to say that there is in *SOTM* material that can provide a partial blueprint for building an artificial mathematician; we see no such material. However, we do see specific examples of human mathematical achievement that are likely to be taken by some as targets for a robot mathematician. One such example involves the proof of an interesting theorem about *hexagonal numbers* and *cubes*; this theorem serves as a pivot around which the formal background for Penrose's new Gödelian case revolves. We turn now to this background, and the theorem in question.

2.4 Background for Penrose's New Gödelian Case

It's uncontroversial that there are certain Turing machines (TMs) which provably never halt. Following one of Penrose's examples, begin

Figure 2.1: Hexagonal Numbers as Arrays

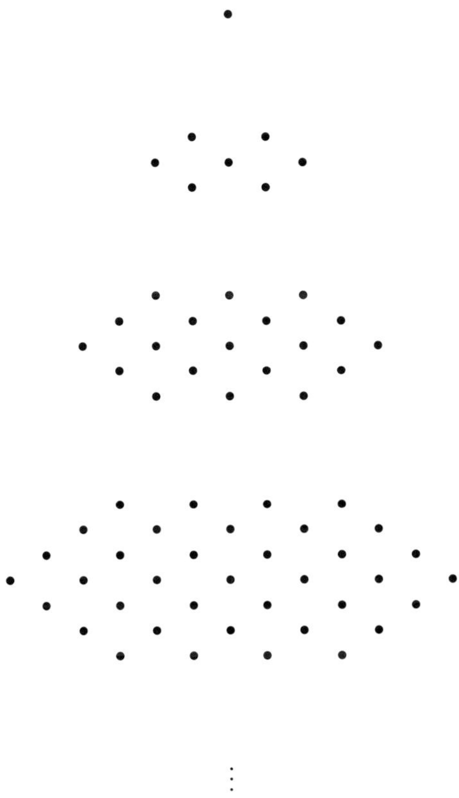

by considering the hexagonal numbers,

$$1, 7, 19, 37, 61, 91, 127, \ldots$$

i.e., the numbers that can be arranged as ever-increasing hexagonal arrays (see Figure 2.1). Now consider the cubes:

$$1 = 1^3, 8 = 2^3, 27 = 3^3, 64 = 4^3, 125 = 5^3, \ldots$$

Let TM $M_{\bar{c}}$ be defined as follows. $M_{\bar{c}}$ adds together the hexagonal numbers successively, starting with 1, checking to see if each sum is a cube. If so, the machine keeps working away; if not, it halts. Does $M_{\bar{c}}$ halt? No, and mathematicians can prove it, for that the pattern

$$1 = 1, 1{+}7 = 8, 1{+}7{+}19 = 27, 1{+}7{+}19{+}37 = 64, 1{+}7{+}19{+}37{+}61 = 125, \ldots$$

continues forever is a theorem.

The basic background idea to be derived from the foregoing is that there is some procedure[6] (let's call it '$\Re$') by virtue of which mathematicians correctly classify some Turing machines (or their "user-friendly" equivalents, e.g., algorithms) as non-halters. Penrose's negative objective in *SOTM* is to establish that $\Re$ is uncomputable.[7] This objective is to be reached, according to his plan, via first demonstrating that

$\mathcal{G}$ For every "knowably sound" algorithm A for classifying Turing machines as non-halters, $\Re \neq A$.

After this attempted demonstration (Chapter 2), Penrose's plan calls for ruling out the other "computationalist" position (Chapter 3; this is the position others seem to regard as quite formidable, (e.g., Chalmers 1995)), namely, that $\Re$ is indeed an algorithm, just not a "knowably sound" one. We first consider Chapter 2's argument against $\mathcal{G}$.

2.5 The Core Diagonal Argument

The heart of the Gödelian Case against $\mathcal{G}$ is a less-than-one-page presentation of a certain diagonal argument designed to show that assuming $\Re$ to be some sound set of computational rules (= some TM) A results in the revelation that $\Re$ cannot be such a set (= such a TM). Here — reproduced verbatim to preclude any inaccuracy — is the diagonal argument for $\mathcal{G}$. It begins with the assumption that

> A is just *any sound* set of computational rules for ascertaining that some computations $C_q(n)$ do not ever halt. Being dependent upon the two numbers q and n, the computation that A performs can be written $A(q, n)$, and we have
>
> **(H)** If $A(q, n)$ stops, then $C_q(n)$ does not stop.
>
> let us consider the particular statements **(H)** for which q is put equal to n... we now have:

---

[6]Not necessarily an *effective* procedure: to assume that the procedure is effective is of course to beg the question against Penrose, since he purports to show that the procedure in question *isn't* effective.

[7]His *positive* objective is to lay the foundation for a science of the mind which accommodates his negative results.

> **(I)** If $A(n, n)$ stops, then $C_n(n)$ does not stop.
>
> We now notice that $A(n, n)$ depends upon just *one* number n, not two, so it must be one of the computations $C_0, C_1, C_2, C_3, \ldots$, since this was supposed to be a listing of *all* the computations that can be performed on a single natural number n. Let us suppose that this is in fact C_k, so we have:
>
> **(J)** $A(n, n) = C_k(n)$.
>
> Now examine the particular value $n = k$. We have, from **(J)**,
>
> **(K)** $A(k, k) = C_k(k)$.
>
> and, from **(I)**, with $n = k$:
>
> **(L)** If $A(k, k)$ stops, then $C_k(k)$ does not stop.
>
> Substituting **(K)** in **(L)** we find:
>
> **(M)** If $C_k(k)$ stops, then $C_k(k)$ does not stop.
>
> From this, we must deduce that the computation $C_k(k)$ does not stop (for if it did then it does not, according to **(M)**!). But $A(k, k)$ cannot stop either, since by **(K)**, it is the *same* as $C_k(k)$. Thus, our procedure A is incapable of ascertaining that this particular computation $C_k(k)$ does not stop even though it does not. Moreover, if we *know* that A is sound, then we *know* that $C_k(k)$ does not stop. Thus, we know something that A is unable to ascertain. It follows that A *cannot* encapsulate our understanding. (Penrose 1994, pp. 74–75)

Immediately after presenting this argument Penrose says

> At this point, the cautious reader might wish to read over the whole argument again ... just to make sure that we have not indulged in any 'sleight of hand'! Admittedly there is an air of the conjuring trick about the argument, but it is perfectly legitimate, and it only gains in strength the more minutely it is examined. (Penrose 1994, pp. 75)

Unfortunately, after taking up Penrose's challenge, having examined the argument minutely, we find that he is simply wrong: the argument, in the end, *is* nothing more than prestidigitation. Parts of it are at best enthymematic, and when the whole thing is rendered precisely, a remarkable number of technical glitches come to light. But over and above these defects, there is a fatal dilemma afflicting the argument: at best, it is an instance of either the fallacy of denying the antecedent, the fallacy of *petitio principii*, or the fallacy of equivocation. In light of this, again, Penrose's new Gödelian Case is a repeat of J.R. Lucas' (Lucas 1964) argument of over three decades ago. Nonetheless, as we've indicated, at the *intuitive* level, Penrose is on target, as we explain later.

2.6 Formal Machinery

In order to expose Penrose's legerdemain, we need only formalize the argument. Our formalization, and the evaluation thereof, will be charitably naive — because they will be carried out without exploiting the fact that Penrose has misdescribed the connection between his diagonal argument and the associated meta-theorems in mathematical logic (not the least of which are Gödel's incompleteness results themselves).[8] We will start by attempting to formalize Penrose's diagonal argument in $\mathcal{L}_I$ (full first-order logic). As will be seen, the formalization will eventually call for $\mathcal{L}_{II}$ (second order logic).[9]

For our formalization we follow the notation of (Ebbinghaus et al. 1984), and hence deploy atomic formulas

$$M_t : u \rightarrow v$$

to denote the fact that TM M_t, starting with u as input on its tape,

[8]Feferman (1995) and Davis (1980) catalogue the inadequacy of Penrose's scholarship when it comes to mathematical logic. However, we don't think the quality of this scholarship, however questionable it may be, creates any fundamental problems for Penrose's core Gödelian arguments against "Strong" AI.

[9]It's important at this point to note that while many of the mathematical issues relating to Penrose's Gödelian arguments are not expressible in $\mathcal{L}_I$, the core arguments themselves must conform to the basic, inviolable principles of deductive reasoning that form the foundation of technical philosophy. *SOTM* is an essay in technical philosophy; it's not a mathematical proof. The present book, of course, is itself technical philosophy. We make reference to logical systems beyond $\mathcal{L}_I$, but our core reasoning is intended to meet standards for deductive reasoning circumscribed in $\mathcal{L}_I$ and $\mathcal{L}_{II}$. See (Ebbinghaus et al. 1984) for a nice introduction to these logical systems, as well as more advanced ones, such as the infinitary systems related to Yablo's Paradox. These systems and this paradox are discussed later in the chapter.

halts and leaves v as output. Similarly,

$$M_t : u \to \text{halt}$$

and

$$M_t : u \to \infty$$

denote, respectively, that the TM in question halts and doesn't halt (on input u). Next, assume that the alphabet with which our TMs work is of the standard sort, specifically $\{|, \bullet\}$, where a natural number n is coded as a string of n |s, and $\bullet$ is used solely for punctuation. Finally, fix some enumeration of all Turing machines and a corresponding Gödel numbering scheme allowing us to reference these machines via their corresponding natural numbers.

With this machinery, humble as it is, it's easy to formalize certified diagonal arguments, like the classic one traditionally used to establish the halting problem, that is, that there is no TM $M_{h \vee \bar{h}}$ which can ascertain whether or not a given TM halts.[10] In the following Fitch-style formalization, with '1' used to signal 'Yes' and '0' 'No,' one starts by assuming that some $M_{h \vee \bar{h}}$ *does* exist, from which a contradiction for *reductio* is derived.[11] (Please note that for the remainder of the chapter we use '$\Rightarrow$' and '$\Leftrightarrow$' to denote the material conditional and biconditional, respectively. We do this because '$\to$' is now being used in formulas about the activity of Turing machines.)

1	$\exists p \forall r \forall s[(M_p : r \bullet s \to 1 \Leftrightarrow M_r : s \to \text{halt}) \wedge$ $(M_p : r \bullet s \to 0 \Leftrightarrow M_r : s \to \infty)] = \phi$	supp.
2	$\phi \Rightarrow \exists m \forall n[M_m : n \to \text{halt} \Leftrightarrow M_n : n \to \infty]$	Lemma 1
3	$\exists m \forall n[M_m : n \to \text{halt} \Leftrightarrow M_n : n \to \infty]$	$1, 2$ MP
4	$\quad \forall n[M_a : n \to \text{halt} \Leftrightarrow M_n : n \to \infty]$	supp.
5	$\quad M_a : a \to \text{halt} \Leftrightarrow M_a : a \to \infty$	$\forall$E
6	$Z \wedge \neg Z$	$3, 4-5$ $\exists$E & RAA
7	$\neg \phi$	$1-6$ RAA

One comment on this proof,[12] an important one given our coming formalization of Penrose's diagonal argument: Note that once an

[10] The form of the halting problem we use is specifically that given a TM M and input to this machine, no TM can ascertain whether or not M halts on this input.

[11] Note that we use '$\forall$E' to denote the inference rule of universal quantifier elimination, etc.

[12] We make no comments about what we've called 'Lemma 1.' For an elegant proof of this lemma (which, if formalized, would require more lines centered around the rules $\exists$I, $\forall$E, $\exists$I, $\forall$I), see (Boolos & Jeffrey 1989).

implicit contradiction $p \Leftrightarrow \neg p$ is obtained in line 5, a contradictory formula, devoid of the constant a, is obtained (via the fact that everything follows from a contradiction). That the instantiating constant a not occur in line 6 is required by one of the standard restrictions on the rule $\exists$E existential elimination. This requirement, as is well-known, ensures that the instantiating constant plays only an intermediary role. To violate it is to allow for absurdities such as, "that from the fact that there is a negative number it follows that two is a negative number."[13]

2.7 Formalizing Penrose's Diagonal Argument

Now, what does Penrose's diagonal argument look like once it's formalized using the machinery at our disposal? The initial part of the formalization is straightforward. Penrose begins by assuming that there is some set A of computational rules (we use 'M_a' to refer to A as TM; this is an identification Penrose himself, following standard mathematical practice, explicitly sanctions in Appendix A of *SOTM*) such that: if A yields a verdict that some TM M fails to halt on input n, then M *does* fail to halt on n. He then moves, via quantifier manipulation, through **(H)** to **(I)**. Here's how this initial reasoning runs:

1′	$\exists m \forall q \forall n[M_m : q \bullet n \rightarrow \text{halt} \Rightarrow M_q : n \rightarrow \infty]$	supp.
2′	$\forall q \forall n[M_a : q \bullet n \rightarrow \text{halt} \Rightarrow M_q : n \rightarrow \infty] = \mathbf{(H)}$	supp.
3′	$\forall n[M_a : b \bullet n \rightarrow \text{halt} \Rightarrow M_b : n \rightarrow \infty]$	2′ $\forall$E
4′	$M_a : b \bullet b \rightarrow \text{halt} \Rightarrow M_b : b \rightarrow \infty$	3′ $\forall$E
5′	$\forall n[M_a : n \bullet n \rightarrow \text{halt} \Rightarrow M_n : n \rightarrow \infty] = \mathbf{(I)}$	4′ $\forall$I

At this point we reach the reasoning from **(I)** to **(J)**, and things begin to turn a bit murky. The reasoning, recall, is (from p. 75 of Penrose 1994):

> We now notice that $A(n, n)$ depends upon just *one* number n, not two, so it must be one of the computations $C_0, C_1, C_2, C_3, \ldots$, since this was supposed to be a listing of *all* the computations that can be performed on a single natural number n. Let us suppose that this is in fact C_k, so we have:
>
> **(J)** $A(n, n) = C_k(n)$.

[13] The proof would be: $\exists x Nx$ as supposition, Nt as supposition, Nt by reiteration, and Nt with second supposition discharged by application of the (here erroneous) rule of $\exists$E.

What, formally speaking, sanctions this reasoning? What entitles Penrose to infer that a TM operating on input $n \bullet n$ is identical to one operating on just n? One possibility that comes to mind is a pair of elementary theorems like

$$(\mathbf{T}_1)\ \forall m \forall n[M_m : n \bullet n \rightarrow \text{halt} \Rightarrow \exists q(M_q : n \rightarrow \text{halt})]$$

$$(\mathbf{T}_2)\ \forall m \forall n \forall o[M_m : n \bullet n \rightarrow o \Rightarrow \exists q(M_q : n \rightarrow o)]$$

$(\mathbf{T}_1)$ is easily established by construction. Given a TM M_1 that operates on input $n \bullet n$ and eventually halts, it's easy to build a TM M_2 which starts with just n on its tape, calls a TM M_2 which copies n so that $n \bullet n$ is written on the tape, and then proceeds to simulate M_1 step for step. The same sort of simple trick verifies $(\mathbf{T}_2)$.

Neither of these two theorems, however, can be what Penrose presupposes, for he needs the *identity* $(\mathbf{J})$. Moreover, that which might do the trick for him is false. Specifically, these two propositions aren't theorems (and are, in fact, easily counter-exampled):

$$(\mathbf{T}_3)\ \forall m \forall n[M_m : n \bullet n \rightarrow \text{halt} \Rightarrow M_m : n \rightarrow \text{halt}]$$

$$(\mathbf{T}_4)\ \forall m \forall n[M_m : n \bullet n \rightarrow \text{halt} \Rightarrow \exists q(M_q = M_m \wedge M_q : n \rightarrow \text{halt})]$$

Charity suggests devising something to rescue Penrose's reasoning. What did he have in mind? We think Penrose, during his moves from $(\mathbf{I})$ to $(\mathbf{J})$, had in mind a rationale like that behind the likes of $(\mathbf{T}_1)$ and $(\mathbf{T}_2)$. His use of identity may stem from an erroneous but tempting conflation of *Turing machine* with Turing machine *computation*, where the latter (following the textbook view; recall that we presented this view, in connection with the programming language $\mathcal{L}$, in Chapter 1.) is a sequence of configurations of a TM. In order to grasp our exegesis, think back to the machines M_1, M_2, and M_3 in the justification given above for $(\mathbf{T}_1)$. Here, it's no doubt safe to say that though M_1, strictly speaking, is diverse from the composite machine M_3 composed of M_2 and M_1, M_1 and M_3 are virtually identical — because the computations involved differ only in regard to the trivial duplication of n.[14] So let us say that in cases like this the machines in

[14]This is as good a place as any to point out that though Penrose's use of the term 'computation' is, to say the least, relaxed, a little analysis reveals that by it he can invariably be read as ultimately making reference to a Turing machine. For example, Penrose sometimes refers to "computations" by way of imperatives, as when he writes:

question are *approximately identical*, written (in this case) $M_1 \approx M_3$. And let us affirm, on Penrose's behalf, the appropriate sentence (see $6'$ in the next portion of the derivation), as well as inference rules for $\approx$ paralleling those for $=$. Now we can continue our careful rendition of Penrose's diagonal argument. Indeed, we can make it beyond **(M)**:

$6'$	$\forall n[(M_n : n \bullet n \rightarrow \text{halt} \Rightarrow M_n : n \rightarrow \infty) \Rightarrow$ $\exists q(M_q \approx M_n \wedge (M_q : q \rightarrow \text{halt} \Rightarrow M_q : q \rightarrow \infty))]$	Lemma
$7'$	$M_a : a \bullet a \rightarrow \text{halt} \Rightarrow M_a : a \rightarrow \infty$	$5'\,\forall$E
8	$(M_a : a \bullet a \rightarrow \text{halt} \Rightarrow M_a : a \rightarrow \infty) \Rightarrow$ $\exists q(M_q \approx M_a \wedge (M_q : q \rightarrow \text{halt} \Rightarrow M_q : q \rightarrow \infty))$	$6'\,\forall$E
9	$\exists q(M_q \approx M_a \wedge (M_q : q \rightarrow \text{halt} \Rightarrow M_q : q \rightarrow \infty))$	$7', 8$ MP
10	$M_k \approx M_a \wedge (M_k : k \rightarrow \text{halt} \Rightarrow M_k : k \rightarrow \infty)$	supp.
11	$M_k : k \rightarrow \text{halt}$	supp.
12	$M_k : k \rightarrow \text{halt} \Rightarrow M_k : k \rightarrow \infty = \mathbf{(M)}$	$10 \wedge$E
13	$M_k : k \rightarrow \infty$	$11, 12$ MP
14	$M_k : k \rightarrow \text{halt}$	11 R
15	$M_k : k \rightarrow \infty$	$11-14$ RAA

At this point it may be thought that things are beginning to look up for Penrose. For not only have we reached his **(M)**, but we have also achieved, in line 15, the formal equivalent to his assertion that "$C_k(k)$ does *not* in fact stop." Two problems, however, stare us in the face.

Problem 1 is that everything to this point is based on two *undischarged* suppositions, $2'$ and 10, in which existentially quantified variables are instantiated to what are supposed to be arbitrary and intermediary constants, a in line 2 and k in line 10. There is no indication whatsoever from the text in question that Penrose intends to discharge these assumptions. In fact, the text clearly indicates that Penrose intends to rest his diagonal argument with the constants a (for his A) and k undischarged. Though we don't understand how this could be

---

Suppose we had tried, instead, the computation

(B) Find a number that is not the sum of four square numbers.

Now when we reach 7 we find that it *is* the sum of *four* squares: $7 = 1^2 + 1^2 + 1^2 + 2^2, \ldots$

It's clear from the commentary here after the presentation of **(B)** that Penrose's use of imperatives is elliptical for a description of an algorithm — the algorithm to be used in the attempt to meet the imperative.

(given the traditional mathematical use of, and restrictions on, instantiating arbitrary constants), we will assume for the sake of argument that this defect can be remedied.

Problem 2 is that it's impossible to derive that which would coincide in our formalization with "But $A(k, k)$ cannot stop either, since by **(K)**, it is the *same* as $C_k(k)$." What would suffice to validate this prose is a derivation of $M_a : k \bullet k \rightarrow \infty$; but this formula can't be derived. (We *can* derive $M_a : k \rightarrow \infty$, by first isolating the "identity" $M_k \approx M_a$ from line 10 and then using this "identity" with line 15 and the indiscernibility of identity.) We see no way to rectify this problem (after burning more than a few grey cells in the attempt to extend and/or modify the proof), but, once again, for the sake of argument we're prepared to assume that somehow Penrose can survive, that is, that he can continue the proof. So we then have:

1′		
⋮		
15	$M_k : k \rightarrow \infty$	11−14 RAA
⋮	⋮	⋮
n	$M_a : k \bullet k \rightarrow \infty$	
$n+1$	$(M_k : k \rightarrow \infty) \wedge (M_a : k \bullet k \rightarrow \infty)$	$15, n \wedge$ I

2.8 Penrose's Dilemma: Either Way a Fallacy

With the diagonal argument done, Penrose now purports to wrap things up:

> Thus, our procedure A is incapable of ascertaining that this particular computation $C_k(k)$ does not stop even though it does not. Moreover, if we *know* that A is sound, then we *know* that $C_k(k)$ does not stop. Thus, we know something that A is unable to ascertain. It follows that A *cannot* encapsulate our understanding. (Penrose 1994, p. 75)

The first sentence in perhaps a bit misleading. What we *can* conclude from line $n+1$ is that M_a doesn't yield a verdict on whether M_k halts on input k, and M_k doesn't halt on k. This, combined with a relevant reinstantiation of $2'$, viz.,

n+2 $M_a : k \bullet k \rightarrow \text{halt} \Rightarrow M_k : k \rightarrow \infty$

gives nothing helpful. Though we admit it's a rather unflattering view, Penrose may fall prey here to the fallacy of denying the antecedent, for with it invoked on lines $n+2$ and n he would have the negation of the consequent in $n+2$, $M_k : k \rightarrow$ halt, which would contradict line 15 — which would in turn give him, by *reductio ad absurdum*, the denial of line 1′. He would then have proved that there is no Turing machine (algorithm, set of computational rules, etc.) that can do what $\Re$ does (viz., give correct verdicts on non-haltingness), which is certainly what he wants to ultimately establish.

The other possibility would seem to be that Penrose's imprecision has led him to confuse the conditional 2′ with a conditional running in the opposite direction, i.e.,

2″ $\forall q \forall n [M_a : q \bullet n \rightarrow \text{halt} \Leftarrow M_q : n \rightarrow \infty]$

For note that 2″, once instantiated (with both q and n to k), yields, by *modus tollens* with line n, a contradiction with line 15; this in turn allows for the *reductio* that would, by the reasoning just explained, make Penrose's day.

Unfortunately, 2″ taken as premise[15] begs the entire question. The reason is as follows. First, 2″ and 2′ combine to produce

2‴ $\forall q \forall n [M_a : q \bullet n \rightarrow \text{halt} \Leftrightarrow M_q : n \rightarrow \infty]$,

Second, (as is well-known) there exists a TM M_h[16] such that

$$\forall r \forall s (M_h : r \bullet s \rightarrow 1 \Leftrightarrow M_r : s \rightarrow \text{halt}).$$

Third, any TM which halts on input m can be effortlessly adapted to print a string u on its tape and *then* halt. It follows that 2‴ implies that there is some machine $M_{a'}$, adapted from M_a, that can solve the halting problem! Since no Turing machine can solve this problem, and since (under the setup Penrose has erected) what implies that A *can* solve it is that A can match $\Re$, it follows that A cannot match $\Re$; and this, again, is precisely what Penrose wants to show. The problem, of course, is that this reasoning has as a premise that $\Re$ can solve the halting problem, and whether or not some human cognition is capable of this is precisely what's at issue in the debate of which

[15]To put it more precisely, 2″ would follow from taking as premise a sentence like 1″, which would be 1′ with $\Rightarrow$ changed to $\Leftrightarrow$.

[16]Which simply simulates, step for step, the TM M_r.

SOTM and *ENM* are a part! So on this reading, Penrose begs the question. His dilemma is hence that either way, so far, the Gödelian Case is fallacious.

But what of the final moves in Penrose's argument? That is, what of

> Moreover, if we *know* that A is sound, then we *know* that $C_k(k)$ does not stop. Thus, we know something that A is unable to ascertain. It follows that A *cannot* encapsulate our understanding. (Penrose 1994, p. 75)

It is easy to render this precise, given our analysis. And what this analysis reveals, alas, is that Penrose is once again perched atop the same dilemma, the horns of which are the fallacies cited above: That "if we *know* that A is sound, then we *know* that $C_k(k)$ does not stop" amounts (courtesy of the proof above from $1'$ to $n+1$, and concessions that both $6'$ is separately provable and failing to discharge 10 is somehow surmountable) to knowledge of

$$\{1'\} \vdash 15.$$

All that remains now is to cash out the final two sentences in the previous quote. These two sentences seem to suggest that "our understanding" is capable of ascertaining whether or not $\Gamma \vdash \psi$, where Γ is some set of first-order formulas, and ψ is one such formula. But of course ascertaining whether or not $\Gamma \vdash \psi$ is provably equivalent to ascertaining whether or not a given Turing machine halts (e.g., see the elegant proof in Boolos & Jeffrey 1989). Hence, A obviously cannot solve the problem of ascertaining whether or not such implications hold. Moreover, to assume that $\Re$, our understanding, can, is to beg the question in the manner discussed above (since $\Re$ would here be capable of solving the halting problem). The other horn, again, is that if Penrose retreats to the circumspect view he started with, that A simply cannot yield a verdict on whether or not $\{1'\} \vdash 15$ (because A is only a *sound* procedure, in the sense that if A says "Yes" then $\{1'\} \vdash 15$, and nothing more), he needs to invoke the fallacious rule $\{\psi \Rightarrow \gamma, \neg\psi\} \vdash \neg\gamma$ in order to get the contradiction $\{1'\} \vdash 15$ and $\{1'\} \nvdash 15$.

2.9 Possible Replies

What might Penrose say for himself? How might he try to dodge the dilemma? We suspect he might say that the diagonal argument is included in another, wider argument which we have mangled; he might claim, more specifically, that he never intended to generate a contradiction from assuming $1'$. In fact, as alert readers will doubtless have already noticed, this *must* be his claim, for $1'$ is in fact an easily proved theorem. For the (trivial) proof, note that all Turing machines can be recast as flow diagrams. Note, in particular, that any TM represented by a flow diagram having as part the fragment shown in Figure 2.2 would be a non-halting TM (because if started in state 1 with its read/write head scanning a block of |s, it will loop forever in this fragment). So we obviously have near at hand a Turing machine M_4 which, upon being given as input a TM M and a string of |s as the input to M, does the following:

- converts M into a flow diagram;
- checks to see if the fragment of Figure 2.2 is in the diagram;
- if so, it outputs a '0' (to indicate that M is a non-halter);
- if not, it goes into an infinite loop of its own.

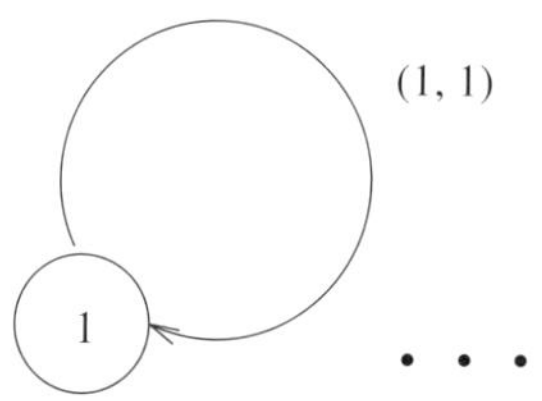

(The node here reflects the start state.)

Figure 2.2: Flow-Diagram Fragment That Entails Non-Halting.

Proposition $1'$ follows from the existence of M_4 by existential introduction. Penrose might also complain that we went wrong in working from the previous quote to the general case of whether or not $\Gamma \vdash \psi$, for all first-order formulas ψ and sets Γ thereof.

What, now, is the wider argument that Penrose can be read as giving? In order to construct it on his behalf, we return to the observation that should Penrose establish that

For every TM M, if M is "knowably" sound in the sense of $1'$, then $\Re \neq M$

he will have succeeded. The next thing to note is that this proposition can of course be established by conditional proof and universal introduction. That is, one could start by assuming that some arbitrary TM M' is sound (in the sense of $1'$ or $2'$; abbreviate this property by 'SOUND'), derive from this somehow that $\Re \neq M'$, move to the conditional, and supplant the constant denoting M' with a variable within the scope of a universal quantifier. If we once again pretend that somehow Penrose can work a miracle on discharging assumptions, we can capitalize on our earlier analysis to quickly build a candidate version of his reasoning:

α_1	$2'$	
$\vdots$		
α_n	$(M_k : k \rightarrow \infty) \wedge (M_a : k \bullet k \rightarrow \infty)$	$15, n \wedge$I
α_{n+1}	$\Re$ yields a verdict of *doesn't halt* w.r.t. $k \bullet k$ by virtue of proving $\{\alpha_1\} \vdash \alpha_n$.	
α_{n+2}	$\forall x \forall y (x = y \Leftrightarrow \forall X (Xx \Leftrightarrow Xy))$	Leibniz's Law
α_{n+3}	$\Re \neq M_a$	$\alpha_n, \alpha_{n+1}, \alpha_{n+2}$
α_{n+4}	$2' \Rightarrow \Re \neq M_a$	$\alpha_{n+1} - \alpha_{n+3}$ CP
α_{n+5}	$\forall q (\text{SOUND}(M_q) \Rightarrow \Re \neq M_q)$	α_{n+4} $\forall$I

Have we finally arrived, then, at a victorious version of Penrose's new Gödelian Case? Hardly. In fact, yet another fallacy rears up here — the fallacy of equivocation. An instance of this fallacy is found in the currently enthymematic presentation of

$$\{\alpha_n, \alpha_{n+1}, \alpha_{n+2}\} \vdash \alpha_{n+3}.$$

The concept of *yielding a verdict* is used equivocally in these inferences: in connection with the TM M_a this concept is used in the straightforward, well-understood sense of a TM doing some work and then printing out '0'; in connection with $\Re$, however, the concept means something quite different — it means carrying out a meta-proof. In order to verify our diagnosis, you have only to look more carefully at how LL is used. In order to use this proposition, the variable X must obviously be instantiated. Let's say that it's instantiated to V. What are V's truth conditions? On the one hand we have the normal, well-understood sense applicable to M_a: V is true of some triple (m, n, o) iff M_m halts upon taking $n \bullet o$ as input. On the other hand, we have the sense according to which $\Re$ is supposed to have this property: V is true of some triple (m, n, o) iff m can deliver a meta-proof showing that a TM M_n goes to infinity on o.[17]

The problem can be made painfully clear if we spotlight the implicit inferences, which run as follows. First, let the meaning of V be the well-understood sense. Now instantiate LL to get

$$\Re = M_a \Leftrightarrow \forall X(X\Re \Leftrightarrow XM_a).$$

Next, suppose for *reductio* that $\Re = M_a$. Then by *modus ponens* we have

$$\forall X(X\Re \Leftrightarrow XM_a).$$

We now observe that $\neg VM_a$, from which it follows (by universal instantiation and biconditional elimination) that $\neg V\Re$. But from α_{n+1} we have $V\Re$ — contradiction.

The problem, of course, is the invalid inference from α_{n+1} to $V\Re$.[18] And the problem arising from an equivocal use of V is unavoidable: If one starts the reasoning we've just gone through with the "meta-proof" sense of V, then we can no longer count on knowing $\neg VM_a$.

Penrose's last chance, at this point, is to somehow define V disjunctively, taking account of both definitions. For example, perhaps he could stipulate that $\Re$ be "operationalized," so that its verdict is delivered by way of something suitably mechanical, perhaps the checking of a box tagged with `DH` for *doesn't halt.* Unfortunately, this doesn't help solve the problem in the least. This is so because what we *know* of M_a is that it fails to halt; for all we know, over and above this, this TM is capable of checking boxes, writing stories, ... carrying out astounding meta-proofs. In searching for a way to straddle the equivocation, Penrose catapults himself back to square one, for as Bringsjord (1992) — and others as well (e.g., Slezak 1982, Webb 1980) — has pointed out

[17] Notice that other technical issues arise here — issues a good deal more subtle than the sort with which Penrose customarily deals. For example, the domain behind deployment of LL can't be the set of all TMs, as it has been to this point — because $\Re$, for all we know at this stage in the proof, isn't a TM. The solution here would perhaps be to adopt as domain not only those machines equivalent to TMs or less, but this set union the set of hypercomputers.

[18] We leave aside issues arising from the self-referentiality of the situation: $\Re$ apparently carries out a proof in which it itself figures.

elsewhere, there is no reason whatever to think that Turing machines (of which, of course, M_a is one) can't deliver the meta-proofs (concerning number theory, recall) with which Penrose framed the entire investigation.

2.10 Given $\mathcal{G}$, The Other Possibilities

Now let's assume for the sake of argument that Chapter 2's (*SOTM*) argument for $\mathcal{G}$, contrary to what we have seen in the foregoing, succeeds. This leaves the following four computationalist possibilities (which Penrose isolates on pages 130–131) with respect to $\mathfrak{R}$:

P1 $\mathfrak{R}$ is unknowable and sound.

P2 $\mathfrak{R}$ is sound and knowable, but not *knowably* sound.

P3 $\mathfrak{R}$ is unsound (i.e., mathematicians unwittingly use an unsound algorithm).

P4 There are different algorithms for different mathematicians (so we cannot speak univocally of $\mathfrak{R}$).

It seems to us that nearly all of the arguments Penrose gives against P1–P4 are rather sloppy. (This is not to say that these arguments fail to fascinate. His treatment of P1 — including as it does the attempt to show that "$\mathfrak{R}$ is unknowable" implies the mystical $\mathcal{A}/\mathcal{D}$ notion that $\mathfrak{R}$ is the result of divine intervention — is quite ingenious.) We don't have the space to treat each possibility and each argument; we focus on P3, and on Penrose's attempt (pp. 137–141) to rule out this possibility. Similar analysis, with similar results, could be given for the remaining trio.

As readers will recall, and as Penrose well knows, if $\mathfrak{R}$ is unsound, then any argument against computationalism from Gödel's first incompleteness theorem will necessarily fail — because the hypothesis of this theorem is that the axioms in question are consistent.[19] What

[19]Here, for reference, is a fairly careful statement of **Gödel I**:

> Let Φ be a consistent, Turing-decidable set of first-order formulas built from the symbol set $\{+, \times, 0, 1\}$, with $+$ and $\times$ binary function symbols (intepreted as addition and multiplication, resp.) and 0 and 1 constants denoting the numbers zero and one, and where Φ is *representable*. (For details on representability, see Ebbinghaus et al. 1984.) Then there is a sentence ϕ built from the same symbol set such that: $\Phi \nvdash \phi$ and $\Phi \nvdash \neg\phi$.

some readers may not know, and what some others may have forgotten, is something Penrose (tendentiously?) fails to even mention, let alone discuss, in *SOTM*: viz., Gödel's *second* incompleteness theorem, and the connection between this result and P3. Here's the theorem in question:

Gödel II: Where Φ is a set of Turing-decidable first-order formulas built from the symbol set $\{+, \times, 0, 1\}$, where $+$ and $\times$ are binary function symbols (intepreted as addition and multiplication, resp.) and 0 and 1 are constants denoting the numbers zero and one, and where $\Phi \subset \Phi_{PA}$, i.e., Φ is a subset of the Peano axioms for arithmetic, then it's not the case that $\Phi \vdash \text{Consis}_\Phi$, where Consis_Φ abbreviates a formula expressing the proposition that from Φ one cannot derive '$0 = 1$,' i.e., that Φ is consistent.[20]

Since classical mathematics includes ZFC set theory[21] or some formal equivalent, it follows immediately from Gödel II that Penrose's belief that $\Re$ is not unsound cannot be the product of the sort of (to use Penrose's respectful phrase) "unassailable reasoning" at the heart of classical mathematics. Given this, why does Penrose reject P3? At first, it appears that his rationale is extraordinarily weak, for he asks at the outset of his discussion of P3:

> But is it really plausible that our unassailable mathematical beliefs might rest on an unsound system — so unsound, indeed, that '1=2' is in principle part of those beliefs? (Penrose 1994, p. 138)

This question is anemic. To say that a system is inconsistent (e.g., that the set Φ of Gödel I is inconsistent) is to say that from it a contradiction can be derived, from which it does indeed follow in standard first-order logic that *any* proposition, including, then, '1=2,' can be derived. *But*

Cognoscenti will note that here and herein we drop Gödel's original concept of ω-consistency in favor of "modern" versions. Penrose has conceded (e.g., see section 2, "Some Technical Slips in *Shadows*," in Penrose 1996) that in *SOTM* he erred in his use ω-consistency. (The errors in question were pointed out by Feferman (1995) and others.) We believe these errors are complete red herrings.

[20] For details on how to construct Consis_Φ, see Chapter X in (Ebbinghaus et al. 1984).

[21] That is, Zermelo-Fraenkel axiomatic set theory with the axiom of choice. ZFC plays a more important role in Chapter 4 than in the present chapter, and is partially described in Chapter 4 (where five of the nine axioms composing ZFC are presented). Readers wanting a fuller treatment are directed to (Ebbinghaus et al. 1984).

the question is whether the contradiction can be found; only if it can, can an absurdity be produced. Finding the contradiction is the issue!

Penrose himself seems to realize that his question (reproduced in the block quote immediately above) is little more than a rhetorical trick: he explicitly considers the possibility that the contradiction could be a "hidden" one (p. 138); and he offers Russell's paradox as an example. As Penrose says: "Without the contradiction having been perceived, the methods of reasoning might well have been trusted and perhaps followed by mathematicians for a good long while" (p. 139). So the obvious question is: Why isn't it possible that P3 is true, and therefore that $\mathfrak{R}$ is unsound, because there is a contradiction hidden in classical mathematics that no one has yet found? For that matter, why isn't it possible that there is a contradiction that will *never* be found?

Penrose's response to these questions is first to say that Russell's paradox could not have gone undetected for any great length of time (p. 139). Because Russell's paradox is so simple we concede for the sake of argument that this response is cogent. This possibility still remains, however: there could be *extraordinarily complicated* contradictions buried within classical mathematics. Penrose himself seems to clearly recognize that this is at least a conceptual possibility, for he writes as follows.

> One might imagine some much more subtle paradox, even lying implicit in what we believe to be unassailable mathematical procedures that we allow ourselves today — a paradox which might not come to light for centuries to come. (Penrose 1994, p. 139)

But immediately after reading this we face an exceedingly peculiar part of *SOTM*: Penrose promptly proceeds to identify the objection based upon the possibility of a "more subtle paradox" with the objection that there is no fixed $\mathfrak{R}$ underlying present-day mathematical understanding, but rather a series of algorithms in constant flux (p. 139). But these are two different objections. The first objection is that $\mathfrak{R}$ and classical mathematics may well be unsound (in which case, as Penrose is forced to admit, his Gödelian case is derailed). The second objection is that talk of the singular and determinate $\mathfrak{R}$ is unjustified. We are not concerned with the second; the first, however, is ours. And Penrose does nothing to disarm it.

Penrose and his supporters might at this point ask: "Well, what paradox do you have in mind? If Russell's paradox doesn't do the trick for you, what does?" It's hard to see how this question can help Penrose. If $\Re$ is unsound because of a hidden contradiction, then the contradiction is just that: hidden. So we can hardly be obligated to display it. At most, the challenge to us is to say what *sort* of paradox might point the way toward the hidden contradiction. This challenge is easy to meet. First, the kind of paradoxes we have in mind are purely logico-mathematical; "physics-related" paradoxes, such as those attributed to Zeno (e.g., Salmon 1975), are irrelevant. Second, the paradoxes we see as supporting the notion that there may for all we know be hidden contradictions in $\Re$ are ones that aren't solved. Our current favorite (somewhat unsurprisingly, perhaps) is Bringsjord's (n.d.) infinitary version of the originally finitary Yablo Paradox (Yablo 1993). The version runs as follows:[22]

> Recall the familiar natural numbers $\mathbf{N} = \{0, 1, 2, \ldots\}$. With each $n \in \mathbf{N}$ associate a sentence as follows, using a truth predicate, T:
>
> $$s(0) = \forall k(k > 0 \rightarrow \neg T(s(k)))$$
>
> $$s(1) = \forall k(k > 1 \rightarrow \neg T(s(k)))$$
>
> $$s(2) = \forall k(k > 2 \rightarrow \neg T(s(k)))$$
>
> $$s(3) = \forall k(k > 3 \rightarrow \neg T(s(k)))$$
>
> $$\vdots$$
>
> Expressed with help from the infinitary system $\mathcal{L}_{\omega_1\omega}$, we can say that
>
> $$s(0) = \bigwedge_{k>0} \neg T(s(k)),\, s(1) = \bigwedge_{k>1} \neg T(s(k)),\, s(2) = \bigwedge_{k>2} \neg T(s(k)) \ldots$$

[22] In the articles discussing Yablo's paradox, writers refer to an infinitary version of Yablo's paradox. For example, both Priest (1997) and Beall (2001) refer to such a formulation in an unpublished note by Forster (n.d.), but the formulation simply isn't there. Forster (n.d.) provides only the one-line kernel of an $\mathcal{L}_{\omega_1\omega}$-based definition of our function s. Likewise, Priest (1997) says that Hardy (1995) provides an infinitary, ω-rule-based version of Yablo's paradox, but actually Hardy only proves, indirectly, that Yablo's paradox entails ω-inconsistency. We specify an infinitary version of Yablo's paradox, expressed in the "background" logic that allows for meta-proofs regarding infinitary logical systems like $\mathcal{L}_{\omega_1\omega}$ (this particular system will turn out to be at the heart of the argument given in the next chapter). This system is presented in encapsulated form in (Ebbinghaus et al. 1984), from which the student interested in infinitary logic can move to (Karp 1964), then to (Keisler 1971), and then to (Dickmann 1975).

Next, suppose that $T(s(0))$. From this it follows immediately that $\neg T(s(1)) \wedge \neg T(s(2)) \ldots$, which in turn implies by conjunction elimination in $\mathcal{L}_{\omega_1\omega}$ that $\neg T(s(1))$. But in addition, if $T(s(0))$ is true, it follows again that $\neg T(s(1)) \wedge \neg T(s(2)) \ldots$, and hence that

$$\neg T(s(2)) \wedge \neg T(s(3)) \ldots,$$

which implies that $T(s(1))$. By *reductio* we can infer $\neg T(s(0))$. The same indirect proof can be given to show

$$\neg T(s(1)), \neg T(s(2)), \neg T(s(3)), \ldots.$$

Hence we can infer by the ω-rule

$$\frac{\alpha(1), \alpha(2), \ldots}{\alpha(n)}$$

that

$$(\star) \quad \bigwedge_{k \in \mathbf{N}} \neg T(s(k)).$$

Hence $\neg T(s(1)), \neg T(s(2)), \neg T(s(3)), \ldots$, that is, $T(s(0))$. But $\neg T(s(0))$ follows from $(\star)$ — contradiction.

Our point is not that Bringsjord's infinitary version of Yablo's Paradox in insoluble. The point is that we seem to have met the challenge to characterize the sort of paradox that should give Penrose pause. $\mathfrak{R}$, for all we know, involves some hidden contradiction of the sort exposed by the infinitary version of Yablo's Paradox, a contradiction that is much more subtle than that produced by Russell's simple paradox.[23]

While Graham Priest (1997) has arguably resolved the original version of Yablo's Paradox by showing that it's self-referential, the

[23]In personal conversation Penrose seemed to be inclined to complain that the original version of Yablo's paradox appears to go beyond what can be derived from axiomatic set theory (e.g., ZFC). This reply is odd, for *SOTM* is quite literally filled with reasoning that appears to go beyond first-order logic. (E.g., consider the diagrammatic "proof" concerning hexagonal numbers we visited above. Such diagrams seem to move beyond first-order logic (Bringsjord & Bringsjord 1996).)

Penrose's arguments in the end any improvement over Lucas' (1964) tiny-by-comparison primogenitor? By our lights, the answer is both "Yes" and "No." The answer is "Yes" because certainly Penrose has fleshed out the line of thought only adumbrated by Lucas. After all, *SOTM* is a big, beautifully written tour, the guide for which is an engaging polymath. Those who read it ought not to be thereby convinced that minds aren't machines (they should be convinced by the likes of the arguments listed in Table 1), but they will learn lots of things about computability theory, mathematics, and physics. On the other hand, the answer is "No" because, alas, Penrose's core Gödelian arguments are no better than the infamously vague one given by Lucas. This we have painstakingly shown. Now, this raises the obvious question: Should we conclude that a denial of the computational conception of mind simply cannot be deduced from Gödelian results? To infer an affirmative answer on the basis of what we have shown would be a non sequitur.

On the other hand, LaForte, Hayes, and Ford have recently argued that Gödel's theorem (and related results) cannot refute computationalism, period (LaForte et al. 1998). Here is what they say:

> *Any* attempt to utilize the undecidability and non-termination results to attack the computationalist thesis is bound to be illegitimate ..., since these results are quite consistent with the computationalist thesis. Theorems of the Gödel and Turing kind are not at odds with the computationalist vision, but with a kind of grandiose self-confidence that human thought has some kind of magical quality which resists rational description. The picture of the human mind sketched by the computationalist thesis accepts the limitations placed on us by Gödel, and predicts that human abilities are limited by computational restrictions of the kind that Penrose and others find so unacceptable. The tension which Penrose and others perceive arises only if one adds further assumptions, often about the nature of truth, human insight, or computation itself, which are already incompatible with the computationalist hypothesis, and indeed often have been explicitly rejected by those working in these areas. (LaForte, Hayes, and Ford 1998, p. 285)

This is marvelously ironic. The L-H-F trio reason here exactly as Penrose does: informally. Identify the computationalist hypothesis L-H-F have in mind with our $\mathcal{A}$. Now, where is the *proof* that Gödel I and $\mathcal{A}$ and propositions expressing the powers of the human mind are consistent? L-H-F don't provide this proof; they don't even sketch it; they just baldly assert this consistency. At least Penrose has *tried* to offer arguments. What would happen if L-H-F did try to prove what they merely assert? They would quickly learn that the proof is rather hard to come by. To see this, return to a scheme introduced in Chapter 1: let the set $\mathcal{P} = \{P_1, P_2, \ldots, P_n\}$ enumerate the familiar properties often offered as candidates for capturing, together, the essence of personhood. As before, if $P_i \in \mathcal{P}$, then let $P_i^\star$ denote a technical, precise correlate for P_i suitable for use in careful argumentation. What L-H-F need to provide is a proof that for *all* $P_i^\star$, these propositions are consistent:

1. Persons have $P_i^\star$
2. Gödel I (or non-termination theorem)
3. $\mathcal{A}$

This proof would consist in specifying a model on which these propositions are at once demonstrably true, and would probably be in the running for one of the most complicated proofs ever conceived!

The upshot is that the future, with respect to Penrose's goal, is still open. Bringsjord is in complete and utter agreement with Penrose that it *is* possible to formally derive the denial of "Strong" AI from Gödelian results, and will soon produce the demonstration (again, see Table 1). This demonstration is based on a *modalized* Gödelian refutation of computationalism anticipated in "Chapter VII: Gödel" of *What Robots Can and Can't Be* (Bringsjord 1992), wherein the kernel of a modal argument against computationalism is presented:

> I should point out that my [just-given] Gödelian argument against the proposition that persons are automata can apparently be modalized ... In this modal argument the central claim would be only that Ralf is such that it's *logically possible* that he act as I have described him acting above via the Fixed Point Theorem, whereas no Turing machine is such that it is logically possible that it act in this way. (Bringsjord 1992, p. 264)

The full specification of the modal Gödelian argument against computationalism will soon be finished, and published.

2.13 Distilling Penrose's Promising Intuitions

Despite succumbing to errors like the ones we have exposed earlier in this chapter, it seems to us that Penrose's *intuitions* about the hypercomputational nature of human mathematical intuition haven't been fully appreciated. Could it be that Penrose has apprehended the hypercomputational nature of human mathematical reasoning, but is unable to convince others in a rigorous third-person scheme (logic and technical philosophy) that this is the case? That Penrose's attempts to articulate a Gödelian case against computationalism are fatally flawed, while his intuitions are sound, is a view that Turing himself would have found palatable, at least in general. In his (1938) dissertation at Princeton University (later published as Turing 1939), Turing distinguished between "intuition" and "ingenuity" in logic and mathematics. He wrote:

> Mathematical reasoning may be regarded rather schematically as the exercise of a combination of two faculties, which we may call *intuition* and *ingenuity*. The activity of the intuition consists in making spontaneous judgments which are not the result of conscious trains of reasoning ... The exercise of ingenuity in mathematics consists in aiding the intuition through suitable arrangements of propositions, and perhaps geometrical figures or drawings. (Turing 1939, p. 214–215)

We believe that while Penrose's ingenuity, so preeminently crisp in mathematical physics, has failed him in the Gödelian sphere, his *intuition* in this sphere *is* in fact exactly right. In order show that this position is more than charity, we begin by turning to some of Penrose's examples of mathematical thinking.

The first example of such thinking given in *SOTM* is in section 1.19, and involves a visualization-based grasping of the arithmetical fact that

$$a \times b = b \times a, \text{where } a, b \in \{0, 1, 2, \ldots\}.$$

As Penrose explains, in the particular case where $a = 3$ and $b = 5$ we have, for $a \times b$:

$$(\bullet\bullet\bullet\bullet\bullet)(\bullet\bullet\bullet\bullet\bullet)(\bullet\bullet\bullet\bullet\bullet)$$

whereas for $b \times a$ we have

$$(\bullet\bullet\bullet)(\bullet\bullet\bullet)(\bullet\bullet\bullet)(\bullet\bullet\bullet)(\bullet\bullet\bullet).$$

But he reminds us that the fact in question can be grasped by visualizing the array

● ● ● ● ●
● ● ● ● ●
● ● ● ● ●

and "rotating the image through a right angle in one's mind's eye, to see that the array representing 5×3 has the same number of elements as that representing 3×5" (Penrose 1994, p. 55). Now we know that it seems to most that this kind of operation is easy enough for computers to carry out. The reason is that computers can process what one of us (Bringsjord & Bringsjord 1996) has called 'simple diagrams,' or S-D's for short. An S-D is a diagram that can be fully represented in standard first-order logic (FOL); and standard FOL is comfortably processed by standard computers (e.g., see Russell & Norvig 1994). Obviously, arrays can be represented via tuples in FOL. Suppose that # is a function mapping $n \times m$ arrays of dots into the natural number corresponding to the number of dots in the array. Let the array above be denoted by the constant a. Then $\#(a) = 15$. Next, suppose that function r^{90c} captures the rotation of an array "through a right angle." Hence

$$\#(a) = \#(r^{90c}(a)).$$

It is also easy enough to give a procedural representation of the objects and operations Penrose describes. For example, a k-tape Turing machine (with multiple tapes) can be programmed to carry out a direct analogue for the mental rotation in question.[26] All of this isn't likely to worry Penrose, for it's the *general* truths that are at issue for him — truths like (where we are assumed to be quantifying over arrays)

$$\forall x \#(x) = \#(r^{90c}(x)).$$

Notice in this truth the quantifier ranges over an infinite number of arrays; it will customarily be established via mathematical induction.

[26]The array Penrose presents can be positioned on three tapes of a five-tape Turing machine, and the operation of rotation can result in three dots being placed contiguously on all five tapes. For a nice introduction to Turing machines with k tapes, see the textbook: (Lewis & Papadimitriou 1981). For a discussion of much more sophisticated image-processing AI systems than this hypothetical Turing machine, in the context of imagistic reasoning seemingly more exotic than what Penrose describes here, see (Bringsjord & Bringsjord 1996).

But note as well that, courtesy of the imagistic cognition Penrose describes, this "infinitary" proposition can be instantly grasped. After all, just imagine the arrays growing incrementally in size, and imagine the rotation performed on these ever-increasing arrays. It's obvious that rotation will preserve equality in each case, is it not? In Turing's terms, intuition suffices to see what's true, and ingenuity can be brought to bear to specify the (in this case tedious) inductive proof.

Here's a second, more interesting example of mathematical thinking in *SOTM*, one involving the hexagonal numbers we visited earlier in the chapter,

$$1, 7, 19, 37, 61, 91, 127, \ldots$$

i.e., the numbers that can be arranged as ever-increasing hexagonal arrays (recall Figure 2.1). Consider again the cubes:

$$1 = 1^3, 8 = 2^3, 27 = 3^3, 64 = 4^3, 125 = 5^3, \ldots$$

And, as before, let Turing machine $M_{\bar{c}}$ be defined as follows. $M_{\bar{c}}$ adds together the hexagonal numbers successively, starting with 1, checking to see if each sum is a cube. If so, the machine keeps working away; if not, it halts. As you now know, $M_{\bar{c}}$ doesn't halt: that the pattern

$$1 = 1, 1 + 7 = 8, 1 + 7 + 19 = 27, 1 + 7 + 19 + 37 = 64, 1 + 7 + 19 + 37 + 61 = 125, \ldots$$

continues forever is a conventionally established theorem. But as Penrose points out, this theorem can be grasped by a human mathematician courtesy of another, more elaborate imagistic trick. He writes:

> First of all, a cube is called a cube because it is a number that can be represented as a cubic array of points as depicted in Figure [2.3]. I want you to try to think of such an array as built up successively, starting at one corner and then adding a succession of three-faced arrangements each consisting of a back wall, side wall, and a ceiling, as depicted in Figure [2.4]. Now view this three-faced arrangement from a long way out, along the directions of the corner common to all three faces. What do we see? A *hexagon* as in Figure [2.5]. The marks that constitute these hexagons, successively increasing in size, when taken together, correspond to the marks that constitute the entire cube. This, then, establishes the fact that adding together successive hexagonal numbers, starting with 1, will always give a cube. Accordingly, we have indeed ascertained that [$M_{\bar{c}}$] will never stop. (Penrose 1994, p. 70)

We suspect that once again Penrose's critics will almost by reflex maintain that the operations he describes here are things that computers can be (and routinely are) programmed to do. This time the demonstration that the diagrams in question are S-D's would be a tad more involved. We would need to represent, in FOL, three-dimensional grids and mechanical operations on them. Then a traditional-style proof would be crafted on the basis of these representations; once again mathematical induction would presumably be used. But such a proof is without question *not* what Penrose has in mind. He has in mind three-dimensional objects, objects that once again are operated upon and inspected, in thought-experimental fashion. The key is *seeing*, with one's mind's eye, that the three-faced objects, as they grow *ad infinitum*, will always be, from a certain perspective, hexagons. Once one visualizes the situation "stretching to infinity," a conventional proof is superfluous, epistemically speaking. A journal may insist upon receiving the proof, but *you* know the truth without even bothering to write the first syllable of a conventional inference.

In the next chapter, to which we now turn, we endeavor to demonstrate that infinitary reasoning of the sort that Penrose may be charitably understood to be pointing toward requires information processing at the level of hypercomputation.

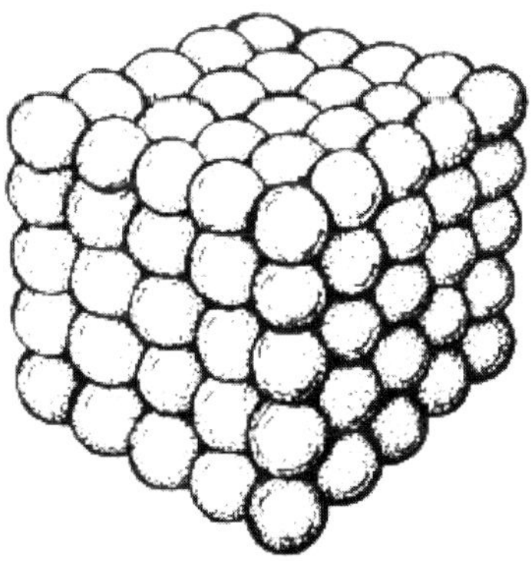

Figure 2.3: Cubic Array of Spheres

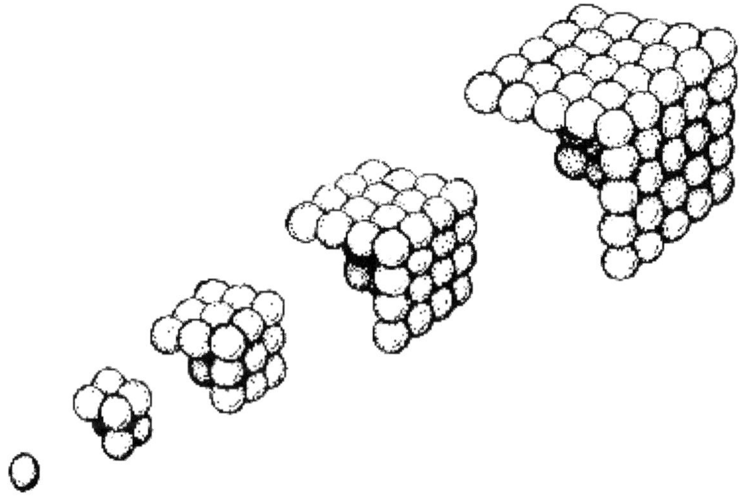

Figure 2.4: Each With a Back Wall, Side Wall, and Ceiling

Figure 2.5: The Points Together Form a Hexagon

Chapter 3

The Argument from Infinitary Reasoning

3.1 Introduction

We ended the last chapter by noting that though Roger Penrose's case against computationalism fails, his *intuitions* about the hypercomputational nature of mathematical reasoning may nonetheless be correct. Specifically, as we show in the present chapter, the *infinitary* nature of some of this reasoning may be part of what makes us superminds. As such, this chapter provides a much more direct route than the Gödelian one Penrose has long been fighting to successfully take.

Of course, it must be immediately admitted that much of the reasoning carried out by human persons is clearly computable. For example, suppose Tal is an expert at taking arbitrary natural numbers, n and m, and quickly returning the sum $n + m$. Many (including, no doubt, some of our readers) can sit down and promptly build an artificial agent that captures Tal's expertise.[1] Of course, Tal is a bit limited, to put it mildly. Yet Tal does appear to share a certain feature with the cognizers AI has traditionally been concerned with "capturing" — diagnosticians, pilots, auditors, and so on. Tal's reasoning, like the

[1]In Common Lisp, presto: `(defun tal (n m) (+ n m))`. This is an **artificial agent** in the sense of that phrase used in AI; e.g., see (Russell & Norvig 1994). Purists demanding to see the artificial agent here cast as reasoning in first-order logic (FOL) can rest assured (as explained in the present chapter) that (i) Common Lisp functions can be expressed as Turing machines, and (ii) Turing machines can be expressed as deduction in FOL.

reasoning possessed by such people, can apparently be captured, at least to a significant degree, by present-day AI technology — technology which is based on computation, and as such is based, in turn, on finitary logic, first-order logic (FOL) to be exact. (The computation-logic equivalence was proved by those — e.g., Alonzo Church and Alan Turing — who gave us computation. We review the equivalence later in this chapter.) But what if we try to cast a broader, braver net in an attempt to catch cognition which doesn't, at least first glance, look like it *can* be rendered in computational terms? Specifically, what about cognition in domains that from the outset put a premium on "infinitary" reasoning? Is such cognition also capturable by computation? Or is *hyper*computation needed for this cognition? In this chapter we focus on certain elementary infinitary reasoning in the domain of mathematical logic. We argue that this reasoning is indeed uncomputable. We end this chapter by discussing briefly the implications of this argument for the practice of AI, and by so doing bridge to the fuller discussion of this issue in the final chapter of the present volume.

Though the argument given in this chapter stands or falls on its own merits, it may be helpful if we contextualize things a bit. Accordingly, let us explain at the outset that in our view there is a genuine difference between truly understanding an infinitary concept and not understanding it. We're concerned herein with those persons who truly understand infinitary concepts (we give some key examples below). We imagine that the understanding achieved by an expert in *many* fields differs from our dim understanding *in kind*.[2] We think students often become aware that the researchers they study (in the area of mathematics, physics, etc.) have a different understanding *in kind* than they do. This chapter is an attempt to put in precise, concrete terms the notion that mathematical reasoning can be quite a "different beast" — that it's possible to obtain a level of mathematical understanding which is uncomputable. In other words, we will show that some mathematical reasoning can only be carried out by superminds.

---

[2] A nice paper on this issue is (Hoffman 1993), wherein (put roughly) the authors argue for the view that experts simply "see" things novices don't. In addition, there seems to be an emerging consensus in education that there is a genuine distinction to be made between "shallow" versus "deep" learning; see (Graesser, VanLehn, Rose, Jordan & Harter 2001, Aleven, Koedinger & Cross 1999).

3.2 Discarding Some Initial Objections

There is a certain objection (which we promised in Chapter 1 to discuss here) that's worth considering even before we move in earnest toward articulating the argument of this chapter, viz.,

> "If only mathematical 'experts who truly understand infinitary concepts' have, or (to stick more closely to your views) *harness* hypercomputation, then so what? Most people aren't such experts. Hypercomputation thus seems inessential to personhood. Your results in this chapter would put up no bar to robots being *persons*. You would merely have shown that robots would be well-advised not to pursue careers in mathematical fields requiring understanding of infinitary things. To put the point another way, maybe that blow to the head I took as a child knocked out my hypercomputational machinery. So much for my career in infinitary mathematics; but I'm still a person."

This objection, as you may recall, grew out of an objection discussed in Chapter 1 — the objection concerning the *general* relationship between personhood and hypercomputation: that hypercomputation isn't essential for personhood. What about the specific point pressed against us here? The answer is really quite simple. The fact is that our focus in this chapter on infinitary reasoning reflects our own intellectual activity. Predictably, it's this activity with which we are intimately familiar. Penrose (e.g., Penrose 1994) has admitted basically the same thing in connection with his repeated attempts to show that mathematical reasoning is beyond the reach of Turing machines and the like: he talks about mathematical reasoning because that is what he knows. He is, after all, a *mathematical* physicist. Likewise, what we know is reasoning; and so we isolate and write *about* reasoning. Presumably a parallel version of the present chapter could be composed for many other aspects of cognition.[3]

[3]We should mention that the example given at end of the preceding chapter as a possible example of infinitary/hypercomputational mathematical reasoning is but one out of many that could be given. For example, it strikes us as entirely possible that the development of infinitesmals happened on the strength of such reasoning. For coverage of this history, see (McLaughlin 1994, Davis & Hersh 1972, Nelson 1977). Leibniz seems to us to have engaged in infinitary reasoning (at the intuitive level; recall the distinction between intuition and ingenuity made in the previous chapter) when discussing infinitesmals.

3.3 The Need for Open-Mindedness

Readers who come to this chapter determined to forever *assume* that all persons are capable of only finite reasoning are encouraged to simply put this book down at this point. If you don't have an open mind about the possibility that some human persons engage in genuinely infinitary (hypercomputational) reasoning, your reading this chapter will be a complete and utter waste of time. This chapter, after all, presents an argument that such an assumption is false.

This is probably the right spot to point out that some well-known computationalists have suppressed careful consideration of mathematical reasoning when setting out and defending computationalism, even when their main topic is reasoning. Johnson-Laird is a perfect case in point. In his highly influential *Mental Models: Towards a Cognitive Science of Language, Inference, and Consciousness* (1983) he writes:

> Mental models can contain only a finite number of entities, but we can reason about infinite quantities and sets of infinite size such as the natural numbers. There is accordingly a distinction between naive or intuitive reasoning, which is directly based on mental models, and mathematical reasoning, which relies on other mechanisms. It is not my task in this book to account for mathematical reasoning ... (Johnson-Laird 1983, p. 444)

The astonishing thing about this quote is that three pages later Johnson-Laird says this:

> But are there other forms of mental representation [than those I have covered in this book]? Apart from the sensory precursors to models produced by the perceptual system, there are no grounds for supposing that higher cognitive processes depend on other forms of representation. Thus what we remember consists of images, models, propositions, and procedures for carrying out actions. (Johnson-Laird 1983, p. 447)

But, pray tell, how can Johnson-Laird know that certain new forms of mental representation are required in mathematical reasoning? After all, he has just told us that his book is only concerned with naive reasoning! The present chapter can be viewed as a focussed challenge to Johnson-Laird and others who have never carefully examined the nature of mathematical reasoning (which is almost from the first moment infinitary in nature). As you will see, the examination quickly reveals that people can develop a mode of reasoning that is beyond computation at or below the Turing Limit. (Later in the chapter we present a new theory of reasoning, **mental metalogic**, which owes a great deal to Johnson-Laird's theory of of reasoning: mental models theory.)

Sometimes those who simply refuse to admit the possibility that human persons reason in infinitary fashion offer "arguments" for their position. For example, here's an argument a number of thinkers (e.g., Eric Steinhart, via personal communication) have presented to us:

$\mathbf{Arg}_1$

	(1)	Human brains are finitely complex.
	(2)	All human mental representations are physical patterns in their brains.
∴	(3)	All human mental representations are finitely complex.

In order to express the core of this argument in such a way that the present chapter is threatened, we would need to expand it to:

$\mathbf{Arg}_{1'}$

	(1)	Human brains are finitely complex.
	(2′)	All mental representations exploited by persons in their reasoning are physical patterns in their brains.
∴	(3)	All mental representations exploited by persons in their reasoning are finitely complex.
	(3a)	If all mental representations exploited by persons in their reasoning are finitely complex, then all reasoning carried out by persons is finitary.
∴	(4)	All reasoning carried out by persons is finitary.

But this argument is not compelling, for these and other reasons: Premise (3a) is inconsistent with the undeniable fact that proofs can be infinitely complex, while each step is finite. Premise (1) assumes something that Bringsjord (2001*b*) has suggested might well be false: that human brains are not incarnations of analog chaotic neural networks (which are infinitely complex). Finally, though Bringsjord believes with Gödel that the brain is probably, at bottom, an ordinary digital computer falling at and below the Turing Limit (the brain is *certainly* such that it can be profitably — scientifically speaking — *treated* as if it is such a thing), it hardly follows from this that *persons* are in this category. Indeed, the penultimate chapter of this book explains why it is that persons may be — to use a phrase with which Bolzano and Descartes and Chisholm are comfortable — **simple souls**.

Let us conclude this section by pointing out that there is an abundance of closed-mindedness with regard to infinitary reasoning. As another confirmatory specimen, we offer a recent "argument" from Priest (1997) against the ω-rule, which for reasons to be explained later in the chapter is central to our case for infinitary reasoning. You will recall that this rule was critical to the infinitary version of Yablo's Paradox we presented in the previous chapter. This rule allows one to infer that n has property α when 1 has this property, 2 has this property, and so on *ad infinitum*. The rule, as you saw in the previous chapter, can be schematized like this:

$$\frac{\alpha(1), \alpha(2), \ldots}{\alpha(n)}$$

Even before we reveal the centrality of this rule to our case for human infinitary reasoning, it should be clear that it *is* in some determinate sense infinitary. While we think humans do genuinely use this rule, and while whether or not they do is something we seek to place upon the table courtesy of the present chapter, Priest (1997) asserts that human persons obviously cannot apply the rule:

> [The suggestion that we can use the ω-rule] would be disingenuous, though. As a matter of fact, we did not apply the ω-rule, and could not have. ... Moreover, no finite reasoner ever really applies the ω-rule. The only way that they can know that there is a proof of each $\alpha(i)$ is because they have a uniform method of constructing such proofs. And it is this finite information that grounds the conclusion [in this case]. (Priest 1997, p. 239).

This is not a strong argument, to put it mildly. What premises and inferences therefrom support the proposition that we didn't and can't use the ω-rule? Priest simply says that "as a matter of fact" this proposition is true; he thus gives us a bald *petitio*. In the reasoning given in the infinitary version of Yablo's Paradox presented in the previous chapter, we did not reason over variables for natural numbers, as can be seen by inspection. However, if we're charitable, we can view the second part of the quote as an argument, viz., an enthymematic version of

(5) No finite reasoner ever really applies the ω-rule.

(6) We are finite reasoners.

(7) ∴ We don't ever really apply the ω-rule.

This argument would appear to be formally valid; at any rate, we concede that it is. But are the premises true? Well, what is a finite reasoner, exactly? One answer, flowing from a computationalist conception of mind, is to say that a finite reasoner is essentially a Turing machine or an equivalent, or a computing machine of lesser power. We can then appeal to well-known theorems to support (5): a Turing machine, rendered in declarative form, corresponds to no more than finite, standard deduction from finite axiom sets, all expressed and carried out in $\mathcal{L}_I$, first-order logic (FOL); indeed, we are about to present this formal material. But why, then, are we supposed to affirm the second premise, (6)? After all, the pivot around which this very book revolves is whether or not we *are* Turing machines (or "less")! Furthermore, as we've noted earlier, there are many arguments against the view that human persons are Turing machines (recall Table 1). If anything is obvious here, it's that Priest is in no position to simply assume (6).

3.4 Plan of the Chapter

Our plan for this chapter runs as follows. In section 3.5 we review the reasons why computationalism implies that reasoning as computation is fundamentally reasoning as deduction in FOL. In section 3.6 we isolate and explicate one aspect of elementary logical reasoning (reasoning with the infinitary system $\mathcal{L}_{\omega_1\omega}$) which can serve as a paradigmatic instance of the sort of infinitary reasoning one regularly sees in the domain of logic and mathematics. In section 3.7 we present the argument that the reasoning explained in section 3.6, in light of the mathematical facts presented in section 3.5, is uncomputable, and hence isn't capturable by ordinary computation. Section 3.8 is a dialectic arising from objections to the argument of section 3.5. In section 3.9, in anticipation of our final chapter, we present an encapsulated version of what we see as the implications of our argument for the practice of AI in the area of reasoning.

3.5 Reasoning as Computation in First-Order Logic

The view that cognition — and, therefore, even sophisticated mathematical reasoning — is computation is one we of course discussed in Chapter 1, and needs no re-introduction here. Computationalism immediately implies that cognition, and hence specifically reasoning, is deduction in FOL. This implication is based on well-known theorems (e.g., the undecidability of FOL: Ebbinghaus et al. 1984, Boolos & Jeffrey 1989) which show that for every computation there is an equivalent deduction in FOL. So that there is no misunderstanding here, we'll review the relevant facts.

As we have noted earlier in this volume, computationalism relies upon automata like Turing machines (TMs) and other equivalent formalisms to render the concept of computation sufficiently rigorous for relevant science and engineering pursuits. But what exactly is a Turing machine?

Put intuitively, TMs include a **two-way infinite tape** divided into squares, a **read/write head** for writing and erasing **symbols** (from some finite, fixed **alphabet**) on and off this tape, a **finite control unit** which at any step in a computation is in one particular state from among a finite number of possible states, and a set of **instructions** (= program) telling the machine what to do, depending upon what state it's in and what (if anything) is written on the square currently scanned by it's head.

There are many readily understandable ways to capture the full set-theoretic description of TMs. One such method is the state diagram approach, which is used in Figures 3.1, 3.2, 3.3, and 3.5. The first of these Turing machines, dubbed "Gordon's 19 in 186," is designed to start on a 0–filled infinite tape and produce, after 186 steps, 19 1's. There is a bit of a story in connection with this machine, which was found by Gordon Greene using a Genetic Algorithm-powered search; this story sets a nice context not only for introducing TMs, but also for introducing a challenge to try to "compute" the uncomputable — a challenge we suggest in the last chapter that AI try in the future to meet — so we share the story here. Gordon Greene, one of Bringsjord's students a few years back, set out to find a machine that could beat the one known as "Greg's Challenge," the machine shown in Figure 3.3. When we say 'beat' here we refer to the 6-state busy beaver competition, in the 4-tuple framework.[4] This competition revolves around the "Busy Beaver" function, $f_{bb} : \mathbf{N} \longrightarrow \mathbf{N}$ ($\mathbf{N}$ here denotes the natural numbers), where $f_{bb}(n)$ yields the greatest number of 1's an n-state TM, starting on a black tape, can leave after halting. This function, and the corresponding proof that it's uncomputable, is due to Rado (1963), who used '$\Sigma(n)$' for '$f_{bb}(n)$.' Busy Beaver candidates (for a particular n) are those n-state TMs which may well produce $f_{bb}(n)$ 1's. When the book *Turing's World* (Barwise & Etchemendy 1993) was first published, Greg's Challenge was apparently the most productive 6-state machine known (again, in the 4-tuple framework). It produced 13 $*$'s; Gordon's 19 in 186 produced 19, a significant jump which established that $f_{bb}(6) \geq 19$. The next chapter was written by Chris Nielsen of the University of Washington, who showed, courtesy of a machine pictured in Figure 3.5, that $f_{bb}(6) \geq 21$. Even Chris' machine has been beaten, as we explain in the final chapter.

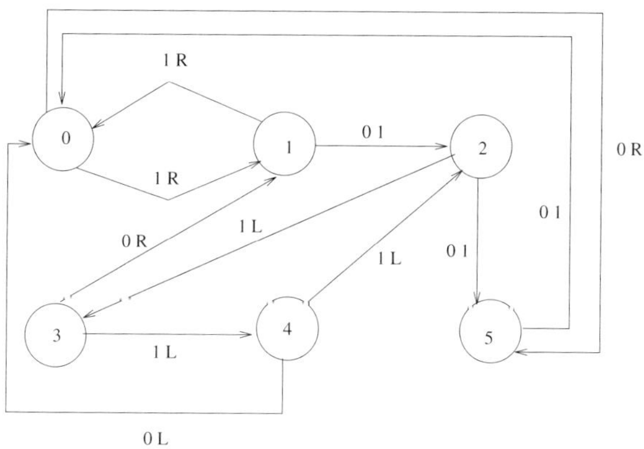

Figure 3.1: Gordon's 19 in 186, Shown in "Pencil-and-Paper" Flow-Graph Form

Now, let's "hand simulate" an initial segment of the computation of Gordon's TM — let's label the machine G — so that we completely fix the core mathematical concepts. The alphabet used is simply {0, 1}. The initial state of G is 0 (represented by the node labeled 0), and at the outset we'll assume that the tape is filled with 0's. The first thing G does is check to see what symbol it finds under its read/write

[4]In the 5-tuple framework, TMs, given what state they're in and what symbol they're currently scanning, can perform not just one action, but two, before entering the next state.

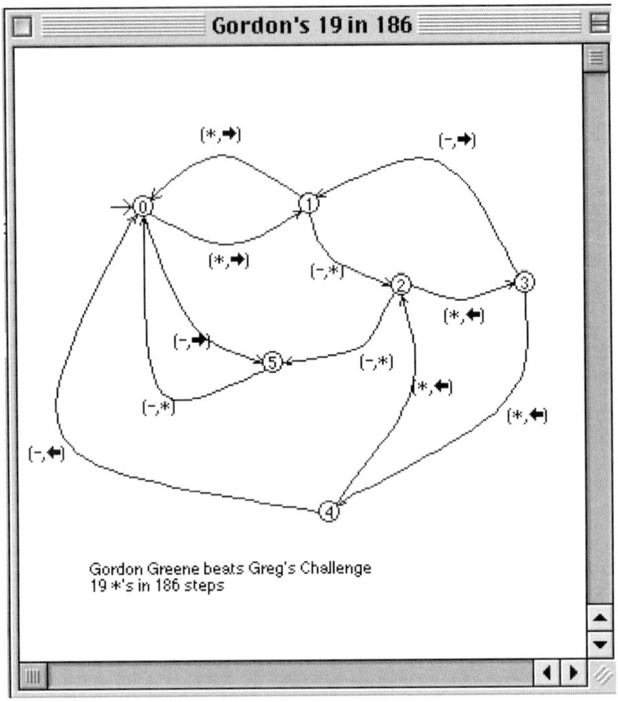

Figure 3.2: Gordon's 19 in 186, Shown in *Turing's World* Flow-Graph Form *Note that instead of the alphabet {0, 1} we use {–, ∗}. Movement to the left and right is indicated by arrows.*

head. In this case it initially finds a 0, so the arc labeled with 0 R[5] is taken, which means that the head moves one square to the right and the machine enters state 5. At this point, since there is another 0 found beneath the head, the 0 is changed to a 1, and the machine reenters state 0. It now finds a 1, and hence takes the arc labeled 1 R to state 1 (i.e., the machine moves its head one square to the right, and then enters state 1) — etc. The machine's activity can be perfectly captured by a tedious catalogue of its configurations from start to finish (Figure 3.4).

If this is your first exposure to TMs, you will probably be struck by how primitive and unassuming they are. But the surprising thing, noted in Chapter 1, is that TMs apparently capture standard compu-

[5]Note that this is an ordered pair composed of zero and R for 'right,' not the word 'OR'.

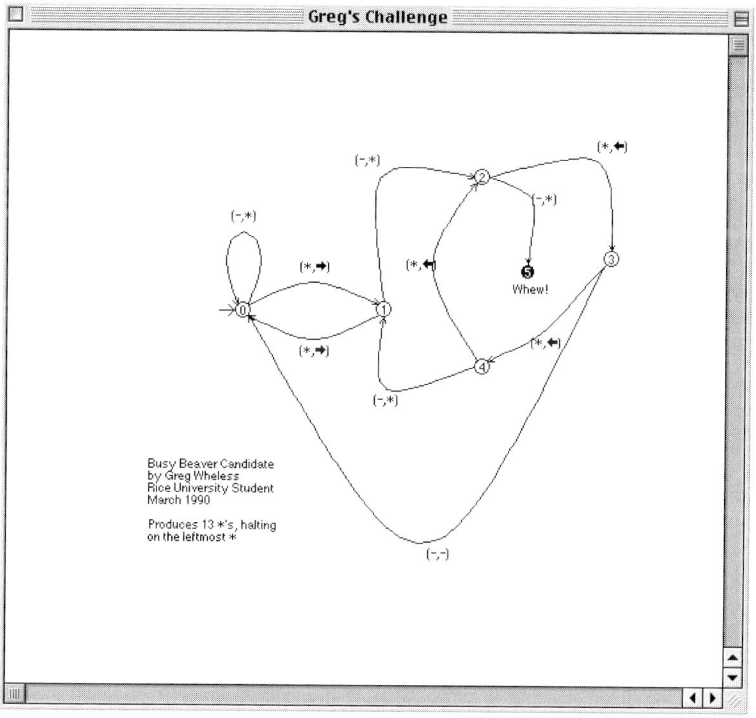

Figure 3.3: Greg Wheless' Original Candidate for a 6-state Busy, Shown in *Turing's World* Flow-Graph Form

tation *in all its guises*. More precisely, whatever can be accomplished by way of an algorithm, by way of a programmed supercomputer, by way of a neural network, a cellular automaton, etc. — whatever can be achieved by any of these can be matched by a TM.[6] Furthermore, we know that adding (non-hypercomputational) capabilities to our TMs doesn't give them any additional power.[7] For example, if we give a TM *two* tapes rather than one, nothing that was impossible for the one-tape machine becomes doable for the two-tape creature.[8]

---

[6]See (Bringsjord 1991) for a discussion of the consequences of this fact for AI.

[7]The capabilities added must, however, be expressible in the language of set theory. If one considers a *physical* Turing machine, then there are perhaps ways of "souping up" such machines so that they can process uncomputable functions. See Chapter 4 in the present volume, and (Bringsjord 2001*b*).

[8]The interested reader can consult an octet of books we find useful: For broad coverage of the basic material, see (Lewis & Papadimitriou 1981, Ebbinghaus et al. 1984, Boolos & Jeffrey 1989, Hopcroft & Ullman 1979). For a nice com-

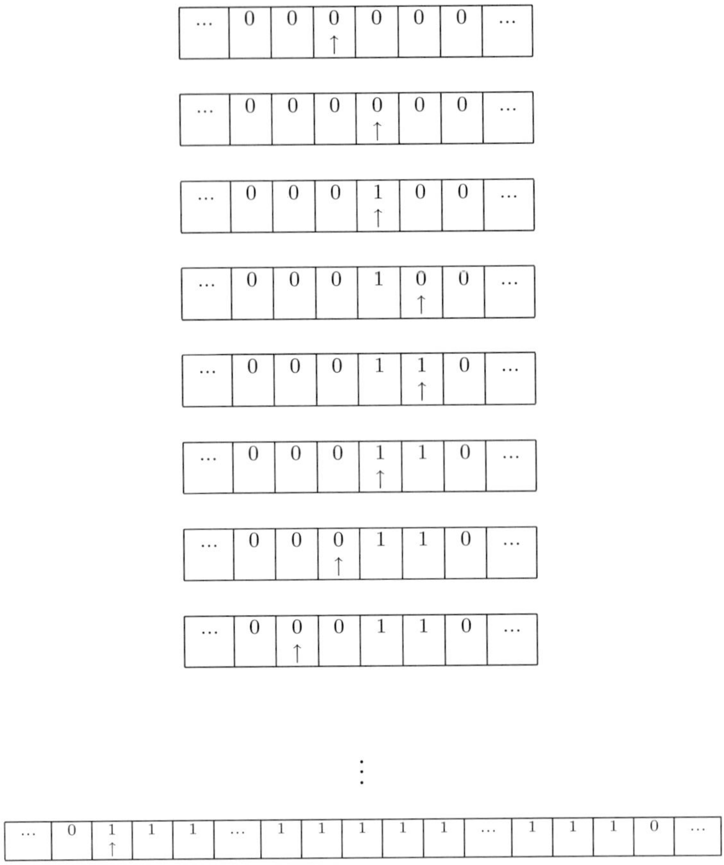

Figure 3.4: Tedious Catalogue of Gordon's TM

The next ingredient we need in order to appreciate the fact that computation is equivalent to deduction in FOL is a quick review of FOL itself: Given an **alphabet** (of **variables** $x, y, \ldots$, **constants** $c_1, c_2, \ldots$, n-ary **relation symbols** $R, G, \ldots$, **functors** $f_1, f_2, \ldots$, **quantifiers** $\exists, \forall$, and the familiar **truth-functional connectives** $(\neg, \vee, \wedge, \rightarrow$

---

prehensive discussion of computability theory that includes succinct coverage of uncomputability, including the Arithmetic Hierarchy, see the book from which we drew heavily in Chapter 1: (Davis et al. 1994); see also the difficult but rewarding (Soare 1980). (Partee et al. 1990), as mentioned in Chapter 1 (when we discussed the Chomsky Hierarchy), contains a very nice discussion of the Chomsky Hierarchy. And, of course, there's always the classic (Rogers 1967).

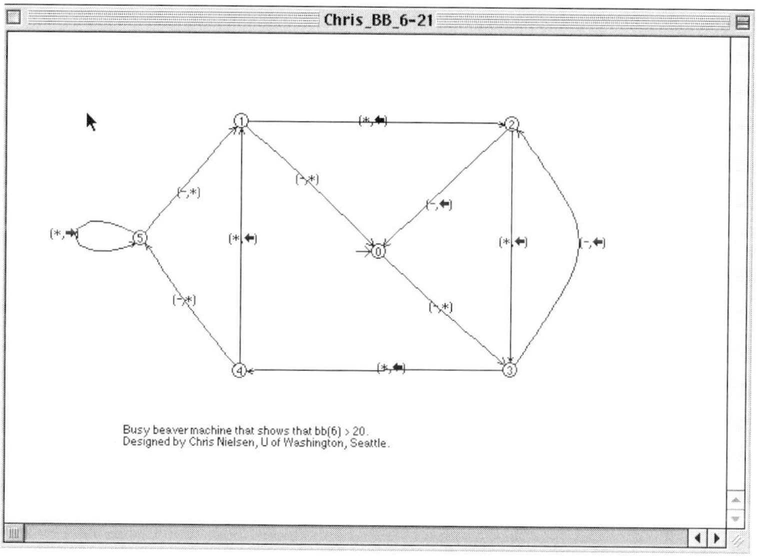

Figure 3.5: Chris Nielsen's TM Showing $f_{bb}(6) \geq 21$, Shown in *Turing's World* Flow-Graph Form; From Chris Nielsen of the University of Washington

, $\leftrightarrow$) one uses standard **formation rules** (e.g., if ϕ and ψ are well-formed formulas, then $\phi \wedge \psi$ is a wff as well) to build "atomic" formulas, and then more complicated "molecular" formulas. Sets of these formulas (say Φ), given certain **rules of inference** (e.g., *modus ponens*: from ϕ and $\phi \rightarrow \psi$ infer to ψ), can lead to individual formulas (say ϕ); such a situation is expressed by meta-expressions like $\Phi \vdash \phi$. First-order logic, like all **logical systems**, includes a semantic side which systematically provides meaning for formulas involved. In FOL, formulas are said to be true (or false) on an **interpretation**, often written as $\mathcal{I} \models \phi$. (This is often read, "$\mathcal{I}$ satisfies, or models, ϕ.") For example, the formula $\forall x \exists y Gyx$ might mean, on the standard interpretation $\mathcal{R}$ for arithmetic, that for every natural number n, there is a natural number m such that $m > n$. In this case, the domain of $\mathcal{R}$ is $\mathbf{N}$, the natural numbers, and G is the binary relation $> \subset \mathbf{N} \times \mathbf{N}$, i.e., $>$ is a set of ordered pairs (i, j) where $i, j \in \mathbf{N}$ and i is greater than j.

In order to concretize things a bit, consider an expert system designed to play the role of a guidance counselor in advising a high school student about applying to college. Suppose that we want a rule in such a system which says "If a student has low SATs, and a low GPA, then none of the top twenty-five national universities ought to be applied to by this student." Assume that we have the following interpreted predicates: Sx iff x is a student, L_sx for x has low SATs, L_gx for x has a low GPA, Tx for x is a top twenty-five national university, Axy for x ought to apply to y. Then the rule in question, in FOL, becomes

$$\forall x \forall y[(Sx \wedge L_sx \wedge L_gx \wedge Ty) \rightarrow \neg Axy].$$

Let's suppose, in addition, that Steve is a student denoted by the constant s in the system, and that he, alas, has low SATs and a low GPA. Assume also that v is a constant denoting Vanderbilt University (which happens to have been a top twenty-five national university according to some *U.S. News and World Reports'* annual rankings). These facts are represented in the system by

$$Ss \wedge L_ss \wedge L_gs \text{ and } Tv.$$

Let's label these three facts, in the order in which they were presented, (1), (2), and (3). Our system, based as it is on FOL, can verify

$$\{(1), (2), (3)\} \vdash \neg Asv,$$

that is, it can deduce that Steve ought not to apply to Vanderbilt.

The point of this example is but to ground our encapsulated presentation of FOL. Our point is *not* that the practice of AI is married to nothing more than FOL. We gladly concede here for the sake of argument that there is a way of doing AI which (unlike the use of probabilistic inference, frames, scripts, forward- and backward-chaining, standard neural nets, etc.) really does depart from FOL *qua* tool. In order to articulate our argument we need only the unassailable fact that standard computation and first-order deduction are equivalent and wholly interchangeable.[9] In order to make this fact clear for readers who may be unfamiliar with it, we return to the TM of Figures 3.1 and 3.2.

It's possible to recast every aspect of the TM "Gordon's 19 in 186" (called G earlier) as an assertion in FOL. We know that this machine,

[9]Of course, whether two formalisms are interchangeable *in practice* is a different question. As we pointed out in the Preface, it's well-known that an abacus (under the assumption that it can be expanded upon demand) is equivalent in power to a digital computer, but someone unskilled in the use of the ancient device certainly could not exchange it for a Macintosh. The bird shown in Figure 2 is skilled, in that it can implement the relevant instructions.

upon starting on a blank tape, will halt with its head over the leftmost 1 of a string of 19 1's on an otherwise blank tape. What we would like (and what Alonzo Church showed could be systematically obtained) is a set Γ of first-order formulas and an individual formula ϕ such that $\Gamma \vdash \phi$ if and only if Gordon's machine operates exactly as we by now know it does. In order to obtain this recasting of G, let's imagine that the squares of the tape upon which TMs operate are numbered in parallel with the integers (Figure 3.6).

...	-3	-2	-1	0	1	2	3	...

Figure 3.6: Numbered TM Tape

Next, let's assume that time works the same way: TMs perform exactly one operation at each "click of the clock," and there is a 0 moment at which the action starts (in this case, with G scanning square 0), with each action taking one more click, time marching on in step with the positive integers.

Now let's fix a first-order language, and interpret it: Let variables range over the integers; let 0 denote (naturally enough) the integer of that name; and let s denote the successor function defined according to tradition: $s(n) = n+1$. What we need now are three relation symbols (predicates):

- $Q_i tx$ if and only if at time t the TM in question is in state i scanning square number x;
- $S_j tx$ if and only if at time t the symbol j is in square number x; and
- $x < y$ if and only if x is less than y.

With the machinery now at our disposal, let's focus on one arc in "Gordon's 19 in 186," namely the one corresponding to the quadruple (0 0 R 5). The general form of such an arc is $(ijRm)$. We're now able to capture such an arc completely in FOL. Let's express it first in English:

> If the machine is in state i at time t and is then scanning square number x on which symbol j occurs, then at time $t+1$ the machine is in state m scanning square number $x+1$, and for every square y, whatever was written there at time t is still written there at $t+1$.

In FOL, using our machinery, this becomes

$$\forall t \forall x \forall y\{(Q_i tx \wedge S_j tx) \rightarrow (Q_m s(t)s(x) \wedge (S_0 ty \rightarrow S_0 s(t)y) \wedge \ldots \wedge (S_r ty \rightarrow S_r s(t)y))\}.$$

In the case of G, the tuple (0 0 R 5) becomes

$$\forall t \forall x \forall y\{(Q_0 tx \wedge S_0 tx) \rightarrow (Q_5 s(t)s(x) \wedge (S_0 ty \rightarrow S_0 s(t)y) \wedge (S_1 ty \rightarrow S_1 s(t)y))\}.$$

This of course isn't the place to present all the details (we're covering herein the minimal number required for accurate exposition). Suffice it to say that every aspect of a given computation (or set or sequence of computations) can be captured according to the procedure we've adumbrated. The upshot of this is that computation is provably equivalent to deduction in appropriately configured FOL. Under the assumption of computationalism, it follows that reasoning, one form or type of cognition, is in turn also provably equivalent to deduction in appropriately configured FOL. This fact has nothing to do with methodological issues involving the *use* of FOL to represent knowledge, reasoning etc. It's a brute fact that many AI practitioners reject FOL on principle (and *a fortiori* other more powerful logics) as a tool for reaching their scientific and engineering objectives. (This rejection not only runs counter to the central role logic continues to play in AI (e.g., see Russell & Norvig 1994), but, as we explain in in the final section of the final chapter, is demonstrably unwise.)

3.6 Sharpening Infinitary Reasoning

Mathematical logic deals with infinite concepts from the outset; such concepts permeate the field. For example, a first course in mathematical logic often begins with Cantor's famous theorem that there are sets (e.g., the reals, or the power set of the set of natural numbers) which are "larger" than the infinite set of natural numbers. But for our purposes herein we need to bring "infinitary" reasoning into sharp focus, otherwise the argument of this chapter will be hopelessly vague.

If this argument is to have any force, it can't be based simply on a hand-wavy reference to all of the infinitary aspects of logic.[10]

The best way we know of to make the notion of infinitary reasoning explicit is to isolate those systems studied by logicians that are *by definition* infinitary. We will focus here specifically on one infinitary system, the "smallest" and simplest one we know of, viz., $\mathcal{L}_{\omega_1\omega}$. The nice thing about this system is that it builds naturally upon FOL, which we've already discussed. In fact, one way to motivate the study of $\mathcal{L}_{\omega_1\omega}$ is to consider the fact that most mathematical concepts cannot be expressed in FOL (a system traditionally denoted by $\mathcal{L}_I$). (For a list of some of those ordinary mathematical concepts beyond the reach of FOL, see (Keisler 1971).) Two such concepts are that of a "finite world (model)," and "ordinary arithmetic." Let's look at these two concepts, and the inability of $\mathcal{L}_I$ to capture them.

In order to understand that it's not possible to capture the ordinary notion of finitude in FOL, consider the $\mathcal{L}_I$-sentence

$$\psi_{\geq 2} = \exists x \exists y \; x \neq y.$$

Any interpretation on which $\psi_{\geq 2}$ is true must have a domain that contains at least two elements, since $\psi_{\geq 2}$ "says" that there exist two distinct things x and y. Put in terms of the standard notation reviewed above, this becomes $\mathcal{I} \models \psi_{\geq 2}$ if and only if $\mathcal{I}$'s domain contains at least two elements. The technique here can be generalized. The sentence

$$\psi_{\geq 3} = \exists x \exists y \exists z \; (x \neq y \land x \neq z \land y \neq z)$$

can only be true in a world with an at-least-three-element domain, etc. Now suppose that we collect together the set of all such formulas, with n getting larger and larger forever. Formally, this set — call it 'Ω' — is $\{\psi_{\geq n} : n \geq 2\}$. Since any interpretation on which all of the members of Ω is true must be an interpretation with at least 2 members, 3 members, 4 members, *ad infinitum*, it follows that such an interpretation must be infinite. In addition, any infinite interpretation will satisfy Ω. So we have assembled a recipe for expressing, in FOL, the notion of infinitude. However, and this is the present point,

[10]We should probably point out at this juncture that by rendering things a bit more precise we're in no way retreating to the view that infinitary mathematical reasoning is constituted by merely dealing with infinite expressions. The "conceptual" take on Cantor's Theorem is, as will be seen, the sort of thing our argument requires: see Objection 5 below.

there is no set of first-order formulas which can express the concept of finitude. (Formally, there is no set of first-order formulas such that an interpretation models this set if and only if it's finite.)

Now to the second limitation of FOL that we want to bring to your attention, namely that it can't capture ordinary arithmetic. In arithmetic, the domain is the set that usually marks a student's first direct contact with the infinite, viz., **N**, the natural numbers $\{0, 1, 2, \ldots\}$, and our operators include the successor function (seen above when we discussed the equivalence of computation and FOL), addition, subtraction, multiplication, and so on. Constants consist of names for the first two natural numbers, 0 and 1. Using these constructs, it's possible to axiomatize all of arithmetic by way of the following three statements (Peano's axioms):

P1 0 is not the value of the successor function (i.e., there is no way to add 1 to a natural number and get back 0).

P2 The successor function is one-to-one (i.e., if you plug two distinct natural numbers in as input, you get back as value two different natural numbers).

P3 For every subset X of the set **N** of natural numbers, if zero is in X, and if the successor of n is in X whenever n is in X, then X is **N** itself.

In other words, all truths of arithmetic, from those we learned starting in Kindergarten to those which call for supercomputers, can be derived from P1–P3 using standard rules of inference. However, there is a small problem: axiom P3 cannot be expressed in FOL! The Peano axiom system is therefore often defined to be the so-called first-order "induction axioms." These axioms include these six

$$\begin{array}{ll} \forall x(x+1 \neq 0) & \forall x, y((x+1 = y+1) \rightarrow x = y) \\ \forall x(x+0 = x) & \forall x, y(x + (y+1) = (x+y)+1) \\ \forall x(x \times 0 = x) & \forall x, y(x \times (y+1) = (x \times y) + x) \end{array}$$

and, for all $x_1, \ldots, x_n, y$ and all formulas ϕ built from the appropriate alphabet (about simple arithmetic; see L^{ar} in Ebbinghaus et al. 1984) such that the free variables in ϕ are together a subset of $\{x_1, \ldots, x_n, y\}$,

$$\forall x_1, \ldots, x_n \left(\left(\phi \frac{0}{y} \wedge \forall y \left(\phi \rightarrow \phi \frac{y+1}{y} \right) \right) \rightarrow \forall y \phi \right)$$

(Note that $\phi\frac{x}{y}$ is the result of replacing x in ϕ with y.)

Because $\mathcal{L}_I$ is so limited, logicans have studied systems like $\mathcal{L}_{\omega_1\omega}$, which we now proceed to define. As we shall see, it turns out that the two limits on FOL just isolated can be handled easily by $\mathcal{L}_{\omega_1\omega}$.

The basic idea behind $\mathcal{L}_{\omega_1\omega}$ is straightforward. This system allows for infinite disjunctions and conjunctions,[11] where these disjunctions and conjunctions are no longer than the size of the set of natural numbers (let's use ω to denote the size of the set of natural numbers).[12] This fundamental idea is effortlessly regimented: First we simply add to the customary alphabet for FOL the symbols $\bigvee$ and $\bigwedge$. To the ordinary formation rules for building grammatically correct first-order formulas, we add

- If Φ is a set of well-formed formulas $\{\phi_1, \phi_2, \ldots\}$ no larger than ω, then $\bigvee \Phi$ ($\bigwedge \Phi$) is also a well-formed formula, viz., the disjunction (conjunction) of the formulas in Φ.

The conditions under which an infinite formula is true is fixed by extending the notion of truth in ordinary FOL:

- A possibly infinite disjunction, $\bigvee \Phi$, is true on an interpretation $\mathcal{I}$ (written $\mathcal{I} \models \bigvee \Phi$) if and only if there is a formula ϕ in Φ which is true on $\mathcal{I}$.
- A possibly infinite conjunction, $\bigwedge \Phi$, is true on an interpretation $\mathcal{I}$ (written $\mathcal{I} \models \bigwedge \Phi$) if and only if every formula ϕ in Φ is true on $\mathcal{I}$.

Proofs (= derivations) in $\mathcal{L}_{\omega_1\omega}$ can, as the relevant literature states, be "infinitely long" (Ebbinghaus et al. 1984). This is because in addition to classical cornerstones like *modus ponens*,

$$\text{from } \phi \rightarrow \psi \text{ and } \phi \text{ infer to } \psi,$$

$\mathcal{L}_{\omega_1\omega}$ allows rules of inference like

$$\text{from } \phi \rightarrow \psi \text{ for all } \psi \in \Phi, \text{ infer to } \phi \rightarrow \bigwedge \Psi.$$

[11]Of course, even finitary logics have underlying alphabets that are infinite in size (the propositional calculus comes with an infinite supply of propositional variables). $\mathcal{L}_{\omega_1\omega}$, however, allows for formulas of infinite length — and hence allows for infinitely long derivations. More about such derivations in a moment.

[12]This chapter is aimed at a multidisciplinary audience assumed to have familiarity with but the rudiments of logic and AI. So this isn't the place to baptize readers into the world of cardinal numbers. Hence we leave the size implications of the subscripts in $\mathcal{L}_{\omega_1\omega}$, and other related niceties, such as the precise meaning of ω, to the side. For a comprehensive array of the possibilities arising from varying the subscripts, see (Dickmann 1975).

This rule says that if in a derivation you have an infinite list of if-thens (i.e., formulas of the form $\phi \rightarrow \psi$) where each consequent (ψ) in each if-then is an element of some infinite set Φ, then you can infer to an if-then whose consequent is the infinite conjunction obtained by conjoining all the elements of Φ. It may be worth pausing a bit to create a picture of the sort of derivation which is here permitted: Suppose that Γ is an infinite set of the same size as $\mathbf{N}$, the natural numbers. So Γ is $\{\gamma_1, \gamma_2, \ldots, \gamma_n, \gamma_{n+1}, \gamma_{n+2}, \ldots\}$. Then here is one possible picture of an infinite derivation:

$\phi \rightarrow \gamma_1$ $\phi \rightarrow \gamma_2$ $\phi \rightarrow \gamma_3$ $\vdots$ $\phi \rightarrow \gamma_n$ $\phi \rightarrow \gamma_{n+1}$ $\vdots$
$\phi \rightarrow \gamma_1 \wedge \gamma_2 \wedge \ldots \wedge \gamma_n \wedge \gamma_{n+1} \wedge \gamma_{n+2} \cdots$

It should be clear from this that derivations in $\mathcal{L}_{\omega_1\omega}$ can indeed be infinitely long.

Now, let's return to the two limitations we saw in the case of FOL in order to see how $\mathcal{L}_{\omega_1\omega}$ surmounts them. The first limitation was that the concept of finitude couldn't be captured by any set of $\mathcal{L}_I$ formulas, let alone by one such formula. But here is one simple formula in $\mathcal{L}_{\omega_1\omega}$ which is such that every interpretation that satisfies it is finite, and every finite interpretation satisfies it:

$$\bigvee_{n<\omega} \exists x_1 \ldots \exists x_n \forall y (y = x_1 \vee \ldots \vee y = x_n).$$

We think it's worth pausing to make sure we understand this formula (and thereby understand some of the power of $\mathcal{L}_{\omega_1\omega}$). This formula is an infinite disjunction; each disjunct has a different value for n. One such disjunct is

$$\exists x_1 \exists x_2 \forall y (y = x_1 \vee y = x_2),$$

which says, put informally, there exist at most two things x_1 and x_2 with which everything in the domain is identical, or there are at most two things in the domain. Obviously, any interpretation which satisfies this formula is finite, since it can at most have a two-element domain.

Another disjunct in the infinite disjunction is the one generated by setting n to 4, i.e.,

$$\exists x_1 \exists x_2 \exists x_3 \exists x_4 \forall y(y = x_1 \vee y = x_2 \vee y = x_3 \vee y = x_4),$$

which says that there are at most four things. Here again, any interpretation which satisfies this formula is finite. But how do we go in the other direction? How do we ensure that any interpretation which is finite satisfies the selected formula? This is where the infinite disjunction does its job. For notice that every finite domain will have a certain size k, where k is some natural number. This domain will make true the disjunct in the infinite disjunction where $n = k$; and since a disjunction, no matter how big, is true if but one of its disjuncts is true, this k-sized domain will make the entire infinite disjunction true.

The second limitation on FOL we isolated above was that arithmetic cannot be formalized in this system, because the third of Peano's axioms cannot be formalized in it. The situation is instantly remedied in $\mathcal{L}_{\omega_1\omega}$, because (where, recall, $s(n)$ gives the successor of n) we can supplant P3 with the infinite formula

$$\forall x(x = 0 \vee x = s(0) \vee x = s(s(0)) \vee \ldots).$$

This formula says that everything in the domain is either 0, or the successor of 0, or the successor of the successor of 0, and so on. In other words, the domain in question is $\mathbf{N}$, the set of natural numbers, exactly as we want it to be. Compare this elegant formula with the rigamarole given a bit earlier to specify the first-order induction axioms.

We end this section by noting and explaining one simple, compelling reason why mathematical reasoning concerning $\mathcal{L}_{\omega_1\omega}$ cannot be recast as reasoning concerning ordinary FOL: The rationcination in question, as we've just witnessed, includes representing both the finitude of interpretations and Peano arithmetic in $\mathcal{L}_{\omega_1\omega}$. (Of course, in "real life," the thinking enjoyed by mathematicians and logicians goes well beyond what we've put on display.) So, if this reasoning *could* be recast in FOL, it would follow that the two limitations on FOL we noted above would not in fact be limitations, for the recasting of the reasoning here would constitute a rendering in FOL of precisely that which we know, on independent and indisputable grounds, to be beyond the reach of FOL.

3.7 The Argument from Infinitary Reasoning

Because we've taken pains to set the stage, articulating the argument is easy and direct:

(1) All reasoning is computable. [assump. for contradiction]

(2) For every case of reasoning R there exists a Turing machine (or other equivalent creature) M such that some computation C of M is such that $R = C$. [from 1]

(3) For every computation C of every TM M there is an equivalent deduction D in some instantiation of the first-order system $\mathcal{L}_I$. [sect. 3.5]

(4) For every case of reasoning R there exists a deduction D in some instantiation of the first-order system $\mathcal{L}_I$ such that $R = D$. [from 2, 3; universal elimination, hypothetical syllogism, universal introduction]

(5) There exists a case of reasoning R^* — viz., reasoning with $\mathcal{L}_{\omega_1\omega}$ — which is such that for every deduction D in some instantiation of the first-order system $\mathcal{L}_I$, $R^* \neq D$. [sect. 3.6]

(6) It's not the case that all reasoning is computable. [*reductio ad absurdum*; 4, 5 contradictory]

This argument is valid in the sense that the inferences are formally correct. This is so because the reasoning in question can be completely formalized using natural deduction rules. We will spare the reader the completely formal proof.

3.8 Dialectic

There are a number of objections which have been brought against this argument. We counter some of them in this section.

3.8.1 Objection 1: It All Hinges on Prior Beliefs

The first objection we consider seeks to reject the entire enterprise of ascertaining whether computationalism, and computationalist approaches to reasoning, are true or false: "As an argument that human thinking is beyond the bounds of computation, yours, guys, is convincing if, but only if, one already accepts that humans are capable of noncomputable feats. But if one starts by believing the opposite, then the argument seems fallacious. If you believe that people are super-mechanistic, the argument may support your belief, but if you believe that they can only compute, it becomes vacuous."

The astonishing thing about this objection is that it itself commits the fallacy of *petitio principii* (begging the question). In this chapter we have put on the table a rigorous case for rejecting computationalism. When someone puts on the table a rigorous, sustained case for not-p, it begs the question to say, "I believe p, so your argument is no good."[13]

The next problem with the present objection is that it seems to require that our argument (and presumably others like it) be *convincing*. But there's a big difference between a convincing argument and a sound argument. (Recall that we discussed this issue at length in Chapter 1. See, specifically, section 1.2.) Many proofs in the history of mathematics were for years after their publication almost universally rejected because acceptance necessitated a realignment of one's beliefs (Cantor's Theorem, mentioned above, is a perfect case in point). We can certainly see why a computationalist, a "true believer," would be disinclined to accept our argument. But such inclinations count for nothing in open-minded debate; logic, emotionless logic, counts for everything. If our argument is to be rejected, then we must be informed as to which premise(s) is (are) false, since the inferences therein are certifiable in accordance with principles affirmed by all computationalists.

On the other hand, the present objection at least has the virtue of being falsifiable. This is so because if some thinkers *are* convinced (or even *partly* convinced) by the argument, the objection here evaporates. And, fortunately (as we see it), many people have been convinced by the argument (and others that preceded it, such as those seen in (Bringsjord 1992)) that computationalism, as it stands, is inadequate. If you aren't yet convinced, wait: you may resist because you affirm one or more of the objections we have yet to consider.

[13]Interestingly enough, the only reason, or at least the main reason, we, and others like us, reject (or recast) computationalism is that we understand and are moved by arguments like the one we've just given! Before Bringsjord learned a thing or two about the math underlying what he was doing when trying to get a computer to do snazzy, intelligent things, he was not only a computationalist, he was a rabid, evangelistic computationalist.

3.8.2 Objection 2: It's All Just Manipulation of Finite Strings

The second objection marks a reaction we've heard from many; it begins as follows: "Your argument, Bringsjord and Zenzen, does not include a specification of what is meant by the terms 'finite,' 'infinite,' and 'infinitary.' Clearly, humans cannot actually manipulate an infinite expression, so to carry out 'infinitary reasoning' with $\mathcal{L}_{\omega_1\omega}$ must mean to reason with the manipulation of finite strings used to represent hypothesized infinite expressions. For example, look at the formula which you made so much of above, viz.,

$$\bigvee_{n<\omega} \exists x_1 \ldots \exists x_n \forall y (y = x_1 \vee \ldots \vee y = x_n).$$

You will notice that this formula is a finite string: it fits nicely on one line of this book. And of course we all know that Turing machines (and the like) have no trouble manipulating finite strings; that, after all, is the essence of what they do, as your own exposition confirms."

The first thing that needs to be said in reply to this objection is that with regard to terminology we are wholly innocent: We've used the terms 'finite,' 'infinite,' and 'infinitary' exactly as they are used in logic and mathematics. Text after text, and article after after, say that $\mathcal{L}_{\omega_1\omega}$ allows for infinite disjunctions and conjunctions; and these creatures are referred to by way of the sorts of finite strings which this objection features. Recall, as well, that we earlier quoted a standard text saying that $\mathcal{L}_{\omega_1\omega}$ allows for "infinite derivations" (a picture for which we offered). And of course it's not just the study of infinitary logic which gives rise to this way of speaking. Turing machines are said to have *infinite* tapes; there is said to be an *infinite* number of natural numbers; FOL is said to have an *infinite* supply of variables; π is said to have an *infinite* decimal expansion (a fact that proves to be strikingly relevant later in this, and in the next, chapter); there is said to be an *infinite* number of real numbers in the interval [0, 1] (a level of infinity greater than that of the natural numbers); certain computing machines are said to be allowed to run for an *infinite* amount of time, and so on, *ad infinitum*. What we have done is to bring the sort of reasoning which is based on this way of speaking and thinking into sharp focus by narrowing things down to $\mathcal{L}_{\omega_1\omega}$. But we have in no way employed an idiosyncratic nomenclature. Quite the contrary: our terminology, whether it pertains to logic or computation, is entirely "off the shelf."

But of course Objection 2 is more than a complaint that certain key terms are ambiguous. The meat of the objection is that those who reason in and about $\mathcal{L}_{\omega_1\omega}$ are simply manipulating finite expressions, and such manipulation is not at all beyond computation.

The first thing worth noting about this aspect of Objection 2 is that it reflects an attitude that seems to be exactly analogous to Hilbert's Gödel-killed finitistic program for mathematics. Hilbert observed that mathematical proofs were invariably presented as finite strings on finite pieces of paper, and he hit upon an idea: proofs were to be entirely mechanical, step-by-step *finite* strings; and all problems in mathematics could be solved by such finitary methods. Demonstrations of consistency were to involve only finite procedures making reference to but a finite number of properties possessed by formulas, and procedures that employed only a finite number of operations over these formulas.

But as we all know by now, Gödel obliterated Hilbert's program. He proved that human mathematical reasoning is *not* always limited to Hilbertian reasoning: some form of infinitistic reasoning must be employed for some proofs of formulas about **N**, the natural numbers — formulas which expert mathematicians and logicians can see to be true. A bit more specifically, Gödel found a sentence of the form $\forall x \phi(x)$ about the natural numbers (i.e., a formula that says that every natural number has a certain property ϕ) which couldn't be proved by finite means, even though each of $\phi(0), \phi(1), \phi(2), \ldots, \phi(n), \phi(n+1), \ldots$ (where each of these formulas says that a particular natural number has the property ϕ) *is* provable by a finite proof from the first-order version of the axioms characterizing the natural numbers! Gödel found a formula which expressed a truth about the natural numbers that couldn't be proved by finite means. What was the reaction? Interestingly enough, many suggested that first-order formalizations of arithmetic be replaced by formalizations in $\mathcal{L}_{\omega_1\omega}$.[14]

So Objection 2, it seems, is just a feeble attempt to exhume Hilbert's

[14] A recent treatment of the issues here can be found in Smullyan's (1992) recent book on Gödel's incompleteness results. Many logicians have a general notion of Gödel's first incompleteness theorem, but few know that Gödel showed that there is a formula $\phi(y)$, with one free variable y, such that $\phi(1), \phi(2), \phi(3), \ldots, \phi(n), \ldots$ are all provable in Peano Arithmetic (PA), while the sentence $\forall y \phi(y)$ *isn't*. This phenomenon — called **ω-incompleteness** by Tarski — can be remedied by invoking the system PA^+, which contains the ω-rule (sometimes also called **Tarski's rule** or **Carnap's rule**) allowing one to infer $\forall y \phi(y)$ from the infinitely many premises $\phi(1), \phi(2), \phi(3), \ldots, \phi(n), \ldots$. This rule of inference, which we discussed earlier, again, is at the heart of the present chapter.

program. But we can nonetheless imagine how one might endeavor to sustain it:

3.8.3 Objection 3: It All Flies in the Face of Finitism

"Bringsjord and Zenzen, there is, I concede, an analogy between my position and Hilbert's, but it's far from a perfect analogy. The fact remains that it's evident to me that no human activity can involve the deployment of infinite things, since infinite things cannot actually exist in this universe. You seem to be assuming what seems to me to be ridiculous, namely that the mathematical reasoning in question is skill with infinite mental objects. Do you really believe that when a *human* logician thinks about $\mathcal{L}_{\omega_1\omega}$, he or she actually has the infinitary concepts referred to by the finite strings in mind? Do you really believe, more generally, that when a mathematician or logician thinks of the integers, he or she has an infinite set in mind? If this is your position, gentlemen, you should state it clearly, and preferably defend it against some obvious objections (e.g., that our brains are finite and can only pass through finitely many states, it would seem)."

In response, let us first state the view underlying our argument: Expert mathematicians and logicians routinely use finite strings to point to infinite objects which they genuinely conceive of, ponder, reflect upon, reason about, manipulate, and so on. Since such activity is genuine, it cannot be recast as activity in finitary FOL, for the reasons given in section 3.7 (to repeat, if reasoning with $\mathcal{L}_{\omega_1\omega}$ could be recast as reasoning with $\mathcal{L}_I$, FOL, then those problems proved to be beyond $\mathcal{L}_I$ would *not* be beyond this system — contradiction). We take it that our view is "motherhood and apple pie," but more about that when we see in detail why Objection 3 fails. Overall, the reason it fails is that it affirms a version of a finitistic philosophy of mathematics. The problem with such an affirmation, in the present dialectic, is three-fold:

1. Anyone who rejects finitistic mathematics will not regard Objection 3 to have any force, and most (for reasons discussed below) reject it.

2. The version of finitism suggested here, according to which the denial of this doctrine entails that infinite objects must somehow "fit inside" the brain, is preposterous.

3. Finitism, as Bertrand Russell and others noted long ago, is highly implausible, in large part because it cannot do justice to even a small

part of the ontology that underlies mathematics *as it is practiced in the real world by real mathematicians and logicians.*

We consider these problems in order.

Problem 1. Suppose that we concede for the sake of argument that a finitistic philosophy of mathematics entails that premise 5 in the argument is false. (This is the only premise which seems to be vulnerable to Objection 3.) Since proponents of such a view are in the minority (amongst philosophers, logicians, and mathematicians, anyway), why should this worry us?

In order to make the point more than a rhetorical one, it may help to consider an analogous exchange, one involving the issue of defining computation. John Searle (1992) believes that 'computation' is a meaningless term from the standpoint of science, because he thinks it can be shown that every physical object can be said to be engaged in computation at any time. This view of Searle's is not one that most affirm, though a number of thinkers do. (Putnam, one of the founders of computation, agrees with Searle. We think Searle does a good job of defending the view, but we're not one of those who affirm it; see Bringsjord 1994.) Now suppose that Smith offers a careful argument purporting to show that AI and CogSci can succeed in creating computer programs having an ability to process visual images. In reply to Smith, Jones, armed with Searle's view, says, "Your argument doesn't convince *me*. After all, central to your argument is the concept of computation, and we think this concept is laughably vague — so vague that it's not suitable for use in scientific discourse." The point, of course, is that this objection is anemic. In light of the fact that most reject the underlying Searlean view, this objection must include a separate defense of this view. The same thing goes for a finitistic view of mathematics as an objection to our argument: we need to see a careful (if not a victorious) argument for this view. Absent that, the objection is but a curiosity.

Problem 2. The second problem with Objection 3 runs as follows. Suppose that one of us thinks about the planet Jupiter — its size, color, distance from Earth, and many other properties it has. Does it follow that Jupiter in some sense fits inside one of our finite brains (if in fact those brains *are* finite)? We doubt it. Suppose that one of us is a theoretical physicist who thinks about the entire universe (as certain such physicists do). Does it follow that the universe is somehow inside a brain that is, after all, smaller than a basketball? (Perhaps it's worth noting that a number of physicists regard the universe to itself be infinite.) Hardly. The problem is that Objection 3 is based on a clear *non sequitur*: it doesn't follow from the fact that a human conceives of (or thinks about, etc.) a that a must in any non-metaphorical sense fit inside this human's brain. What a human presumably *does* need in order to conceive of a is some sort of *representation* of a. And we see no reason why a human can't have a representation scheme by virtue of which he or she conceives of infinite objects. Indeed, such a view is the dominant one in classical mathematics, and it's the one we affirmed above. (Recall Arg_1 and Arg_2, presented at the outset of the present chapter.)

Moreover, this second problem with Objection 3 can be refuted by a rather simple *reductio*: Suppose for the sake of argument that the fact that the brain is finite does rule out a human's ability to genuinely conceive of and think about an infinite object like the set $\mathbf{Z}$ of integers. Since the brain is finite, it follows that it is of size k, and is able to pass through only $f(k)$ states, where both k and $f(k)$ are themselves positive integers. (What it means for something to be finite is that it is of a size that corresponds to some integer.) So, if Objection 3 succeeds, it does so because things like $\mathbf{Z}$ are a heck of a lot bigger than either k or $f(k)$ — too big, so the story here goes, for a brain of this size to genuinely conceive of them. But suppose that there is some object which is known to be finite, but which is also known to be larger than k and $f(k)$. (If you're looking for examples, the field of complexity theory within computer science will provide an endless source.) We are forced to conclude that this object, say the perfect winning strategy for chess — which can be coded in the form of a *finite* state automaton guaranteed to dwarf k and $f(k)$ — cannot be conceived of by a human with a brain of the size assumed here — but that is absurd (just ask those who think about such objects for a living).

Problem 3. The third problem is that finitism is generally thought to be untenable. Bertrand Russell, in his famous essay "The Limits of Empiricism" (Russell 1936), fatally lampooned it long ago; anyone embracing finitistic attitudes toward mathematics is obliged to start by derailing Russell's biting observations. There isn't space to discuss Russell's essay here. We will say only that Russell was no wild-eyed dualist. He calmly asserted, over and over, that he was capable of routinely doing that which we claim herein expert mathematicians and logicians routinely do. In fact, Russell (and others, e.g., Weyl 1949) claimed that they could conceive of infinite objects provably "larger" than the objects at the heart of $\mathcal{L}_{\omega_1\omega}$! For example, one of the finitists of Russell's day, Ambrose, claimed that it wasn't possible for a human to know that there are not three consecutive 7's in the expansion of π. Russell replied — and this is the reply he would no doubt make to those who make Ambrose's finitistic claim about, say, the infinite derivations which are part of $\mathcal{L}_{\omega_1\omega}$ — that it is only *medically* impossible to carry out the expansion in order to check for the three 7's. Russell said that he could easily imagine an Omniscient Diety, and such a being could not only know the answer, but could share it with a human. Russell also claimed that he could imagine an infinite number of operations taking place in a finite amount of time (Russell 1936, p. 144).

At any rate, the fundamental point, and the one Russell made, is that when a mathematician uses the existential quantifier $\exists$ to assert the existence of some infinite set (as for example when such a person proves in axiomatic set theory that there exists a set which we call the integers), this isn't an assertion (*contra* Objection 3) which merely amounts to the claim that there exists a finite string inscribed on pieces of paper which mathematicians and logicians use.

3.8.4 Objection 4: "Moore's Case for Finitism"

"It follows from the foregoing dialectic that the two of you do concede that *if* finitism is true, *if* there is no such thing as the infinite or an infinite set, then your argument against computationalism fails. Hence, if it can be *shown* that finitism is true, your argument is destroyed. Now, you have done a fairly good job of casting doubt upon finitism (although I daresay that Russell deserves most of the credit), but it just so happens that a compelling case has been made for finitism by A. W. Moore (1990) in his seminal book *The Infinite*. Your praising Bertrand Russell's old attack on finitism does nothing to stop Moore's case."

There's really no getting around the fact that if the argument of this chapter is to succeed, we are going to have to overthrow Moore's case. It's clearly insufficient for us to merely applaud Russell's remarks, when Moore's sustained argument for finitism has been placed on the table. We're going to have to overthrow Moore's argument. We do so now.

Moore offers two intertwined arguments for the view that there is literally no such thing as the infinite, and no thing (e.g., a set) that is infinite. His first argument is essentially that if finitism is false, all sorts of unsolvable paradoxes arise. Since an unsolvable paradox implies a contradiction, and a contradiction can't be true, it follows by *modus tollens* that finitism is true. The paradox Moore is most concerned to press against infinitists in this argument is The Second-Order Paradox of Thought About The Infinite. He does have a number of other paradoxes in mind, and many of them will be familiar to many readers: The Achilles, The Liar, The Paradox of the Divided Stick, Hilbert's Hotel, and so on.[15] But the paradox Moore calls "perhaps the most serious of all" (1990, p. 12) is The Second-Order Paradox of Thought About The Infinite (TSOPOTATI).

This paradox is presented on pages 11 and 12 of Moore's book. Moore first says here that "One of the paradoxes of thought about the infinite, then, is that there are reasons both for and against admitting the concept of infinity" (p. 11). This is supposed to be a "first-order" paradox. The kernel of this paradox is that there are good reasons for accepting the existence of a set that contains 0, 1, 2, and so on *ad infinitum*. (This set is of course standardly referred to as **N**, the so-called natural numbers.) On the other hand, there are good reasons for refusing to countenance the existence of any infinite sets — because when one accepts the existence of such sets, paradoxes burst on the scene. As a response to this paradox, Moore reports that one might say that we cannot come to know anything about the infinite, or say anything coherent about it — and that in support of saying this one can cite the various well-known paradoxes (e.g., The Liar, The Achilles, ...). But to say this is to accept line 1 in the argument that follows, and that's the only premise in that argument — an argument which itself constitutes a paradox, viz., TSOPOTATI. Moore calls this paradox a "second order" one, and, as we've indicated, apparently regards it to be the toughest paradox of all. Here, where p^{inf} denotes a proposition about the infinite, is the argument Moore has in mind:

#	**Proposition**	**Justification**
1	$\exists p^{inf} K p^{inf} \rightarrow (\phi \wedge \neg\phi)$	various paradoxes
2	$\neg(\phi \wedge \neg\phi)$	propositional tautology
3	$\neg\exists p^{inf} K p^{inf}$	1, 2 *modus tollens*
4	$K\neg\exists p^{inf} K p^{inf}$	3, (doxastic) necessitation
5	$\exists q^{inf} K q^{inf}$ (Set $q^{inf} = \neg\exists p^{inf} K p^{inf}$)	? – but we grant the inference
6	$3 \wedge 5$ form a contradiction!	$\wedge$ Introduction

[15]The introduction in (Moore 1990) is one of the best synoptic accounts of these and other paradoxes.

Moore's solution seems to be to deny that there is anything referred to by such phrases as 'the infinite.' We read:

> The paradox is that I appear to be able to grasp the infinite as that which is ungraspable. The solution is to deny that I can grasp it in any way. There is nothing there to be grasped. (1990, p. 225)

So here we have it: finitism saves the day, and saves us from having to swallow a contradiction. Unfortunately, Moore blunders badly. The reason is that if there is nothing there to be grasped, it presumably follows immediately that

$$\neg\exists p^{inf} K p^{inf},$$

from which line 1 follows by propositional logic![16] Therefore, invoking finitism, *contra* Moore, provides no escape from TSOPOTATI.

The second argument Moore offers for finitism relies on a simpler version of premise 1 in the above argument, and runs as follows. If finitism is false, then various paradoxes arise. In particular, if finitism isn't true, The Paradox of the Staccato Run (PSR) arises. PSR implies that "supertasks" are coherent. But such tasks are incoherent. Ergo, finitism is true. Moore's description of PSR, and the claim that PSR implies the coherence of super-tasks, takes us back to the expansion of π, mentioned above in connection with Russell:

> Suppose that Achilles runs for half a minute, then pauses for half a minute, then runs for a quarter of a minute, then pauses for a quarter of a minute, and so on *ad infinitum*. At the end of two minutes he will have stopped and started in this way infinitely many times. Yet there is something repugnant about admitting this possibility, even as a conceptual — let alone a physical — possibility. For example, suppose that each time he pauses he performs a task of some kind, there being no limit to how quickly he can do this. Then at the end of two minutes he will have performed infinitely many of these tasks. He might, say, have written down the complete decimal expansion of π (3.141592...), for which he needs only a finite sheet of paper and the ability to write down digits that get smaller without limit (see [Figure 3.7]) ... We are loath to admit this as a conceptual possibility, though we seem bound to do so. (1990, p. 4)

---

[16] Recall that from $\neg p$ one can infer $\neg p \lor q$, from which one can infer $\neg p \rightarrow q$.

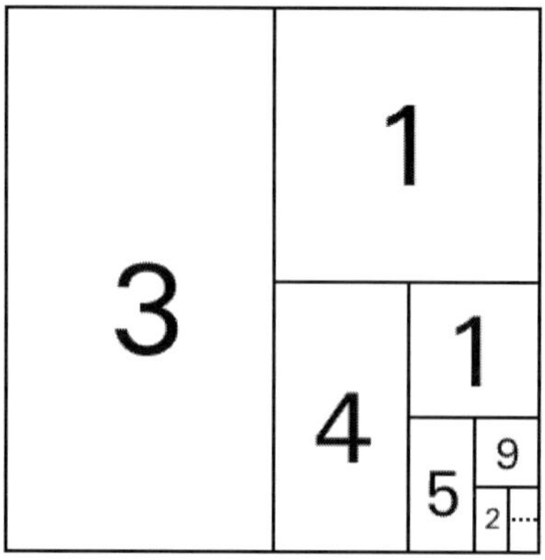

Figure 3.7: Expanding π as a Supertask

Here we see the by-now-familiar supertask: writing down the infinite decimal expansion of π in a finite amount of time. Another example would be to write down (or construct, visit, etc.) all of the natural numbers **N** in a finite amount of time. This is the particular "generic" supertask that Moore focuses on later in his book. (This super-task is also the one famously said by Boolos and Jeffrey (1989) to be possible for a superhuman character they call 'Zeus.' You will recall that Zeus machines were discussed in Chapter 1.) The idea seems to be that it isn't just PSR that implies the coherence of supertasks: even the seemingly innocent acceptance of the existence of **N** seems to imply the coherence of supertasks.

Moore's idea here should be promptly discarded. After all, following the relevant part of classical theoretical computer science (as we did in Chapter 1), we are fully entitled to presuppose the coherence of supertasks. Evidently Moore is completely unaware of this uncontroversial branch of mathematics. Even in elementary calculus the notion of a limit — as explained, for example, by Salmon (1975) — presupposes the coherence of supertasks. (For other explanations of this point, see Steinhart forthcoming and Koetsier & Allis forthcoming.) Even children are frequently taught that supertasks are perfectly coherent, because they are prepared early on, in mathematics, for calculus down the road. For example, see Figure 3.8, which is taken from page 268 of (Eicholz et al. 1995).

Bringsjord's son, Alexander, in the 7th grade, was asked to determine the "percent pattern" of the outer square consumed by the ever-decreasing shaded squares. The pattern, obviously, starts at $\frac{1}{4}$, and then continues as $\frac{1}{16}, \frac{1}{64}, \frac{1}{256}, \ldots$. When asked what percent "in the limit" the shaded square consumes of the original square, Alexander was expected to say "Zero" — but the notion of a limit was a bit tricky for him (perhaps understandably). When asked what percentage the shaded square would "get down to" if someone could work faster and faster, and smaller and smaller, at drawing the up-down and left-right lines that make each quartet of smaller squares, Alexander said zero. This is anecdotal, yes, but what it indicates is something in keeping with what we discussed earlier: some humans may start with an intuitive picture of a supertask, and move from there to the formalisms in question (i.e., those seen in elementary calculus). It would be interesting to systematically poll students about these matters.

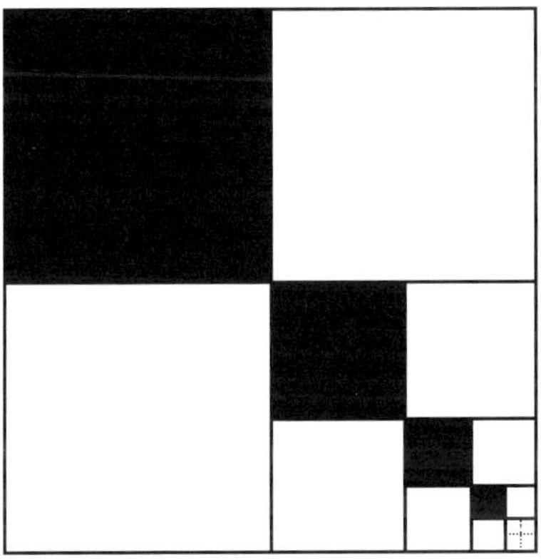

Figure 3.8: Picture of Supertask from Seventh Grade Math

At any rate, Moore does have an argument for the view that supertasks aren't coherent, and if it's any good it would certainly trump an anecdotal report about a child's reasoning. Here's his argument:

> If it did make sense to say that I had just constructed all of the natural numbers in a minute, by the Zenonian procedure, then it would also make sense to say this: while I was constructing them, my constantly increasing speed of performance meant that time seemed to be going more and more slowly to me; it seemed that I was constructing them at a steady rate. Yet there is nothing that could count for me as a retrospective grasp of such an experience, in its apparent endlessness. (I could not have an apparently endless experience, apparently followed by further experience.) I must, subsequently, have forgotten all but an initial segment of it. How can this be? Surely what we have

> here is symptomatic of the fact that nothing could ever count, for anyone, as a grasp of an infinite reality. The grammar of 'infinity' is not geared to this. The special problems that arise when we envisage time seeming to go more slowly merely serve to make graphic an incoherence that is there to be acknowledged anyway — an incoherence that crept in at the very beginning of the story. It does not make sense to say that I have just performed infinitely many tasks of any kind, nor to say that anything is infinite in any respect. (Moore 1990, p. 213)

Put in a more explicit form that can be evaluated, the argument runs like this:

Moore's Argument for the Incoherence of Super-Tasks

(7) If super-tasks are coherent, then it could seem to someone who (conceptually speaking) performs such a task that each action along the way took the same amount of time (say n seconds).

(8) If it could seem to someone who performs a super-task that each action along the way took n seconds, then someone could retrospectively experience, or relive, each action for n seconds — and then proceed to have other, normal life experiences.

(9) No one could retrospectively experience, or relive, each action in a super-task for n seconds — and then proceed to have other, normal life experiences.

(10) Super-tasks are incoherent. [(7), (8), (9)]

Is this argument any good? Well, the argument appears to be formally valid, an instance of hypothetical syllogism and *modus tollens*; at any rate we're prepared to concede that the argument is valid. Premise (9) certainly seems to be true. After all, $n + n + n + \ldots$ will eventually exceed the amount of time a human has to operate with (in this lifetime, anyway), whereas a supertask's duration is some such sum as

$$\sum_{1}^{\infty} \frac{1}{2^n}.$$

Premise (7) seems extremely plausible. For suppose that Jones is a sprinter who runs 100 meter races in 9 seconds. Couldn't it nonetheless seem to Jones that he runs the race in 15 seconds? Couldn't it in fact seem to Jones that he runs the race in *30* seconds? And why not a minute? After all, we're talking here about Jones' subjective perspective. He could hallucinate during his sprint, or go into some kind of wild dream that seems to span 10 years. In general, then, when a human performs an action having property F, it may not seem to the human that that action has F; the human may perceive the action to have a radically different property G.

So where does this leave the argument? It leaves it hinging on premise (8). But this premise is at best controversial; at worst, the premise begs the question against us. The problem is that it doesn't follow from the fact that a series of actions can *seem* to someone to have certain properties, that that person can in any way genuinely experience actions having these properties. It may seem to me that a moment ago I leaped over a tall building in a single bound (perhaps I had a dream in which I wore a skin-tight bodysuit with a big **S** emblazoned at chest level, etc.), but it doesn't follow from this that I can experience leaping a tall building in a single bound. Likewise, it may seem to someone that they just squared the circle, but they cannot experience squaring the circle. (If one experiences squaring the circle, squaring the circle can be pulled off — but it *can't* be pulled off.)

Finally, why do we say that premise (8) may make the argument in question circular? We imagine that once one has (conceptually) performed a supertask, the only way to retrospectively experience this task is for the retrospective experience to itself be a supertask. To the extent that premise (8) rules out by fiat this method of retrospective experience, it preemptively rules on precisely what's at issue: Moore's argument becomes a *petitio*.

3.8.5 Objection 5: "Reasoning About" vs. "Reasoning With"

"Though you offered it merely for the record, gentlemen, and not as part of your argumentation, your statement that

> Expert mathematicians and logicians routinely use finite strings to point to infinite objects which they genuinely conceive of, ponder, reflect upon, reason about, manipulate, and so on. Since such activity is genuine, it cannot be recast as activity in finitary FOL ...

reflects your fundamental error. Notice that you slip from 'conceive of, reflect upon, reason about' (with which I have no problem) to 'manipulate.' Here you switch from describing infinite sentences to somehow *using* them. That's precisely your mistake.

"However, I confess that I am puzzled by your 'second problem' to Objection 3. There you seem to be quite happy with the idea that we can use a finite representation to describe or refer to something infinite. I heartily agree. But now we seem to have switched horses. I thought you were denying this — via a claim that human mathematicians and logicians are *using* $\mathcal{L}_{\omega_1\omega}$. Surely it is quite possible that a human expert mathematician or logician uses some finite mental representation to reason *about* $\mathcal{L}_{\omega_1\omega}$. Not even you two can be reasoning *with* $\mathcal{L}_{\omega_1\omega}$: it can't be a proper description of your language of thought."

This objection is better than its predecessors, but it too ultimately fails. We readily admit that sometimes mathematicians and logicians (and, for that matter, non-mathematicians) merely reason *with* a finite representation and reasoning system. (We suppose the paradigmatic case of this would be the carrying out of derivations in some natural deduction system for $\mathcal{L}_I$.) But it doesn't follow from this fact that our argument founders on the distinction at the heart of Objection 5. The key is that we've chosen $\mathcal{L}_{\omega_1\omega}$ for good reason: some of the reasoning *about* this logical system is clearly reasoning *with* a representation and reasoning system having *at least* the infinitary grade of $\mathcal{L}_{\omega_1\omega}$. In order to see this we have but to look a bit at what goes on when a relevant theorem about $\mathcal{L}_{\omega_1\omega}$ is pondered and proved. Take, for example, the following simple theorem, which Bringsjord has often asked students learning about $\mathcal{L}_{\omega_1\omega}$ to prove:

> *Scott's Isomorphism Theorem.* Let $\mathcal{I}$ be an interpretation for $\mathcal{L}_I$. Then there is a sentence ϕ of $\mathcal{L}_{\omega_1\omega}$ such that for all countable interpretations $\mathcal{I}^*$ for $\mathcal{L}_I$, $\mathcal{I}^* \models \phi$ iff $\mathcal{I}^*$ is isomorphic to $\mathcal{I}$.

Intuitively, this theorem says that a single infinitary sentence can perfectly characterize a countable interpretation for $\mathcal{L}_I$. The customary proof involves (among other things) constructing infinitely long conjunctions (*outside of* $\mathcal{L}_{\omega_1\omega}$), each conjunct of which is an atomic formula capturing a truth about the elements in the domain of $\mathcal{I}$. For example, if the domain of $\mathcal{I}$ is $\mathbf{N}$, the natural numbers, and $\mathcal{I}$ includes $>$ (ordinary greater than), then the following are elements of $>$: (3,2), (4,3), (5,4),.... Hence, if we are to capture $\mathcal{I}$, there must be an atomic formula corresponding to each such fact, and the conjunction of these formulas (which is still only a part of the construction at the heart of Scott's Theorem) becomes (with the relation symbol G interpreted as $>$, and c_i as constants):

$$Gc_3c_2 \wedge Gc_4c_3 \wedge Gc_5c_4 \wedge \dots ,$$

or, in the notation for infinitely long conjunctions in $\mathcal{L}_{\omega_1\omega}$,

$$\bigwedge\{Gc_ic_j : c_i, c_j \text{ are constants } \& \ \mathcal{I} \models Gc_ic_j\}.$$

The point is that, *contra* Objection 5, the sort of mathematical reasoning needed for carrying out such proofs requires that one reason *with* a "language of thought" that parallels $\mathcal{L}_{\omega_1\omega}$ itself.[17]

3.8.6 Objection 6: But Mathematicians Themselves Don't Agree!

Here's the final objection: "Gentlemen, but mathematicians themselves don't agree with you, as you must realize. For example, in his highly regarded *The Math Gene*, Keith Devlin (2000) defends the view that mathematical thinking is reducible to language use, and originally emerged from evolutionary reinforcement of what he calls 'offline' thinking — thinking about situations represented in an abstract form removed from their flesh-and-blood concretizations in the physical world."

Devlin's attempt to give an evolutionary explanation for mathematical thinking is one we discuss in Chapter 7, when considering objections to our argument therein that such thinking (at least sophisticated strains thereof) can't in fact be explained from the standpoint of evolutionary psychology (at least as this psychology is expressed in Pinker's 1997 *How The Mind Works*). At any rate, this attempt is tangential to the objection given here, for that objection amounts to

(11) If members of field F, or at least the vast majority of such members, don't affirm ϕ, and ϕ makes an assertion about

[17] Please note that our rebuttal doesn't in the least conflate object theory with metatheory. We are in fact invoking this very distinction — but we're pointing out that the metatheory in question (unsurprisingly) deploys some of the very same infinitary constructions as seen *in* $\mathcal{L}_{\omega_1\omega}$. This seems utterly undeniable — and so much the worse for those (e.g., Ebbinghaus et al. 1984) who believe that "background" logic/mathematics is fundamentally first-order.

the nature of the cognition of members of F, then ϕ is false (or: ought to be rejected by theorists).

(12) The vast majority of mathematicians don't affirm the claim (ϕ^{inf}) that some mathematical reasoning is genuinely and irreducibly infinitary.

(13) $\therefore$ It's not the case that some mathematical reasoning is genuinely and irreducibly infinitary.

As should be obvious, this is a ... well, a miserable argument. For starters, the argument runs afoul of the possibility that some members of a field F might not even entertain some ϕ, and might for that reason fail to affirm it. Presumably the underlying idea is that the practitioners in some discipline *reject* some proposition ϕ — and that this rejection, because these practioners are presumed to have some special insight into matters relating to their field, ought to count against ϕ itself. Accordingly, let's be charitable and assume for the sake of argument that Devlin explicitly rejects ϕ^{inf}, and that all mathematicians follow suit. (The truth of the matter is that in *The Math Gene*, Devlin (2000) doesn't discuss mathematical or philosophical logic of any kind, and nowhere discusses infinitary reasoning. Let's assume that his failure to discuss these matters implies that he rejects ϕ^{inf}.) In addition, let's supplant (11) with the more plausible

(11′) If members of field F, or at least the vast majority of such members, reject ϕ, and ϕ makes an assertion about the nature of the cognition of members of F, then ϕ is false.

What now are we to say about this modified version? Well, the problem is that there is really no reason to affirm (11′). This is particularly easy to see if one is a computationalist, for consider the following scenario. Assume that all painters on the planet reject the proposition that the process of painting, top to bottom and start to finish, is computational; that is, they reject the idea that the process can be carried out by a Turing machine. If the argument here is sound, then computationalists, faced with such collective disdain for the view that painting is at bottom computation, must reject what they believe to be true, namely, that painting, like all human activity, is at or below the Turing Limit. Clearly, computationalists would be positively irrational to allow such painters to overthrow computationalism. It would likewise be irrational for us to agree that because Devlin (and, by our charitable hypothesis, others) rejects ϕ^{inf}, we ought to join him.

3.9 Simon's Dream and Mental Metalogic

Our principal concern to this point has been to evaluate the philosophical, logical and mathematical status of the computationalist approach to human reasoning. As we've tried to show, this approach, when confronted with mathematical reasoning, is bound to fail. At bottom, this is because mathematicians and logicians, from the moment they step on the job, must directly contemplate, reason about, manipulate, represent ... *cognize* infinitude. If we're right, where does this leave us? What are the consequences of what we've uncovered for the practice of AI aimed at capturing "high-powered" human reasoning? We don't have complete answers to these rather far-reaching questions, but we can say a few words, some of them in connection with what Bringsjord has dubbed 'Simon's Dream,' a goal Herb Simon set out to reach in the summer before he died.

3.9.1 Logical Illusions

3.9.1.1 Some Warm-up Problems

Due in no small part to ingenious experimentation carried out by Johnson-Laird and Yingrui Yang, one of the dominant themes in current psychology of reasoning is the notion of a cognitive or logical illusion. In a visual illusion of the sort with which you're almost certainly familiar, one seems to see something that, as a matter of objective fact, simply isn't there. (In the Sahara, what seems to be a lovely pool of water is just the same dry-as-a-bone expanse of sand — and so on.) In illusions of the sort that Johnson-Laird and Yang have brought to our attention, what seems to hold on the basis of inference doesn't. To orient you, here's a specimen (from Johnson-Laird & Savary 1995; variations are presented and discussed in Johnson-Laird 1997*a*):

Illusion 1

(14) If there is a king in the hand, then there is an ace, or else if there isn't a king in the hand, then there is an ace.

(15) There is a king in the hand.

Given these premises, what can you infer?

Johnson-Laird has recently reported that

> Only one person among the many distinguished cognitive scientists to whom we have given [Illusion 1] got the right answer; and we have observed it in public lectures — several hundred individuals from Stockholm to Seattle have drawn it, and no one has ever offered any other conclusion (Johnson-Laird 1997*b*, p. 430).

The conclusion that nearly everyone draws (including, perhaps, you) is that there is an ace in the hand. Bringsjord (and, doubtless, many others) has time and time again, in public lectures, replicated Johnson-Laird's numbers — presented in (Johnson-Laird & Savary 1995) — *among those not formally trained in logic.* (For those so trained, the story is different — though this isn't the place to tell it.) This is wrong because 'or else' is to be understood as exclusive disjunction, so (using obvious symbolization) the two premises become

(14′) $((K \rightarrow A) \vee (\neg K \rightarrow A)) \wedge \neg((K \rightarrow A) \wedge (\neg K \rightarrow A))$

(15′) K

It follows from the right (main) conjunct in (14′) by DeMorgan's Laws that either $K \rightarrow A$ or $\neg K \rightarrow A$ is false. But by the truth-table (of course, a fully syntactic proof is easy enough to give) for '$\rightarrow$,' in which a conditional is false only when its antecedent is true while it's consequent is false, it then follows that either way A is false, i.e., there is *not* an ace in the hand. By the way, even when you make the exclusive disjunction explicit, the results are the same. For example, you still have an illusion if you use

Illusion 1′

(14″) If there is a king in the hand then there is an ace, or if there isn't a king in the hand then there is an ace, but not both.

(15″) There is a king in the hand.

Given these premises, what can you infer?

3.9.1.2 A New Illusion

Now, consider the following logical illusion, devised by Bringsjord.

Illusion 2

(16) The following three assertions are either all true or all false:

- If Billy helped, Doreen helped.
- If Doreen helped, Frank did as well.
- If Frank helped, so did Emma.

(17) The following assertion is definitely true: Billy helped.

Can it be inferrred from (16) and (17) that Emma helped?

(If you're interested in pondering this illusion without knowing the answer, pause before reading the rest of this paragraph.) Sure enough, in preliminary results obtained by Bringsjord and Yang, most students, and even most professors, often including, in both groups, even those who have had a course in logic, answer with "No." The rationale they give is that, "Well, if all three of the if-thens are true, then you can chain through them, starting with 'Billy helped,' to prove that Doreen helped. But you can't do that if the three are false, and we're told they might be." But this is an illusion. In the case of reports like these, the premises are represented correctly, but there is an invalid argument for the belief that E can't be deduced from the conjoined negated conditionals (while they do have in mind a *valid* argument from the true conditionals and 'Billy helped' to 'Emma helped'). This argument is invalid because if the conditionals are negated, then we have

$$\neg(B \rightarrow D)$$
$$\neg(D \rightarrow F)$$
$$\neg(F \rightarrow E)$$

But the first two of these yields by propositional logic the contradiction $D \wedge \neg D$. Given this, if we assume that Emma failed to help, that is, $\neg E$, we can conclude by *reductio ad absurdum* that she did help, that is, E.[18] Notice that this solution makes use of both the rule that Lance Rips (1994) says humans don't have (a negated conditional to the truth of its antecedent) and the rule that most psychologists of reasoning say humans don't have (*reductio*). In light of this, it's noteworthy that a number of those to whom we presented Illusion 2 *did* produce the full solution we just explained.

We next draw your attention to an interesting feature of Illusion 2, and use it as a stepping stone to a new theory of human reasoning devised by Yang & Bringsjord (2001): mental metalogic.

[18] A fully formal proof in $\mathcal{F}$ (the system of natural deduction introduced by Barwise & Etchemendy 1999) is shown in Figure 3.9. The predicate letter `H` is of course used here for 'helped.'

	✓ Given
$((H(b) \to H(d)) \land (H(d) \to H(f)) \land (H(f) \to H(e))) \lor (\neg(H(b) \to H(d)) \land \neg(H(d) \to H(f)) \land \neg(H(f) \to H(e)))$	✓ Given
$H(b)$	✓ Given
$(H(b) \to H(d)) \land (H(d) \to H(f)) \land (H(f) \to H(e))$	✓ Assume
$H(b) \to H(d)$	✓ ∧ Elim
$H(d)$	✓ → Elim
$H(d) \to H(f)$	✓ ∧ Elim
$H(f)$	✓ → Elim
$H(f) \to H(e)$	✓ ∧ Elim
$H(e)$	✓ → Elim
$(\neg(H(b) \to H(d)) \land \neg(H(d) \to H(f)) \land \neg(H(f) \to H(e)))$	✓ Assume
$\neg(H(b) \to H(d))$	✓ ∧ Elim
$H(b) \land \neg H(d)$	✓ Taut Con
$\neg(H(d) \to H(f))$	✓ ∧ Elim
$H(d) \land \neg H(f)$	✓ Taut Con
$\neg H(e)$	✓ Assume
$H(d) \land \neg H(d)$	✓ Taut Con
$H(e)$	✓ ¬ Intro
$H(e)$	✓ ∨ Elim

Figure 3.9: A Proof That `Emma helped` in $\mathcal{F}$

3.10 Mental MetaLogic: A Glimpse

A cognizer fooled by Illusion 2 imagines an argument for the view that there is no way to derive 'Emma helped' from the negated trio of conditionals and 'Bill helped.' Such arguments cannot be expressed in standard proof theories for FOL, such as the theory $\mathcal{F}$.[19] And when we look to psychology of reasoning for help in making sense of such arguments we draw a blank. The reason for this is that psychology of reasoning is dominated by two competing models of human reasoning (especially deductive reasoning), and neither one allows for the expression of disproofs. The two theories are mental logic (ML) and mental models (MM). ML is championed most prominently by Braine (1998*b*) and Rips (1994), who hold that human reasoning is based on syn-

[19] $\mathcal{F}$ is the natural deduction system presented in in Barwise and Etchemendy's *Language, Proof, and Logic* (1999). In such a system, there is always a rule for introducing a logical connective, and a rule for eliminating such a connective.

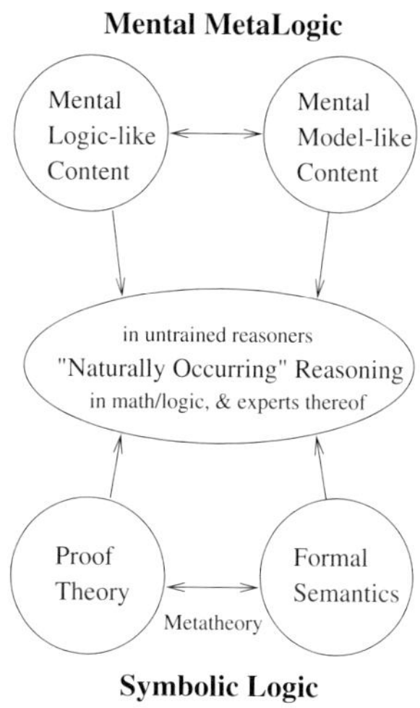

Figure 3.10: Overview of Mental MetaLogic Symmetry with Symbolic Logic

tactic rules like *modus ponens*. MM is championed by Johnson-Laird (1983), who holds that humans reason on the basis not of sequential inferences over sentence-like objects, but rather over possibilities or scenarios which they in some sense imagine. As many readers will know, the meta-theoretical component of modern symbolic logic covers formal properties (e.g., soundness and completeness) that bridge the syntactic and semantic components of logical systems. So, in selecting from either but not both of the syntactic or semantic components of logical systems, psychologists of reasoning have, in a sense, broken the bridges built by logicians and mathematicians. Mental logic theory and mental model theory are incompatible from the perspective of symbolic logic: the former is explicitly and uncompromisingly syntactic, while the latter is explicitly and uncompromisingly semantic. The theory of Bringsjord and Yang, mental metalogic (or just 'MML'), bridges between these two theories (see Figure 3.10). In MML, step-by-step proofs (including, specifically, disproofs) can at once be syntactic and semantic, because situations can enter directly into line-by-line proofs. The HYPERPROOF system of Barwise and Etchemendy (1994) can be conveniently viewed as a simple instantiation of part of MML.[20]

[20]We have our own, more sophisticated instantiations of MML, but HYPERPROOF is simpler, is widely used in teaching logic, and in the present context is perfectly sufficient.

In HYPERPROOF, one can prove such things as that $\Psi \not\vdash \psi$. Accordingly, let's suppose that in Illusion 2, a "tricked" cognizer moves from a correct representation of the premises when (16)'s conditionals are all true along with (17), to an incorrect representation when the conditionals in (16) are false. To flesh out this hypothetical reasoning, suppose, specifically, that the negated conditionals give rise to a situation, envisaged by the cognizer, in which four people (objects represented as cubes) `b`, `d`, `f`, and `e` are present, the sentence 'Billy is happy' is explicitly represented by a corresponding formula, and the issue is whether it follows from this given information that Emma is happy. This situation is shown in Figure 3.11. (We have moved from helping to happiness because HYPERPROOF has 'Happy' as a built-in predicate.) Notice that the full logical import of the negated conditionals is nowhere to be found in this figure: this is, after all, by design an incorrect (incomplete) representation of the given information in Illusion 2. Next, given this starting situation, a isproof in HYPERPROOF is shown in Figure 3.12. Notice that a new, more detailed situation has been constructed, one which is consistent with the original given info (hence the CTA rule), and in which it is directly observed that Emma isn't happy. This demonstrates that Emma's being happy can't be deduced from the original given information.

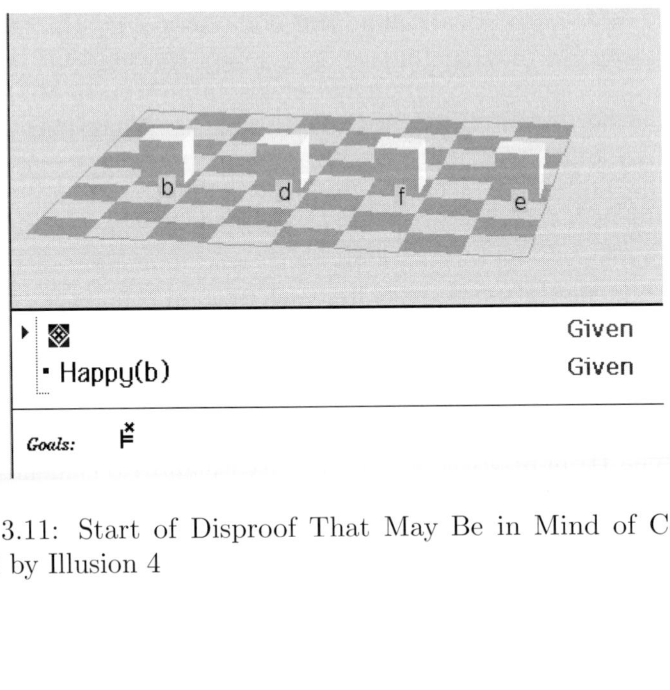

Figure 3.11: Start of Disproof That May Be in Mind of Cognizers Tricked by Illusion 4

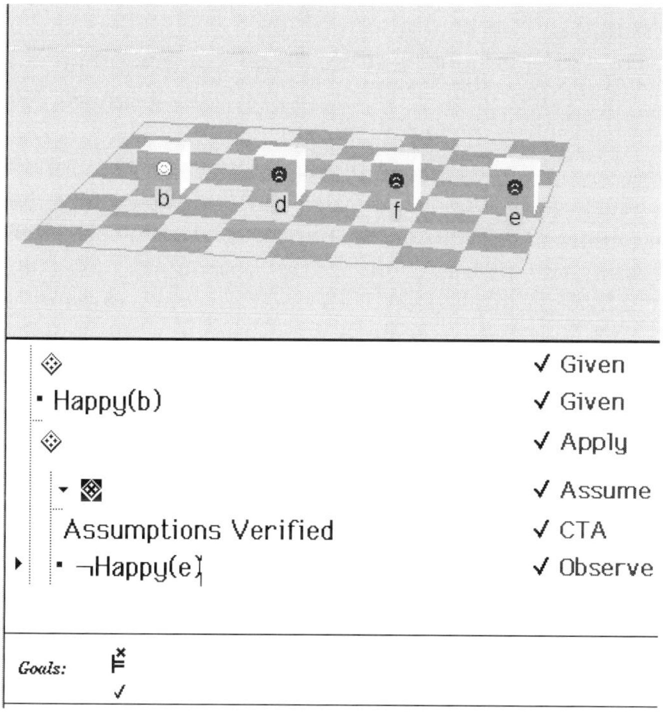

Figure 3.12: Completed Disproof That May Be in Mind of Cognizers Tricked by Illusion 4

3.10.1 Simon's Dream

Bringsjord and Yang hope that mental metalogic can make "Simon's Dream" a reality. But what *is* Simon's Dream? This dream, which, again, Herb Simon, one of the visionary founders of AI, expressed at the annual Computing and Philosophy Conference (Pittsburgh, PA) in the Summer (2000) before he died, is to build an AI system able to produce conjectures, theorems, and proofs of these theorems as intricate, interesting, subtle, robust, profound, etc. as those produced by human mathematicians and logicians. In his talk Simon pointed out something that everybody in the relevant disciplines readily concedes: machine reasoning, when stacked against professional human reasoning, is laughably primitive.[21]

[21]The situation is different than chess — radically so. In chess, as explained in (Bringsjord 1998*a*), we knew that brute force could eventually beat humans. In

Why is machine reasoning still impoverished? This is a topic outside the scope of this book, but we do offer a few pointers for interested readers. First, John Pollock, before delivering a recent talk at RPI on his singularly powerful system for automated reasoning, OSCAR, conceded that all theorem provers, including his own, are indeed primitive when stacked against expert human reasoners, and he expressed a hunch that one of the reasons is that such provers don't use diagrammatic, pictorial, or spatial reasoning. (Wc understand this observation to imply that these unimpressive provers can all be viewed as machine incarnations of mental logic.) Assuming that Pollock is correct, work like that reported in (Glasgow, Narayanan & Chandrasekaran 1995) should be investigated by interested readers. Next, we point out that Simon's "condemnation" of today's machine reasoning has also been carefully expressed in specific connection with Gödel's incompleteness theorems: Bringsjord (1998*b*) has explained why the well-known and highly regarded prover OTTER, despite claims to the contrary, doesn't really prove Gödel's incompleteness results in the least. Finally, Bringsjord & Yingrui (2002) explain that one of the main reasons why machine reasoning systems are no match for human ones may be that humans reasons in groups the members of which reason about the reasoning of members, while machines reason on their own.

It seems to us that the only way to fight toward Simon's Dream is to affirm the primacy of psychology of reasoning. Otherwise we will end up building systems that are anemic. The fact is that first-rate human reasoners use techniques (such as those alluded to in the previous paragraph) that haven't found their way into machine systems (e.g., humans use extremely complicated, temporally extended mental images and associated emotions to reason). No machine, no theorem prover, no cognitive architecture, uses such a thing. We thus take quotes like this one from Dale Jacquette very seriously: "Logic can be understood as descriptive of how some reasoning occurs, at the very least the reasoning of certain logicians." (Jacquette 1997, p. 324) We specifically submit that mental metalogic, used in conjunction with a brand of Jacquette-inspired psychologism, is AI's best bet for reaching Simon's Dream. As we have said, expert reasoners often explicitly

---

reasoning, brute force shows no signs of exceeding human reasoning. Therefore, unlike the case of chess, in reasoning we are going to have to stay with the attempt to understand and replicate in machine terms what the best human reasoners do. See prescription P4 in the final chapter.

work within a system that is purely syntactic (and hence within a system that relates only to mental logic) – but it's also undeniable that such reasoners often work on the semantic side. Roger Penrose, as we saw in Chapter 2, has recently provided us with an interesting example of semantic reasoning (recall the case of the mathematician who is able to see via an image of the sort that is shown in Figure 2.4 that adding together successive hexagonal numbers, starting with 1, will always yield a cube).

MML (Yang & Bringsjord 2001), as we've said, is a theory of reasoning that draws from the proof theoretic side of symbolic logic, the semantic (and therefore diagrammatic) side, and the content in between: metatheory. Furthermore, while theories of reasoning in psychology and cognitive science have to this point been restricted to elementary reasoning (e.g., the propositional and predicate calculi), MML includes psychological correlates to all the different forms of deductive reasoning that we find in relevant human experts; for example, modal, temporal, deontic, conditional, ... reasoning, and combinations thereof. Accordingly, mental metalogic can serve as the basis of an automated reasoning system that realizes Simon's Dream, and we will attempt to demonstrate this in the future.

A few more points are in order here.

First, note that Simon's Dream and supermentalism are a natural fit, given what the present chapter has disclosed. Though Simon believed that even the best of human reasoning is merely computational, we now know better (in light of infinitary reasoning, for starters). AI and the narrower study of (and attempt to replicate) human reasoning needs to be careful not to inadvertently cultivate a warranted perception that certain domains have been conveniently left aside. Even a good many laypeople know that the reasoning underlying medical diagnosis, for example, has been significantly rendered in computational terms. And beginning students of AI and reasoning can no doubt guess that attempts to analyze and replicate the reasoning of, say, auditors and accountants will meet with considerable success. But of course AI has been dogged by accusations that it loves "toy worlds." This accusation will persist as long as the study of reasoning fails to grapple with human performance at least ostensibly beyond computation. The antidote to the lingering "toy world" concern, with respect to reasoning, is none other than setting Simon's Dream as a goal: that is, to tackle head-on some domain which at least looks to be impenetrable from the standpoint of ordinary computation.

No doubt some readers will be puzzled at this point. Some will ask: "How can you advocate an attack on Simon's Dream, given that you purport to have demonstrated that some human reasoning is hypercomputational?" The answer is simply to recapitulate our position on the purely *behavioral* side of computationalism. Suppose that we are correct; suppose that infinitary mathematical reasoning is uncomputable, and that therefore it is reasoning in which no computer can engage. From this is doesn't follow that no system can *appear* to reason in this manner. There are some well-known uncomputable functions which many are doing their best to "solve." Our favorite line of such research — and this will come as no surprise given the TMs we showed earlier in this chapter — is the attack on the uncomputable busy beaver function; see, for example, Buntrock, J. Marxen & Buntrock 1990. (We will have more to say about this function in connection with our recommendations for AI carried out in light of supermentalism. These recommendations are made in the final chapter.) AI, as far as we can see, has never settled the fundamental clash between those who, like Turing, aim at engineering a device whose behavior is indistinguishable from ours, and those who seek not only to create the behavior but also the underlying conscious states which we enjoy. Nothing we've said in this chapter precludes success in the attempt to engineer a computational system which *appears* to carry out infinitary reasoning; and nothing we have said precludes the mathematization of infinitary reasoning. What we purport to have shown, or at least made plausible, is that no implemented system at or below the Turing Limit can *in fact* enjoy such reasoning. We don't know the limits of an approach which resigns itself to engineering behavior only, no one does, but it seems to us to be *the* route for future research and development in AI. It is the avenue, indeed the creed, which one of us, when wearing not the hat of a theoretician, but that of an engineer, has happily followed (e.g., see Bringsjord & Ferrucci 2000). We will return to this approach when, in the final chapter, we discuss the *practice* of AI in light of supermentalism. There we advocate a form of AI expressed in eight prescriptions.

Chapter 4

Supermentalism and the Fall of Church's Thesis

In this chapter we refute one of the core parts of computationalism, as this doctrine is defined in Chapter 1: namely, Church's Thesis. More precisely, we refute the Church-*Turing* Thesis, the view, put roughly for now, that that which is algorithmic in the intuitive sense can be identified with what a (standard) Turing machine can accomplish. As you will see, our refutation makes crucial use of the formal terrain above the Turing Limit (recall, yet again, Figure 1.1). The refutation owes a great debt to Elliot Mendelson, who, in a widely affirmed paper in the *Journal of Philosophy* (1986), challenges what he rightly calls the "standard conception" (Mendelson 1986, p. 230) of Church's Thesis (CT) — the conception being that CT is unprovable. Mendelson got Bringsjord thinking about CT in earnest; he provided the stimulus that eventuated in this chapter. However, our core argument owes much to many other thinkers who have offered critiques, and our case has in the process — by our lights, anyway — grown progressively stronger.

To be more specific about the plan for the present chapter: Once Mendelson's target, and his attack upon it, are rigorously analyzed with help from uncomputability theory, we show that his challenge to the standard conception of CT evaporates, and that the cogent rationale for this conception is revealed. This analysis will in turn constitute a foundation for overthrowing CT on the basis of our hypercomputational ability to assimilate and assess narrative. Needless to say, we rebut a series of objections to our attack on CT. We end by considering some other attacks on CT (some of which, as we shall see, help to paint a fascinating portrait of philosophizing about CT through the years, for some of these attacks are ones Mendelson himself tried to derail over forty years ago), and the relation between these attacks and our own.

4.1 Background

At the heart of CT is the notion of an **algorithm**, characterized in traditional fashion by Mendelson as

> an effective and completely specified procedure for solving a whole class of problems. ... An algorithm does not require ingenuity; its application is prescribed in advance and does not depend upon any empirical or random factors. (Mendelson 1986, p. 225)

An **effectively computable** function is then said to be the computing of a function by an idealized "worker" or "computist" following an algorithm.[1] (Without loss of generality, we can for present purposes view all functions as taking natural numbers into natural numbers; i.e., for some arbitrary f, $f : \mathbf{N} \to \mathbf{N}$). CT also involves a more formal notion, typically that of a so-called **Turing-computable** function (or, alternatively, and equivalently, that of a **recursive** function, or, ...); as you'll recall, we covered these concepts in Chapter 1. Mendelson employs Turing's approach, and Turing machines will by now be familiar to all of our readers; we'll follow him: a function $f : \mathbf{N} \to \mathbf{N}$ is Turing-computable iff there exists a TM M which, starting with n on its tape (perhaps represented by n |s), leaves $f(n)$ on its tape after processing.[2] (The details of the processing are harmlessly left aside for now.) Given this definition, CT amounts to

CT A function is effectively computable if and only if it's Turing-computable.[3]

Now what exactly is Mendelson's aim? He tells us:

---

[1]Recall our remarks, in the Preface, about Turing (1936)'s "computists" and Post (1944)'s "workers," humans whose sole job was to slavishly follow explicit, excruciatingly simple instructions.

[2]Once again, details were provided in Chapter 1, and some readers may wish to refer back to that chapter.

[3]This is often called the Church-*Turing* Thesis for obvious reasons.

> Here is the main conclusion I wish to draw: it is completely unwarranted to say that CT is unprovable just because it states an equivalence between a vague, imprecise notion (effectively computable function) and a precise mathematical notion (partial-recursive function). (Mendelson 1986, p. 232)

From this it follows that Mendelson's target is the traditional *argument* for the unprovability of CT. And the line of reasoning he means to attack runs as follows.

Arg$_1$

	EQU	If some thesis T states an equivalence between a vague, imprecise notion and a precise, mathematical notion, T is unprovable.
∴	(1)	If CT states an equivalence between a vague, imprecise notion and a precise, mathematical notion, CT is unprovable.
	(2)	CT states an equivalence between a vague, imprecise notion and a precise, mathematical notion.
∴	(3)	CT is unprovable.

4.2 Mendelson's Attack

Mendelson's attack on Arg$_1$ is based on "theses" analogous to CT — "Peano's Thesis," "Tarski's Thesis," "Frege's Thesis," and "Weierstrass' Thesis."[4] The first three, respectively, are:

PT f is an intuitive, rule-based function if and only if f is a set of ordered pairs satisfying ($\star$) if $(x, y) \in f$ and $(x, z) \in f$, then $y = z$.

TT Let L be a first-order language, and $\mathcal{I}$ an interpretation based on L. Then a wff ϕ of L is true on $\mathcal{I}$ in the intuitive sense iff $\mathcal{I} \models \phi$, i.e., $\mathcal{I}$ satisfies ϕ, in the Tarskian model-theoretic sense.

FT Again, let L be a first-order language, and $\mathcal{I}$ an interpretation based on L. Then a wff ϕ is valid in Frege's intuitive sense iff $\models \phi$, i.e., ϕ is valid in the model-theoretic sense.

[4] WT identifies the intuitive notion of limit with the standard ϵ-δ definition. Mendelson thinks that these four theses are just some among many such "theses." He mentions "the notion of measure as an explication of area and volume, the definition of dimension in topology, the definition of velocity as a derivative, the definition of logical implication and logical equivalence in first-order logic, and the definitions of circle, triangle, interior of an angle, and many other geometric concepts" (Mendelson 1986, p. 232).

But how does Mendelson use these three theses as ammunition for his three-pronged attack on Arg_1? (The prongs are distinguished on pp. 232–233; he says his "argument" is based on "three points.") Let's look at the three prongs in turn, and blunt each.

The first prong, the most sophisticated and promising of the three, is an attack on Arg_1's premise (2): Mendelson seems to be saying that the equivalence this premise attributes to CT is chimerical:

> The concepts and assumptions that support the notion of partial-recursive function are, in an essential way, no less vague and imprecise than the notion of effectively computable function; the former are just more familiar and are part of a respectable theory with connections to other parts of logic and mathematics. (The notion of effectively computable function could have been incorporated into an axiomatic presentation of classical mathematics, but the acceptance of CT made this unnecessary.) The same point applies to [PT, FT, and TT]. Functions are defined in terms of sets, but the concept of set is no clearer than that of function and a foundation of mathematics can be based on a theory using function as primitive notion instead of set. Tarski's definition of truth is formulated in set-theoretic terms, but the notion of set is no clearer than that of truth. The model-theoretic definition of logical validity is based ultimately on set theory, the foundations of which are no clearer than our intuitive understanding of logical validity. (Mendelson 1986, p. 232)

But how does not-(2) follow from this? What, exactly, is Mendelson's argument? The key idea seems to be that (2) is false because

(4) The notion of Turing-computable function is no clearer than, nor more mathematically useful (foundationally speaking) than, the notion of an effectively computable function.

We can probably all agree that (4) implies not-(2). But is (4) true? Mendelson gives both a direct rationale for (4) and an argument for it based on PT, FT, TT, and WT. Let's consider, first, the argument based on these other theses. Mendelson's hope appears to be that (4) follows from

X is no clearer than, nor ... than Y

when this template, tied to the other "theses," is filled in in the expected way. For example, with respect to TT, the template becomes

'true on some $\mathcal{I}$' is no clearer than, nor ... than, 'intuitive truth'

And with respect to PT the template becomes

> '($\star$)-based function' is no less vague than, nor ... than, 'intuitive function'

But there is a problem: Mendelson doesn't *establish* these statements. He simply asserts them. And that's not enough — especially when the default intuition is likely to be that these statements are false. Now, Mendelson does seem to think that such statements follow from (or are at least supported by) the fact that things like ZFC (Zermelo-Fraenkel axiomatic set theory — with the axiom of choice, its presence indicated here by C —, presented in concise form in Ebbinghaus et al. 1984)[5] can in principle be replaced with foundations that take the concept of function as primitive. But *how* does it follow from this that, e.g.,

> '($\star$)-based function' is no clearer than, nor ... than, 'intuitive function'?

Mendelson doesn't answer this question.

But let's assume for the sake of argument that the template filled in for PT (and FT, TT, WT) is true. We can then ask whether (4) follows from this assumption. Does it? No. In fact, it seems relatively easy to show that (4) is false, once one looks a bit at uncomputability theory. Here's how the demonstration works: Clearly, if

(4) The notion of Turing-computable function is no clearer than, nor more mathematically useful (foundationally speaking) than, the notion of an effectively computable function,

---

[5]Nine axioms are included in ZFC. Here are five.

Axiom of Extensionality

$$\forall x \forall y(\forall z(z \in x \leftrightarrow z \in y) \rightarrow x = y)$$

Axiom Schema of Separation

$$\forall x_0 \ldots \forall x_{n-1} \forall x \exists y \forall z(z \in y \leftrightarrow (z \in x \wedge \phi(z, x_0, \ldots, x_{n-1})))$$

Power Set Axiom

$$\forall x \exists y \forall z(z \in y \leftrightarrow \forall w(w \in z \rightarrow w \in x))$$

Axiom of Infinity

$$\exists x(\emptyset \in x \wedge \forall y(y \in x \rightarrow y \cup \{y\} \in x))$$

Axiom of Choice

$$\forall x((\emptyset \notin x \wedge \forall u \forall v((u \in x \wedge v \in x \wedge u \neq v) \rightarrow$$

$$u \cap v = \emptyset)) \rightarrow \exists y \forall w(w \in x \rightarrow \exists^{=1} z z \in w \cap y))$$

then

(5) The notion of Turing-decidable set is no clearer than, nor more mathematically useful (foundationally speaking) than, the notion of an effectively decidable set,

and

(6) The notion of Turing-enumerable set is no clearer than, nor more mathematically useful (foundationally speaking) than, the notion of an effectively enumerable set,

⋮

Now suppose that (4) is true. From this assumption, and the conditional indicated immediately above, it would seem to follow by *modus ponens* and simplification of conjunction that

(+) The notion of a formally defined program for guiding the operation of a TM is no clearer than, nor more mathematically useful (foundationally speaking) than, the notion of an algorithm.

This proposition, it would then seem, is the very heart of the matter. If (+) is true then Mendelson has made his case; if this proposition is false, then his case is doomed, since we can chain back by *modus tollens* and negate (4). What's the verdict? It's possible to demonstrate the falsity of (+), by way of the following straightforward reasoning.

$\mathbf{Arg_2}$

(7) If (+), then one should be able to construct the Arithmetic Hierarchy by way of the notion of an algorithm.
(8) One cannot construct the Arithmetic Hierarchy by way of the notion of an algorithm.
∴ (9) Not-(+)

This argument is obviously valid. Are the premises true? In order for you to answer this question, you may need to return to Chapter 1 and review the presentation there of the Arithmetic Hierarchy. You will see there a more formal presentation of this quantifier-based way to generate AH:

Σ_n set of all predicates definable in terms of totally computable predicates using at most n quantifiers, the first of which is *existential*

Π_n set of all predicates definable in terms of totally computable predicates using at most n quantifiers, the first of which is *universal*

Δ_n $\Sigma_n \cap \Pi_n$

You will also see that we have the Arithmetic *Hierarchy* because, where $\subset$ is proper subset,

$$\begin{gathered}\Sigma_0 \subset \Sigma_1 \subset \Sigma_2 \ldots \\ \Pi_0 \subset \Pi_1 \subset \Pi_2 \ldots \\ \text{for every } m > 0,\ \Sigma_m \neq \Pi_m \\ \Pi_m \subset \Sigma_{m+1} \\ \Sigma_m \subset \Pi_{m+1}\end{gathered}$$

At any rate, even a cursory look at AH reveals that Arg_2, the counter-argument against Mendelson, is sound. Why? We take it, first, that (7) is uncontroversial.[6] Since Arg_2 is formally valid, if (8), the only other premise in the argument, is true, the argument is sound. Premise (8), again, is

(8) One cannot construct the Arithmetic Hierarchy by way of the notion of an algorithm.

But a quick look at AH verifies this premise, for the simple reason that algorithms correspond to those things (e.g., programs $\mathcal{P}$ in $\mathcal{L}$, as we explain in Chapter 1) which must be significantly modified in order to open up the infinite landscape of AH. Mendelson is of course correct that it's possible to supplant ZF with any number of equivalent constructions which don't take 'set' as primitive. But if one takes 'algorithm' as primitive, one will forever close off AH, since to gain access to this paradise one must *formalize* 'algorithm,' and then begin to "tinker" with the details of this formalization.[7]

It's important to note that the crucial (8) is distinct from

($8^\star$) One cannot construct the Arithmetic Hierarchy without the notion of an algorithm.

Proposition ($8^\star$) is false, because it's possible (and to some, preferable) to develop the theory of relative computability (of which AH is

[6] In fact, it would seem that (7) is itself provable via conditional proof: Start by assuming that (+), and then simply consider building AH by way of the unformalized notion of an algorithm.

[7] Such a project is what Steinhart (forthcoming) carries out in "digital physics-based" fashion not found in textbooks on computability and uncomputability.

but a small part) by way of the notion of a function, with programs, oracles, algorithms and the like left by the wayside. Premise (8), on the other hand, says that AH cannot be constructed by way of the notion of an algorithm — and this is so, to repeat, because such a method gives you nothing like the fine-grained analysis of the "function route" (which requires composition, primitive recursion, unbounded minimalization, etc.).[8]

What about the second prong in Mendelson's attack? This prong is aimed against the EQU principle, which was the first premise in Arg_1, the argument Mendelson believes he has undermined. Here is the relevant quote:

> The assumption that a proof connecting intuitive and precise mathematical notions is impossible is patently false. In fact, half of CT (the "easier" half), the assertion that all partial-recursive functions are effectively computable, is acknowledged to be obvious in all textbooks on recursion theory. (Mendelson 1986, p. 232)

Mendelson proceeds to give the "proof" of the "if" part of CT (Mendelson 1986, p. 231). We readily concede that he does in fact prove the "if" part. But a question remains: How does not-EQU follow from this sub-proof? EQU would be overthrown by a counterexample in which this principle's antecedent is true but its consequent is not. But this is *not* the situation Mendelson creates by way of his sub-proof. At best, he has overthrown *this* principle:

$\text{EQU}^{\rightarrow}$ If some thesis T states a conditional connection between a vague, imprecise notion and a precise, mathematical notion, then T is unprovable.

The obvious question then arises: are any of PT, FT, TT provable? If so, then Mendelson might be able to sustain the fight. But it's hard to see how one would even begin to prove these. (Skeptics are encouraged to embark upon the proofs. Better yet, next time you or a colleague teaches first-order logic, assign one of these proofs on an exam.) At any rate, the proofs would clearly be non-trivial; and since Mendelson has provided neither these proofs, nor sketches for how to carry them out, there is no reason, in the context of the present

[8]Heading off the erroneous conflation of ($8^\star$) and (8) is something Kostas Arkoudas, in objecting to our argument, stimulated us to consider; we're indebted.

dialectic, to suppose that the trio analogous to CT — PT, FT, TT — is provable.

What is Mendelson's third prong? Essentially, it's the claim that "the usual viewpoint concerning CT is that it assumes that the only way to ascertain the truth of the equivalence asserted in CT is to *prove* it" (Mendelson 1986, p. 233). Mendelson goes on to claim that

> ... equivalences between intuitive notions and apparently more precise mathematical notions often are simply "seen" to be true without proof, or are based on arguments that are a mixture of such intuitive perceptions and standard logical and mathematical reasoning. (Mendelson 1986, p. 233)

Here Mendelson seems to commit a bald *non sequitur*. For notice that nothing in this quote threatens Arg_1. Nowhere in Arg_1 is there a hidden premise to the effect that there aren't arguments-short-of-proof for CT. On the contrary, as is well-known, those impressed by *inductive* arguments often affirm CT on the basis of such reasoning. So how does Mendelson's third prong constitute a challenge to the orthodox view on CT? Apparently, it simply doesn't. Even Mendelson himself seems to concede that the third prong is a bit beside the point:

> That CT is true follows, I believe, from Turing's analysis of the essential elements involved in computation. But this is not what I have tried to establish. The point I have attempted to make is the equivalences between intuitive notions and "precise" notions need not always be considered unprovable theses.[9] (Mendelson 1986, p. 233)

The third prong would be relevant if there were good deductive arguments for CT, *and* if what Mendelson calls the "usual viewpoint" ruled them out. But to enlarge the "usual viewpoint concerning CT" this way would be to create a straw man. Besides, we have a formidable deductive argument *against CT*, and so it would be rather double-minded if we went in search of deductive arguments *for* the thesis. Our deductive argument against CT doesn't derive from the view that this thesis connects a vague notion with a mathematical one. It derives from another application of uncomputability theory. But before

[9] The last sentence here is somewhat peculiar, since it could be verified by something that would not necessarily defeat Arg_1. Indeed, this sentence could be verified by an argument which lacked reference to CT.

presenting that argument, we consider Mendelson's response[10] to what we have so far said against him.

4.3 Mendelson's Rebuttal

Mendelson concedes that a significant part of our case succeeds, that is, that (4) is indeed false: He agrees that the formal concepts in question (e.g., 'Turing-computable function') are more useful than their informal partners (e.g., 'effectively computable function'); and he admits that "One could, with some justification, claim that the notion of a Turing-computable function is 'clearer' than that of an effectively computable function because the former is more specific and ties in closely with other well-known mathematical concepts." However, Mendelson goes on to say:

> My point in this case has nothing to do with relative clarity of concepts. Rather, the point is that the notion of an effectively computable function is not essentially different from the notions that underlie the theory of Turing-computable functions, and, more specifically, that the former notion can be used in mathematical proofs just as legitimately as the latter notions. This was illustrated in my paper by the proof that all partial-recursive functions are effectively computable. That proof, which Professor Bringsjord himself accepts, undermines the basis for the traditional belief in the unprovability of the the Church-Turing Thesis, namely, that there is in principle an unbridgeable gap between, on the one hand, arguments that involve 'vague, intuitive' notions, and, on the other hand, 'respectable' proofs that can be formalized within, say, ZF or PA.

Unfortunately, this rebuttal fails. Yes, we did indeed concede that Mendelson's mathematical argument for the so-called "easier" half of CT constitutes a proof (though we find Mendelson's comment following that argument — "This simple argument is as clear a proof as I have seen in mathematics" (233) — rather hard to believe). But Mendelson seems to ignore our observation that his proof doesn't overthrow the premise in Arg_1 called 'EQU:'

EQU If some thesis T states an equivalence between a vague, imprecise notion and a precise, mathematical notion, T is unprovable.

[10] Personal communication, December 14, 1993.

Lest it be thought that we tendentiously cooked up this premise, and set its negation as an unreachable target for Mendelson, we remind the reader that the text in question is clear: there can be little doubt that Mendelson targets EQU. Consider again, for example, what he says on page 232:

> Here is the main conclusion I wish to draw: it is completely unwarranted to say that CT is unprovable just because it states an *equivalence* between a vague, imprecise notion (effectively computable function) and a precise mathematical notion (partial-recursive function). (Mendelson 1986, p. 232; emphasis ours)

And again on page 233 Mendelson says that "*equivalences* between intuitive and 'precise' notions need not always be considered unprovable theses" (emphasis ours).

Now Janet Folina (1993) has suggested that Mendelson's aim be charitably scaled back — so that what he is said to be aiming at is *not* a demonstration that CT is provable (an aim she agrees we have shown Mendelson cannot reach), but rather a demonstration that proofs merely *connecting* intuitive with formal notions are possible. This reconstruction of Mendelson's main argument is, predictably, one that we would gladly accept.[11]

With our foundation laid, we turn now to our narrational attack on CT.

4.4 Attacking Church's Thesis

Our suspicion that CT is false first arose in connection with the concept of **productive** sets, which have two properties:

[11]Incidentally, the reconstruction probably has fatal problems, as Folina (1993) points out. After all, *is* what Mendelson calls a proof here a proof? One common conception — indeed, probably the *dominant* conception — of a proof is of a transformation in some formal system. Yet Mendelson says about his proof: "The fact that it is not a proof in ZF or some other axiomatic system is no drawback; it just shows that there is more to mathematics than appears in ZF" (Mendelson 1986, p. 233). (Remember this quote later when we consider Mendelson's rejection of Kalmár's argument against CT because in part it falls outside any standard formal system.) A lot of thinkers will balk at this. As Folina (1993) notes, many will diagnose the situation by saying that what Mendelson has shown is that there is more to mathematics than proofs. (This is something we've known all along, of course.) Moreover, if Mendelson's reasoning *isn't* a proof, then what is it? If it's merely a *precise, compelling argument* connecting an intuitive notion with a formal one, then it shows something we knew to be true all along.

P1 They are classically undecidable (= no program, Turing machine, etc. can decide such sets).

P2 There is a computable function f from the set of all standard programs to any such set, a function which, when given a candidate program P (for deciding the set in question), yields an element of the set for which P will fail.

Put informally, a set A is productive iff it's not only classically undecidable, but also if any program proposed to decide A can be counter-exampled with some element of A. Clearly, if a set A' has these properties, then $A' \notin \Sigma_0$ and $A' \notin \Sigma_1$. If A' falls somewhere in AH, and is effectively decidable, then CT falls. But what could possibly fit the bill? We have become convinced that the set $\mathcal{S}$ of all interesting stories provides a perfect fit.

This no doubt catches you a bit off guard. Interesting stories? Well, let us first remind you that the view that there are productive sets near at hand is far from unprecedented. Douglas Hofstadter (1982), for example, holds that the set $\mathcal{A}$ of all As is a productive set. In order to satisfy P1, $\mathcal{A}$ must forever resist attempts to write a program for deciding this set; in order to satisfy P2, there must at minimum always be a way to "stump" a program intended to decide $\mathcal{A}$. That $\mathcal{A}$ satisfies both these conditions isn't all that implausible — especially when one faces up to the unpredictable variability seen in this set. For example, take a look at Figure 4.1, taken from (*Graphic Art Materials Reference Manual* 1981).

In order for a program to decide $\mathcal{A}$, it must capitalize on some rules that capture the "essence" of the letter in question. But what sorts of rules could these be? Does the bar in the middle need to touch the sides? Apparently not (see 2 A). Does there have to be a bar that *approximates* connecting the sides? Apparently not (see 7 G). And on and on it goes for other proposed rules.[12]

However, it must be conceded that no *argument* for the productivity of $\mathcal{A}$ has been provided by Hofstadter. For all we know, some company could tomorrow announce a letter recognition system that will work for all As. The situation is a bit different in the case of the mathematician Peter Kugel (1986), who makes clever use of an elementary theorem in unmistakably *arguing* that the set of all beautiful objects is located above Σ_1 in AH:

[12]Relevant here is Hofstadter's LETTER SPIRIT program, which generates fonts from the first few letters in the font in question. For an argument that this program, and others, aren't really creative, see (Bringsjord, Ferrucci & Bello 2001).

Figure 4.1: Various Letter As

> We seem to be able to recognize, as beautiful, pieces of music that we almost certainly could not have composed. There is a theorem about the partially computable sets that says that there is a uniform procedure for turning a procedure for recognizing members of such sets into a procedure for generating them. Since this procedure is uniform — you can use the same one for all computable sets — it does not depend on any specific information about the set in question. So, if the set of all beautiful things were in Σ_1, we should be able to turn our ability to recognize beautiful things into one for generating them ... This suggests that a person who recognizes the Sistine Chapel Ceiling as beautiful knows enough to paint it, [which] strikes me as somewhat implausible. (Kugel 1986, pp. 147–148)

The main problem with this line of reasoning is that it's disturbingly exotic. Beauty is perhaps a promising candidate for what Kugel is after, but it must be conceded that most of those scientists who think seriously about human cognition don't think a lot about beauty. Indeed, they don't seem to think *at all* about beauty.[13] And this isn't (they would insist) because beauty is a daunting concept, one that resists recasting in computational terms. The stance would doubtless be that beauty is left aside because one can exhaustively analyze cognition (and replicate it on a machine) without bothering to grapple in earnest with this concept.

This claim about the irrelevance of beauty may strike some as astonishing, and it certainly isn't a view affirmed by each and every computationalist, but we gladly concede it for the sake of argument: for the record, we grant that ignoring beauty, in the context of attempts to model, simulate, and replicate mentation, is acceptable.[14] However, we think there is another concept that serves our purposes perfectly: namely, the concept of a *story*. Stories are thought by many to be at the very heart of cognition. For example, in their lead target chapter in *Knowledge and Memory: The Real Story* (Wyer 1995), Roger Schank and Robert Abelson, two of the most eminent scientists in the world working in the area of cognition and computation, boldly assert on the first page that "virtually all human knowledge" is based on stories.[15] Schank and Abelson go on to claim that since the essence of cognition inheres in narrative, we can jettison propositional, logic-based, rule-based, formal ... schemes for knowledge representation. Among the 17 commentators who react to the target piece, 13 affirm the story-based view (the remaining four authors are skeptical). Moreover, this book is one of many. For example, Schank has devoted a book to the view that stories are at the very heart of human cognition: (Schank 1995). For another example, note that Dennett's (1991) *Consciousness Explained* can be read as a defense of the view (his "multiple drafts" view of consciousness) that thinking amounts to the spinning out of parallel stories.

---

[13] A search for coverage of this concept in standard texts about cognition — e.g., (Ashcraft 1994) and (Stillings et al. 1995) — turns up nothing whatever.

[14] What argument could be mustered for ignoring beauty in the context of attempts to reduce cognition to computation, or to build an artificial agent capable of behaviors analogous to human ones typically taken to involve beauty? We envisage an argument running parallel to the one John Pollock (1995) gives for ignoring human emotions in his attempt to build an artificial person. Pollock's view, in a nutshell, is that human emotions are in the end just "time savers;" with fast enough hardware, and clever enough algorithms, artificial persons could *compute* the need to quickly flee (say) a lion, whereas we take one look and immediately feel a surge of fear that serves to spark our rapid departure.

[15] An insightful review of this book has been written by Tom Trabasso (1996).

The other nice thing about stories, from our perspective, is that apparently one of us knows a thing or two about them, in connection to computation. For over a decade now, Bringsjord has co-directed an AI research group — *Autopoeisis* — devoted to creating an artificial agent capable of autonomously creating sophisticated fiction. Bringsjord first discussed this project in his *What Robots Can and Can't Be* (1992), in which he specifically discussed the challenge of characterizing, precisely, the class of *interesting* stories. (His main claim was that formal philosophy offers the best hope of supplying this characterization.) For those who seek to build agents capable of creative feats like good storytelling, this is a key challenge. It's easy enough to build systems capable of generating *un*interesting stories. For example, the world's first significant artificial story generator, TALE-SPIN (Meehan 1981), did a good job of that. Here, for example, is one of TALE-SPIN's best stories:

"Hunger"

> Once upon a time John Bear lived in a cave. John knew that John was in his cave. There was a beehive in a maple tree. Tom Bee knew that the beehive was in the maple tree. Tom was in his beehive. Tom knew that Tom was in his beehive. There was some honey in Tom's beehive. Tom knew that the honey was in Tom's beehive. Tom had the honey. Tom knew that Tom had the honey. There was a nest in a cherry tree. Arthur Bird knew that the nest was in the cherry tree. Arthur was in his nest. Arthur knew that John was in his cave. ...

How are things to be improved? How is one to go about building an agent capable of creating interesting stories? It has been the sustained attempt to answer this question, in conjunction with the concept of productivity discussed above, that has persuaded the two of us that CT is indeed false. Let us explain.

First, to ease exposition, let $\mathcal{S}^I$ denote the set of all interesting stories. Now, recall that productive sets must have two properties, P1 and P2; let's take them in turn, in conneciton with $\mathcal{S}^I$. First, $\mathcal{S}^I$ must be classically undecidable; i.e., there is no program (or TM, etc.) which answers the question, for an arbitrary story in $\mathcal{S}$, whether or not it's interesting. Second, there must be some computable function f from the set of all programs to $\mathcal{S}^I$ which, when given as input a program P that purportedly decides $\mathcal{S}^I$, yields an element of $\mathcal{S}^I$ for which P fails. It seems to us that $\mathcal{S}^I$ does have both of these properties —

because, in a nutshell, the *Autopoeisis* research group seems to invariably and continuously turn up these two properties "in action." Every time someone suggests an algorithm-sketch for deciding $\mathcal{S}^I$, it's easily shot down by a counter-example consisting of a certain story which is clearly interesting despite the absence in it of those conditions P regards to be necessary for interestingness. (It has been suggested that interesting stories must have inter-character conflict, but monodramas can involve only onc character. It has been suggested that interesting stories must embody age-old plot structures, but some interesting stories are interesting precisely because they violate such structures, and so on.)

The situation we have arrived at can be crystallized in deductive form as follows.

Arg$_3$

	(9)	If $\mathcal{S}^I \in \Sigma_1$ (or $\mathcal{S}^I \in \Sigma_0$), then there exists a procedure P which adapts programs for deciding members of $\mathcal{S}^I$ so as to yield programs for enumerating members of $\mathcal{S}^I$.	
	(10)	There's no procedure P which adapts programs for deciding members of $\mathcal{S}^I$ so as to yield programs for enumerating members of $\mathcal{S}^I$.	
$\therefore$	(11)	$\mathcal{S}^I \notin \Sigma_1$ (or $\mathcal{S}^I \notin \Sigma_0$).	10, 11
	(12)	$\mathcal{S}^I \in$ AH.	
$\therefore$	(13)	$\mathcal{S}^I \in \Pi_1$ (or above in the AH).	disj syll
	(14)	$\mathcal{S}^I$ is effectively decidable.	
$\therefore$	(15)	CT is false.	*reductio*

Clearly, Arg$_3$ is formally valid. Premise (9) is not only true, but necessarily true, since it's part of the canon of elementary computability theory. What about premise (10)? Well, this is the core idea, the one expressed above by Kugel, but transferred now to a different domain: People who can *decide* $\mathcal{S}^I$, that is, people who can decide whether something is an interesting story, can't necessarily *generate* interesting stories. Students in *Autopoeisis* have been a case in point: with little knowledge of, and skill for, creating interesting stories, they can nonetheless recognize such narrative. That is, students who are, by their own admission, egregious creative writers, are nonetheless discriminating critics. They can decide which stories are interesting (which is why they know that the story generators AI has produced so far are nothing to write home about), but *producing* the set of all such stories (including, as it does, such works as not only *King Lear*, but *War and Peace*) is quite another matter. These would be, necessarily, the *same* matter if the set of all interesting stories, $\mathcal{S}^I$, was in either Σ_0 or Σ_1, the algorithmic portion of AH.

But what's the rationale behind (13), the claim that $\mathcal{S}^I$ is effectively decidable? The rationale is simply the brute fact that a normal, well-adjusted human computist can effectively decide $\mathcal{S}^I$. Try it yourself: First, start with the sort of story commonly discussed in AI; for example:

"Shopping"

> Jack was shopping at the supermarket. He picked up some milk from the shelf. He paid for it and left.[16]

Well? Your judgement? Uninteresting, we wager. Now go back to "Hunger," and come up with a judgement for it, if you haven't done so already. Also uninteresting, right? Now render a verdict on "Betrayal," a story produced by Bringsjord and Ferrucci's (2000) BRUTUS:

"Betrayal"

> Dave Striver loved the university. He loved its ivy-covered clocktowers, its ancient and sturdy brick, and its sun-splashed verdant greens and eager youth. He also loved the fact that the university is free of the stark unforgiving trials of the business world — only this *isn't* a fact: academia has its own tests, and some are as merciless as any in the marketplace. A prime example is the dissertation defense: to earn the PhD, to become a doctor, one must pass an oral examination on one's dissertation. This was a test Professor Edward Hart enjoyed giving.
>
> Dave wanted desperately to be a doctor. But he needed the signatures of three people on the first page of his dissertation, the priceless inscriptions which, together, would certify that he had passed his defense. One of the signatures had to come from Professor Hart, and Hart had often said — to others and to himself — that he was honored to help Dave secure his well-earned dream.

[16]From page 592 of (Charniak & McDermott 1985). The story is studied in the context of attempts to resolve pronouns: How do we know who the first occurrence of 'He' refers to in this story? And how do render the process of resolving the pronoun to Jack as a computational one?

> Well before the defense, Dave gave Hart a penultimate copy of his thesis. Hart read it and told Dave that it was absolutely first-rate, and that he would gladly sign it at the defense. They even shook hands in Hart's book-lined office. Dave noticed that Hart's eyes were bright and trustful, and his bearing paternal.
>
> At the defense, Dave thought that he eloquently summarized Chapter 3 of his dissertation. There were two questions, one from Professor Rodgers and one from Dr. Teer; Dave answered both, apparently to everyone's satisfaction. There were no further objections.
>
> Professor Rogers signed. He slid the tome to Teer; she too signed, and then slid it in front of Hart. Hart didn't move.
>
> "Edward?" Rogers said.
>
> Hart still sat motionless. Dave felt slightly dizzy.
>
> "Edward, are you going to sign?"
>
> Later, Hart sat alone in his office, in his big leather chair, saddened by Dave's failure. He tried to think of ways he could help Dave achieve his dream.

This time, interesting, right?

Now at this point some readers may be thinking: "Now wait a minute. Isn't your position inconsistent? On the one hand you cheerfully opine that 'interesting story' cannot be captured. But on the other you provide an interesting story! — a story that must, if I understand your project, capitalize upon some careful account of interestingness in narrative."

"Betrayal" is based in significant part upon formalizations, in intensional logic, of definitions taking the classic form of necessary and sufficient conditions seen in analytic philosophy. These definitions are given for "immemorial themes;" in "Betrayal" the two themes are self-deception and, of course, betrayal. Here is the definition of betrayal with which BRUTUS works:[17]

D Agent s_r betrays agent s_d at t_b iff there exists some state of affairs p and $\exists t_i, t_k$ $(t_i \leq t_k \leq t_j \leq t_b)$ such that

[17]Note that the variables t_i range over times, and that $\leq$ means "earlier or simultaneous." Note also the following clauses, which appear in clause 3′.

3 s_r agrees with s_d that p ought to occur;

6′ s_d wants that there is some action a which s_r performs in the belief that thereby p *will* occur.

1 s_d at t_i wants p to occur;

2 s_r believes that s_d wants p to occur;

3′ $(3 \wedge 6') \vee$

6″ s_d wants at t_k that there is no action a which s_r performs in the belief that thereby p will not occur;

4″ there is some action a such that:

4″a s_r performs a at t_b in the belief that thereby p will *not* occur; and

4″b it's not the case that there exists a state of affairs q such that q is believed by s_r to be good for s_d and s_r performs a in the belief that q will not occur;

5′ s_r believes at t_j that s_d believes that there is some action a which s_r will perform in the belief that thereby p *will* occur.

All of this sort of work (i.e., the gradual crafting of such definitions in the face of counter-example after counter-example; the crafting in the case of betrayal is described in Chapter 4 of Bringsjord & Ferrucci 2000) is perfectly consistent with the absence of an account of 'interesting story.' *In fact*, this kind of philosophical analysis figures in the observation that proposed accounts of interestingness are invariably vulnerable to counter-example. For example, suppose we try (here, schematically) something Bringsjord and colleagues have tried: Let $c_1, \ldots, c_n$ enumerate the definitions of all the immemorial themes involved in narrative. Now suppose we venture a defintion having the following structure.

D′ A story s is interesting iff

1 …

⋮

k s instantiates (inclusive) either c_1 or c_2 or … or c_n.

$k+1$ …

⋮

p …

The problem — and, alas, Bringsjord has experienced it time and time again — is that along will come a counter-example; in this case, a story which explicitly fails to satisfy k from D′'s definiens will arrive. For example, an author can write a very interesting story about a phenomenon like betrayal as cashed out in definition D, except that instead of clause 4″, the following weaker clause is satisfied.

4′ there is some action a which s_r performs in the belief that thereby p will *not* occur.

The story here might involve a courageous, self-sacrificial mother who assures her addicted son that she will procure drugs to relieve his misery (as he desires), but intends only to confront the pusher and put an end to his destructive dealings. Ironically, clearly some of the interestingness in this story will derive precisely from the fact that the mother is not betraying her son. On the contrary, she plans to save him and others. In short, devising accounts like D' seems to be to fight a battle that can never be won; good narrative cannot be bottled.

At this point, we suspect that many readers are chomping at the bit, raring to tear into our position with additional objections. Let's see if we can't anticipate and disarm them now.

4.5 Objections

4.5.1 Objection 1

"Look, Bringsjord and Zenzen, you must have gone wrong somewhere! Stories are just strings over some finite alphabet. In your case, given the stories you have put on display above, the alphabet in question is { `Aa, Bb, Cc, ..., :, !, ;, ...`}, that is, basically the characters we see before us on our computer keyboard. Let's denote this alphabet by 'E.' Elementary string theory tells us that though E^*, the set of all strings that can be built from E, is infinite, it's *countably* infinite, and that therefore there is a program P which enumerates E^* (P, for example, can resort to lexicographic ordering). From this it follows that your $\mathcal{S}$, the set of all stories, is itself countably infinite. (If we allow, as no doubt we must, all natural languages to be included — French, Chinese, and even Norwegian — the situation doesn't change: the union of a finite (or for that matter a countably infinite) number of countably infinite sets is still just countably infinite.) So what's the problem? You say that your students are able to decide $\mathcal{S}^I$? Fine. Then here's what we do to enumerate $\mathcal{S}^I$: Start P in motion, and for each item S generated by this program, call your students to pass verdict on whether or not S is interesting. This composite program — call it P': P working in conjunction with your students — enumerates $\mathcal{S}^I$. So sooner or later, P' will manage to write *King Lear*, *War and Peace*, and even more recent belletristic narrative produced by Bringsjord's favorite author: Mark Helprin."

There is good reason to think that even if $\mathcal{S}$ is in some sense typographic, it needn't be countably infinite. Is $\mathcal{A}$, the set of all `A`s, countable? (You might at this point want to return to Figure 4.1.) If not, then simply imagine a story associated with every element within $\mathcal{A}$; this provides an immediate refutation of Objection 1. For a parallel route to the same result, think of a story about π, a story about $\sqrt{2}$, indeed a story for every real number!

Now, someone might claim that $\mathcal{S}$ (or $\mathcal{S}^I$) can be partitioned into a countably infinite set of equivalence classes. But without an argument for this claim, we have no reason to take it seriously. Not only is an argument lacking, but of course the equivalence classes must themselves be specified. What might these classes be? As far as we can see, the only candidates correspond to categories used by mass-market editors; for example: mysteries, romance novels, erotica, and so on. But this will get our opponent nowhere: these categories can be broken by original fiction, just as the rules proposed for interestingness can be broken. To assume otherwise for these categories is to simply beg the question. To see the *petitio* from a different angle, consider what we should make of the point that **R** can be partitioned into a countably infinite set of equivalence classes. **R** can be partitioned into intervals on the "real number line," but what defines each class in this case is "extrinsic." For example, "between 0 and 1, with 0," "between 1 and 2, with 1," ... would give rise to a countably infinite set of sets composing **R**, but this approach leverages properties extrinsic to each real. To maintain that something similar can be carried out for $\mathcal{S}$ (or $\mathcal{S}^I$) is, again, to simply beg the question. After all, the issue is whether there *are* relatively simple structures that allow us to categorize stories.

The fact of the matter is that stories, in the real world, are often neither strings nor, more generally, typographic. After all, authors often think about, expand, refine, ... stories without considering anything typographic whatsoever. They may "watch" stories play out before their mind's eye, for example. In fact, it seems plausible to say that strings (and the like) can be used to *represent* stories, as opposed to saying that the relevant strings, strictly speaking, *are* stories.

At this point, a critic might complain that this lattitudinarian view of stories implies that they are not the sort of well-defined things that are in the Arithmetic Hierarchy. This complaint can't be on the right track, for if it were, then even the concepts involved in standard, well-oiled computational work would have to be said to fall outside of AH, which would in turn cut an unacceptable chasm between the theory of computation and computation itself. For example, a good chess move, a sensible medical test to be conducted given the presentation of certain symptoms, an expert judgement about where to drill for oil given certain evidence, and so on — the assumption is always that these concepts, each of which is at the core of concrete work, are ultimately in Arithmetic HierarchyAH, even though it seems at the moment well nigh impossible to express them in terms of quantifiers over computable predicates.

4.5.2 Objection 2

"Ah! You concede then that you have a decision procedure for $\mathcal{S}^I$. But uncountably infinite sets like **R**, the reals, are *not* decidable!"

This objection is anemic (though we have had it earnestly expressed to us). And the reason it is, of course, is that we need only maintain that $\mathcal{S}^I$ is *effectively* decidable, not that there is some program (or Turing machine, etc.) that can decide this set. (CT is the customary justification given for identifying effective decidability with formal decidability, but of course one can hardly invoke CT in the present context without falling prey to yet another *petitio*.)

Though Objection 2 is misguided, it does suggest an interesting parallel for Arg_3:

Arg$'_3$

	(9′)	If $\mathbf{R} \in \Sigma_1$ (or $\mathbf{R} \in \Sigma_0$), then there exists a procedure P which adapts programs for deciding members of **R** so as to yield programs for enumerating members of **R**.	
	(10′)	There's no procedure P which adapts programs for deciding members of **R** so as to yield programs for enumerating members of **R**.	
∴	(11′)	$\mathbf{R} \notin \Sigma_1$ (or $\mathbf{R} \notin \Sigma_0$).	10′, 11′
	(12′)	$\mathbf{R} \in$ AH.	
∴	(13′)	$\mathbf{R} \in \Pi_1$ (or above in the AH).	disj syll
	(14′)	**R** is effectively decidable.	
∴	(15′)	CT is false.	*reductio*

As we know by now, premise (9′) is an instantiation of a simple theorem of elementary computability theory; (11′) and (13′) are simply intermediate conclusions; (15) does indeed follow from (13′) and (14′), since these two propositions counter-example CT's "only if" part; and the other two inferences are unassailable. Everything boils down to (10′) and (14′). But we know that in the case of the reals, (11′) is true (and, of course, so is (10′)), and the technique of getting **R** from **N** (the natural numbers) via (e.g.) Dedekind cuts constitutes a proof of (12′). Of course, it's doubtful that **R** or a subset like $[0, 1]$ is effectively decidable. Such is not the case, as we've explained, with $\mathcal{S}^I$.

4.5.3 Objection 3

"I now see your error, gentlemen: premise (12) in Arg_3. If $\mathcal{S}^I$ is to be in AH, then your key predicate — 'Interesting'; denote it by 'I' — must be a bivalent one. (More precisely, I must be isomorphic to a predicate that is built via quantification out of the totally computable bivalent predicates of Σ_0.) But a moment's reflection reveals that I *isn't* bivalent: different people have radically different opinions about whether certain fixed stories are interesting! Clearly, though Jones and Smith may share the same language, and may thus be able to fully understand 'Shopping,' 'Hunger,' 'Betrayal,' *King Lear*, and *War and Peace*, their judgements may differ. "Shopping" might be downright thrilling to an AInik interested in determining how, upon reading such a story, humans know instantly that the pronoun 'He' refers to Jack."[18]

It is important to realize that we are talking about stories *qua* stories; stories as narrative. Hence a better way to focus the present objection is to note that Jones may find *Kind Lear* to be genuine drama, but monstrously boring drama (because, he says, King Lear, is but a lunatic), while Smith is transfixed. It's undeniable that differences of opinion like those existing between Jones and Smith are common. But this fact is not a threat to our argument. First, note that such differences are present in *all* domains, not just in the domain of narrative. Wittgenstein, remember, teased much out of a clash between someone who says that $2 + 2 = 4$ and someone who flatly denies it — so even the arithmetical realm, if Objection 3 goes through, would lack bivalent properties, and if anything is suffused with bivalence, it's arithmetic. Moreover, there is nothing to prevent us from stipulating that these agents come decked out with some fixed "value system" — for judging stories. In fact, let us heretofore insist that I be read as not just interesting *simpliciter*, but interesting given (what must surely be one of the world's most refined systems for gauging stories)

[18]This intelligent objection is originally due to Michael McMenamin (1992), though a number of thinkers have conveyed its gist to us.

the knowledge and ability of none other than Umberto Eco.[19] Our new predicate, then, can be I_{UE}.

The objection could perhaps be sustained as follows:

> "I seriously doubt that Umberto Eco *has* a fixed effective decision system by which *he* decides. I take it this is an illusion predicated on the fact that Eco has the *authority* to say what interests him (*a la* Wittgenstein on the incorrigibility of 'introspection'). *Whatever* Eco sincerely pronounces 'interesting' *is* interesting for Eco; what he says goes. This seems akin to what you two envision your 'decked out' agents doing (just reading and pronouncing); this seems unlike effective deciding. You might as well say that each of us has an effective procedure for deciding the set of things that will be said by us in our lifetime: just by saying that we do we 'enumerate the set.' You might as well say the U.S. Supreme Court has a rote procedure for deciding cases: in deciding them they 'enumerate the set' of Supreme Court decisions. Eco's own infallibility being a matter of authority, nothing guarantees that identically 'decked out' agents — lacking authority — will decide the same as him (or each other for that matter)."

This is a decidedly weak objection. Clearly, one claim made against us is simply that Eco has no system by which he judges interestingness. But this claim is wrong. The reason is that Eco doesn't rely on mere authority: he presents the system: again, we refer interested readers to: (Eco 1979). (One might say that Eco has become an authority *because* he has described his system.) Given this, the analogies to the Supreme Court, and to what we say in our lifetimes, fail. In neither of these domains is there even the hint of a description of the scheme by which verdicts are produced; the situations are therefore disanalogous. We do suspect that *individual* members of the Supreme Court *would* be analogous to Eco. Indeed, analyses of and careful commentaries on Supreme Court opinions routinely contain descriptions of the scheme deployed by a particular member of the Court.

4.5.4 Objection 4

"At the start of this chapter you affirmed Mendelson's characterization of 'algorithm.' Let me remind you that according to that character-

[19] Those unfamiliar with Eco's non-fiction work, might start with his surprising reasons for finding Ian Fleming's 007 (James Bond) series to be very interesting; see "Chapter Six: Narrative Structures in Fleming," in (Eco 1979).

ization, 'An algorithm does not require ingenuity.' Are you not now bestowing remarkable ingenuity upon the readers/judges you have in mind?"

Recall that in order to parse 'effectively computable,' as we have noted, it's necessary to invoke the generic concept of an agent, either Turing's "computist" or Post's "worker." (At the very least, the standard way to unpack 'effectively computable' is through this generic concept.) The agent in question, as none other than Elliot Mendelson reminded us nearly forty years ago (Mendelson 1963), needn't be a *human* agent, because, following the mantra at the heart of computability theory, we impose no practical restrictions on the length of calculations and computations (we followed this approach, note, in Chapter 1). It follows immediately that the agents we have in mind have enough raw time and energy to process the longest and most complex contenders in $\mathcal{S}$. Furthermore, if we are going to seriously entertain CT, we must, all of us, allow the agents in question to have certain knowledge and ability, for example the knowledge and ability required to grasp the concepts of number, symbol, change, movement, instruction, and so on. The agents we have in mind are outfitted so as to be able to grasp stories, and the constituents of stories. And in deploying I, and in moving to I_{UE}, we assume *less* on the part of agents (workers, computists, etc.) than what even defenders of CT through the years have assumed. This is so because such thinkers freely ascribe to the agents in question the knowledge and ability required to carry out sophisticated proofs — even proofs which cannot be formalized in first-order logic. The agents capable of deciding $\mathcal{S}^I$ need only read the story (and, for good measure, read it n subsequent times — something mathematicians routinely do in order to grasp proofs), and render their decision.

4.5.5 Objection 5

"Yes, but what your computists do is not decomposable into smaller, purely mechanical steps, which is the hallmark of an algorithm. They are supposed to read a story (and, if I understand you, perhaps read it again some finite number of times), and then, just like that, render a judgment. This is more like magic than mechanism."

This objection is a complete non-starter. In order to see this, let's prove, in a thoroughly traditional manner, that a certain well-defined problem is effectively solvable. Recall that all Turing machines can be recast as flow diagrams (e.g., see Boolos & Jeffrey 1989). Next, note that any TM represented by a flow diagram having as part the fragment shown in Figure 4.2 would be a non-halting TM (because if started in state 1 with its read/write head scanning the leftmost 1 in a block of 1s — and we can assume the alphabet in question to be a binary one consisting of {0,1} — it will loop forever in this fragment). Let m be a fixed TM specified for computist Smith in flow diagram form, and let this diagram contain the fragment of Figure 4.2. Suppose that Brown looks for a minute at the diagram, sees the relevant fragment, and declares: "Nonhalter!" In doing this, Brown assuredly decides m, and his performance is effective. And yet what's the difference between what Brown does and what our "Eco-ish" agents do? The activity involved is decomposable in both cases. There are innumerable "subterranean" cognitive processes going on beneath Brown's activity, but they are beside the point: that we don't (or perhaps can't) put them on display does not tell against the effectiveness in question. The fact is that Brown simply looks at the diagram, finds the relevant fragment, assimilates, and returns a verdict.[20] The same is true of our agents in the case of stories.

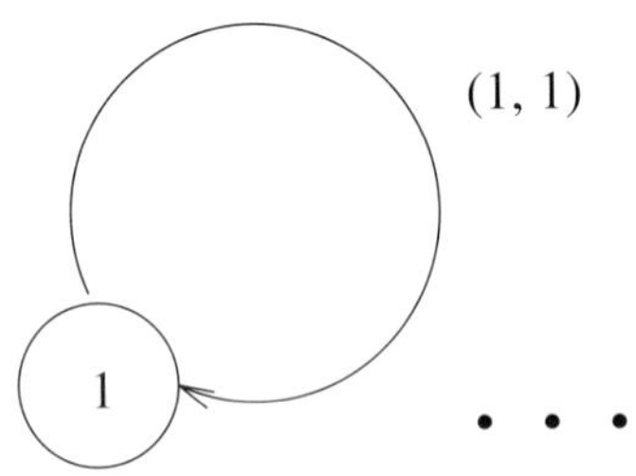

(The node here reflects the start state.)

Figure 4.2: A Flow-Diagram Fragment That Entails Non-Halting

Before turning to consider other attacks on CT, we point out that the predicates I and I_{UE} really aren't exotic, despite appearances to the contrary. All those who try to harness the concepts of theoretical computer science (concepts forming a superset of the formal ones canvassed in this book) in order to get things done end up working with predicates *at least* as murky as these two. A good example is

[20]Our example is perfectly consistent with the fact, reviewed and exploited in Chapter 1, that the set of TMs, with respect to whether or not they halt, is not Turing-decidable.

to be found in the seminal work of John Pollock, which is based on the harnessing of theoretical computer science (including AH) so as to explicate and implement concepts like warrant, defeasibility, *prima facie* plausibility, and so on.[21]

4.6 Our Arg_3 in Context: Other Attacks on CT

Over the past six decades, the possibility of CT's falsity has not only been raised,[22] but CT has been subjected to a number of outright attacks. While we obviously don't have the book-long space it would take to treat each and every attack, we think it's possible to provide a provisional analysis that is somewhat informative, and serves to situate our own attack on CT. What this analysis shows, we think, is that Arg_3 is the best attack going.

Following R.J. Nelson (1987), we partition attacks on CT into three categories:

CAT1 Arguments against the arguments for CT;

CAT2 Arguments against CT itself; and

CAT3 Arguments against doctrines (e.g., the computational conception of mind) which are said to presuppose CT.

Consider CAT3 first. Perhaps the most promising argument in this category runs as follows. Assume for the sake of argument that all human cognition consists in the execution of effective processes (in brains, perhaps). It would then follow by CT that such processes are Turing-computable, i.e., that computationalism is true. However, if computationalism is false, while there remains incontrovertible evidence that

[21] Here is one example from (Pollock 1995): Pollock's OSCAR system is designed so as to constantly update that which it believes in response to the rise and fall of arguments given in support of candidate beliefs. What constitutes correct reasoning in such a scheme? Pollock notes that because a TM with an ordinary program can't decide theorems in first-order logic (the set of such theorems isn't Turing-decidable), answering this question is quite tricky. He ingeniously turns to super-computation for help: the basic idea is that OSCAR's reasoning is correct when it generates successive sets of beliefs that approach the ideal epistemic situation in the limit. This idea involves AH, as Pollock explains.

[22] Boolos and Jeffrey, for example, in their classic textbook *Computability and Logic* (1989), provide a sustained discussion of CT — and take pains to leave the reader with the impression that CT can be overthrown.

human cognition consists in the execution of effective processes, CT is overthrown.

Attacks of this sort strike us as decidedly unpromising. For starters, many people aren't persuaded that computationalism is false (despite the many careful arguments we have ourselves given; recall Table 1). Secondly, this argument silently presupposes some sort of physicalism, because the evidence for the effectiveness of cognition (in the sense that *all* cognition is effective; only this view can support an overthrow of CT in CAT3) no doubt derives from observation and study of processes in the central nervous system. Thirdly, it is certainly at least an open question as to whether the processes involved *are* effective. Indeed, by our lights, some of the processes that constitute cognition aren't effective. We are after all in the business of demonstrating this.

What about CAT1? Well, our refutation of Mendelson falls within it — and yet who would claim that what we have revealed about Mendelson's reasoning constitutes, by itself, a serious attack on CT? The same fundamental question derails even the work of those who intend to attack CT by attacking the time-honored rationales for it. For example, William Thomas (1973) seeks to capitalize on the fact (and it *is* a fact, that much is uncontroversial) that the main rationale behind CT involves empirical induction — a form of reasoning that has little standing in mathematics. Unfortunately, Thomas' observations don't threaten CT in the least, as is easy to see. Most of us believe, *unshakably* believe, that the universe is more than 3 seconds old — but what mathematical rationale have we for this belief? As Russell pointed out, mathematics is quite consistent with the proposition that the universe popped into existence 3 seconds ago, replete not only with stars, but with light here on Earth from stars, and also with minds whose memories include those we have. More generally, of course, from the fact that p doesn't follow deductively from a set of propositions Γ, it hardly follows that p is false; it doesn't even follow that p is the slightest bit implausible.

We are left, then, with CAT2 — the category into which our own attack on CT falls. How does Arg_3 compare with other attacks in this category? To support the view that our own attack is superior, let us consider a notorious argument from four decades back, one due to László Kalmár (1959) (and rejected by none other than Elliott Mendelson 1963), and the only other modern attack on CT that we know of, one given by Carol Cleland (1993, 1995).[23]

---

[23] Perhaps we should mention here something that students of CT and its his-

4.6.1 Kalmár's Argument against CT

Here's how Kalmár's argument runs. First, he draws our attention to a function g that isn't Turing-computable, given that f is:[24]

$$g(x) = \mu_y(f(x, y) = 0) = \begin{cases} \text{the least } y \text{ such that } f(x, y) = 0 \text{ if } y \text{ exists} \\ 0 \text{ if there is no such } y \end{cases}$$

Kalmár proceeds to point out that for any $n \in \mathbf{N}$ for which a natural number y with $f(n, y) = 0$ exists, "an obvious method for the calculation of the least such y ... can be given," namely, calculate in succession the values $f(n, 0), f(n, 1), f(n, 2), \ldots$ (which, by hypothesis, is something a computist or TM can do) until we hit a natural number m such that $f(n, m) = 0$, and set $y = m$.

> On the other hand, for any natural number n for which we can prove, not in the frame of some fixed postulate system but by means of arbitrary — of course, correct — arguments that no natural number y with $f(n, y) = 0$ exists, we have also a method to calculate the value $g(n)$ in a finite number of steps: prove that no natural number y with $f(n, y) = 0$ exists, which requires in any case but a finite number of steps, and gives immediately the value $g(n) = 0$. (Kalmár 1959, p. 74)

Kalmár goes on to argue as follows. The definition of g itself implies the *tertium non datur*, and from it and CT we can infer the existence of a natural number p which is such that

(i) there is no natural number y such that $f(p, y) = 0$; and

(ii) this cannot be proved by any correct means.

Kalmár claims that (i) and (ii) are very strange, and that therefore CT is at the very least implausible.

---

tory will be familiar with, viz., *given an intuitionistic interpretation of 'effectively computable function,'* CT can be disproved. The basic idea is to capitalize on the fact that any subset of $\mathbf{N}$ is intuitionistically enumerable, while many such sets aren't effectively enumerable. [A succinct presentation of the disproof can be found on page 592 of Nelson (1987).] The main problem with such attacks on Church's Thesis, of course, is that they presuppose (certain axioms of — see e.g., Kreisel 1965, 1968) intuitionistic logic, which most reject.

[24]The original proof can be found on page 741 of (Kleene 1983).

This argument is interesting, but really quite hopeless, as a number of thinkers have indicated. For example, as Mendelson (1963) (see also Moschovakis' 1968 review of both Kalmár's paper and Mendelson's reaction) points out, Kalmár's notion of 'correct proof,' for all Kalmár tells us, may fail to be effective, since such proofs are outside the standard logical system (set theory formalized in first-order logic). This is surely historically fascinating, since — as we have seen — it would be Mendelson who, nearly thirty years later, in another defense of CT (the one we examined earlier), would offer a proof of the 'only if' direction of this thesis — a proof that he assumes to be correct but one that he admits to be beyond ZF. Mendelson's proof, however, at least has the virtue of having been presented. The root of Kalmár's problem is that his proofs, on the other hand, are wholly hypothetical: we don't have a single one to ponder. And things get even worse for Kalmár (as Nelson 1987 has pointed out), because even absent the proofs in question, we know enough about them to know that they would vary for each argument to g that necessitates them, which would mean that Kalmár has failed to find a *uniform* procedure, a property usually taken to be a necessary condition for a procedure to qualify as effective.

Though Kalmár does anticipate the problem of lack of uniformity,[25] and though we personally happen to side with him on this issue, it is clear that his argument against CT falls flat: If Kalmár's argument is to succeed, (ii) can be supplanted with

(ii$'$) this cannot be proved by any effective means.

But then how can the argument be deductively valid? It is not, at bottom, a *reductio*, since (i) and (ii$'$) surely are not absurd, and this is the only form a compelling version of the argument could at core be. Kalmár himself, as we have noted, confesses that his argument is designed only to show that CT is implausible, but this conclusion goes

---

[25]He says:

> By the way, [the assumption that the procedure in question] must be uniform seems to have no objective meaning. For a school-boy, the method for the solution of the diverse arithmetical problems he has to solve does not seem uniform until he learns to solve equations; and several methods in algebra, geometry and theory of numbers which are now regarded group-theoretic methods were not consider as uniform before group-theory has (sic) been discovered. (Kalmár 1959, p. 73)

through only if (i) and (ii′), if not absurd, are at least counter-intuitive. But are they? For some, perhaps; for others, definitely not.

Our own take on Kalmár's argument is that it can be rather easily *shown* to be impotent via the machinery we set out above to analyze Mendelson's recent defense of CT: First, let

$$m_1, m_2, \ldots, m_n, m_{n+1}, \ldots$$

enumerate the set of Turing machines. Now substitute for Kalmár's g the following function.

$$h(m_i) = \begin{cases} 1 & \text{if } m_i \text{ halts} \\ 0 & \text{if } m_i \text{ doesn't halt} \end{cases}$$

Recall that if a TM halts, simulating this machine will eventually reveal this fact. This allows us to produce an exact parallel to Kalmár's reasoning: Start with m_1; proceed to simulate this machine. Assuming it halts, return 1, and move on to m_2, and do the same for it; then move to m_3, and so on. While this process is running, stand ready to prove "not in the frame of some fixed postulate system but by means of arbitrary — of course, correct — arguments" that the machine m_i fails to halt, in which case 0 is returned. The parody continues as follows. Given CT, and the law of the excluded middle (which the definition of the function h presupposes), we infer two implausible propositions — propositions so implausible that CT is itself cast into doubt. They are:

(i_h) there exists an m_k such that $h(m_k) = 0$; and

(ii'_h) this cannot be proved by any effectively computable means.

This is a parody, of course, because both of these propositions are fully expected and welcomed by all those who both affirm CT and have at least some familiarity with the formalisms involved.

Now, what about *our* case against CT? Well, it would seem to be free of the defects that plague Kalmár's argument. First, our narrational case is deductive, as Arg_3 makes plain. Second, the process of reading (and possibly rereading a finite number of times) a story, assimilating it, and judging whether or not it's interesting on a fixed evaluation scheme — this process is transparently effective. (Indeed, related processes are routinely requested on standardized tests containing reading comprehension problems, where stories are read, perhaps reread, and judged to express one from among n "main ideas.") Third, the process we're exploiting would seem to be uniform.[26]

4.6.2 Cleland's Doubts about CT

Cleland (1993, 1995) discusses three variants on our CT:

CT_1 Every effectively computable number-theoretic function is Turing-computable.

CT_2 Every effectively computable function is Turing-computable.

CT_3 Every effective procedure is Turing-computable.

Before evaluating Cleland's arguments against this trio, some exegesis is in order. First, each of these three theses is a conditional, whereas our CT is a *bi*conditional. There should be no question that the biconditional is more accurate, given not only Mendelson's authoritative affirmation of the biconditional form, but also given that Church himself originally refers to his thesis as a *definition* of "effectively calculable function" in terms of "recursive function" (Church 1936).[27] However, since we have happily conceded the 'if' direction in CT, there is no reason to worry about this aspect of Cleland's framework. The second point is this: by 'number-theoretic' function Cleland simply means what we have in the present chapter (and in Chapter 1) called a function, that is, a mapping from **N** to **N**. We thus now understand function *simpliciter*, as for example as it's used in CT_2, to allow functions from the reals to reals.[28] There is of course no denying that Church and Turing failed to advocate CT_2, but CT_1 is certainly the "left-to-right" direction of our CT.

Now, what does Cleland say against CT_1–CT_3? She claims, first, that CT_3 can be disproved; the argument is simply this. One type of effective procedure coincides with what Cleland calls "mundane procedures," which are "ordinary, everyday procedures such as recipes

[26] No doubt test designers are correct that a uniform procedure needs to be followed in order to excel in their reading comprehension sections. So why wouldn't the process at the heart of Arg_3 be uniform as well?

[27] On the other hand, Church then immediately proceeds to argue for his "definition," and the reader sees that he is without question urging his readers to affirm a *thesis*.

[28] It will not be necessary to present here the formal extension of computability with number-theoretic functions to computability with functions over the reals. For the formal work, see, e.g., (Grzegorczyk 1955, 1957).

for making Hollandaise sauce and methods for starting camp fires; they are methods for manipulating physical things such as eggs and pieces of wood" (Cleland 1995, p. 11). Turing machine procedures, on the other hand, are "methods for 'manipulating' abstract symbols" (Cleland 1995, p. 11). Since mundane procedures have "causal consequences," and TMs (*qua* mathematical objects) don't, it follows straightaway that mundane procedures aren't Turing-computable, that is, $\neg CT_3$.[29]

Cleland's reasoning, when formalized, is valid; no question about that. The problem is that CT_3 (at least on her reading) has next to nothing to do with those propositions placed in the literature under the title "Church's Thesis"! CT_3 is a variant that no one has ever taken seriously. It may *seem* to some that CT_3 has been taken seriously, but this is only because one construal of it, a construal at odds with Cleland's, has in fact been recognized. On this construal, that a procedure is Turing-computable can be certified by either a relevant design (e.g., a TM flow-graph for making Hollandaise sauce, which is easy to come by; recall Figure 4.2), or by a relevant artifact (e.g., an artificial agent capable of making Hollandaise sauce, which again is easy to come by). At any rate, we're quite willing to concede that CT_3, on Cleland's idiosyncratic reading, is provably false. (Note that we have known for decades that even CT_1, on an intuitionistic (and hence idiosyncratic) reading of "effectively computable function," is provably false. See note 23.) It's worth noting that Cleland herself has sympathy for those who hold that her reading of CT_3 is not a *bona fide* version of Church's Thesis (Cleland 1995, p. 10). What then, about CT_2 and CT_1?

Here Cleland no longer claims to have a refutation in hand; she aims only at casting doubt on these two theses. This doubt is supposed to derive from reflection upon what she calls "genuinely continuous devices" (Cleland 1995, p. 18), which are objects said to "mirror" Turing-uncomputable functions (Cleland 1995, p. 16–17). An object is said to **mirror** a function iff (a) it includes a set of distinct objects which are in one-to-one correspondence with the numbers in the field of the function, and (b) the object pairs each and every object corresponding to a number in the domain of the function with an object corresponding to the appropriate number in the range of the function. Cleland takes

[29] In (Bringsjord & Zenzen forthcoming) we explain why Cleland's placing recipes for such things as cheese balls alongside mathematical accounts of computation is unacceptable.

pains to argue, in intuitive fashion, that there are objects which mirror Turing-uncomputable functions (e.g., an object moving through a 2-dimensional Newtonian universe). She seems completely unaware of the fact that such objects *provably* exist — in the form, for example, of analog chaotic neural nets and, generally, analog chaotic dynamical systems (Siegelmann & Sontag 1994, Siegelmann 1995). (These objects are known to exist in the mathematical sense. Whether they exist in the corporeal world is another question, one everyone — including Cleland — admits to be open.) We will be able to see Cleland's fundamental error (and, indeed, the fundamental error of anyone who attacks CT by taking her general route) if we pause for a moment to get clear about the devices in question. Accordingly, we'll present here an analog dynamical system via the "analog shift map," which is remarkably easy to explain.

First let's get clear on the general framework for the "shift map." Let A be a finite alphabet. A **dotted sequence** over A is a sequence of characters from A^* wherein one dot appears. For example, if A is the set of digits from 0 to 9, then 3.14 is a dotted sequence over A. Set $A^{\cdot}$ to the set of all dotted sequences over A. Dotted sequences can be finite, one-way infinite (as in the decimal expansion of π), or bi-infinite. Now, let $k \in \mathbf{N}$; then the shift map

$$S^k : A^{\cdot} \rightarrow A^{\cdot} : (a)_i \rightarrow (a)_{i+k}$$

shifts the dot k places, negative values for a shift to the left, positive ones a shift to the right. (For example, if $(a)_i$ is 3.14159, then with $k = 2$, $S^2(3.14159) = 314.159$.) Analog shift is then defined as the process of first replacing a dotted substring with another dotted substring of equal length according to a function $g : A^{\cdot} \rightarrow A^{\cdot}$. This new sequence is then shifted an integer number of places left or right as directed by a function $f : A^{\cdot} \rightarrow \mathbf{Z}$. Formally, the analog shift is the map

$$\Phi : a \rightarrow S^{f(a)}(a \oplus g(a)),$$

where $\oplus$ replaces the elements of the first dotted sequence with the corresponding element of the second dotted sequence if that element is in the second sequence, or leaves it untouched otherwise. Formally:

$$(a \oplus g)_i = \begin{cases} g_i & \text{if } g_i \in A \\ a_i & \text{if } g_i \text{ is the empty element} \end{cases}$$

Both f and g have "finite domains of dependence" (DoDs), which is to say that they depend only on a finite dotted substring of the sequence on which they act. The domain of *effect* (DoE) of g, however, may be finite, one-way infinite, or bi-infinite. Here is an example from (Siegelmann 1995 p. 547) which will make things clear, and allow us to see the fatal flaw in Cleland's rationale for doubting CT_2 and CT_1. Assume that the analog shift is defined by (where π^2 is the left-infinite string ...51413 in base 2)

DoD	f	g
0.0	1	π^2
0.1	1	.10
1.0	0	1.0
1.1	1	.0

and that we have a starting sequence of u = 000001.10110; then the following evolution ensues:

$$000001.00110$$

$$0000010.0110$$

$$\pi^2.0110$$

$$\pi^2 0.100$$

$$\pi^2 0.100$$

$$\pi^2 01.00$$

$$\pi^2 1.00$$

$$\pi^2 01.00$$

At this point the DoD is 1.0 and hence no changes occur; this is a **fixed point**. Only the evolution from an initial dotted sequence to a fixed point counts.[30] In this case the input-output map is defined as the transformation of the initial sequence to the final subsequence to the right of the dot (hence in our example u as input leads to 00). The class of functions determined by the analog shift includes as a proper subset the class of Turing-computable functions (the proof is straightforward: Siegelmann 1995). Moreover, the analog shift map is a mathematical model of idealized physical phenomena (e.g., the motion of a billiard

[30]For a nice discussion of the general concept of a fixed point in connection with supertasks, see (Steinhart forthcoming).

ball bouncing among parabolic mirrors). From this it follows that we provably have found exactly what Cleland desires, that is, a genuinely continuous device that mirrors a Turing-uncomputable function. So, if Cleland can establish that

(16) If x mirrors a function, then x computes it,

she will have overthrown both CT_2 and CT_1. Unfortunately, given our analysis of the analog shift map, we can see that Cleland doesn't have a chance; here is how the reasoning runs. Recall, first, the orthodox meaning of 'effectively computable function,' with which we started this chapter: a function f is effectively computable provided that, an agent having essentially our powers, a computist (or worker), can compute f by following an algorithm. So let's suppose that you are to be the computist in the case of the analog shift map. There is nothing impenetrable about the simple math involved; we'll assume that you have assimilated it just fine. So now we would like you to compute the function Φ as defined in our example involving π. To make your job as easy as possible, we will guarantee your immortality, and we will supply you with an endless source of pencils and paper (which is to say, we are "idealizing" you). Now, please set to work, if you will; we will wait and observe your progress ...

What happened? Why did you stop? Of course, you stopped because you hit a brick wall: it's rather challenging to write down and manipulate (or imagine and manipulate mentally) strings like π in base 2! (Note that the special case where the DoE of g is finite in the analog shift map generates a class of functions identical to the class of Turing-computable ones.) Yet this is precisely what needs to be done in order to attack CT_2 and CT_1 in the way Cleland prescribes.

Cleland sees the informal version of the problem, for she writes:

> Is there a difference between mirroring a function and computing a function? From an intuitive standpoint, it seems that there is. Surely, falling rocks don't compute functions, even supposing that they mirror them. That is to say, there seems to be a difference between a mere representation of a function, no matter how detailed, and the computation of a function. [Q:] But what could this difference amount to? (Cleland 1995, p. 20)

She then goes on to venture an answer to this question:

> A natural suggestion is that computation requires not only the mirroring of a function but, also, the *following* of a procedure;

> falling rocks don't compute functions because they don't follow procedures. (Cleland 1995, p. 20)

Cleland then tries to show that this answer is unacceptable. The idea is that since the answer doesn't cut it, she is entitled to conclude that (16) is true, that is, that there *isn't* a difference between mirroring a function and computing a function,[31] which then allows the mere existence of (say) an idealized billiard ball bouncing among parabolic mirrors to kill off CT_2 and CT_1.

What, then, is Cleland's argument for the view that the "natural suggestion" in response to Q fails? It runs as follows:

> Turing machines are frequently construed as *purely* mathematical objects. They are defined in terms of the same kinds of basic entity (viz., sets, functions, relations and constants) as other mathematical structures. A Turing machine is said to *compute* a number-theoretic function if a function can be *defined* on its mathematical structure which has the same detailed structure as the number-theoretic function concerned; there isn't a distinction, in Turing machine theory, between computing a function and defining a function ... If computing a function presupposes following a procedure, then neither Turing machines nor falling rocks can be said to compute functions. (Cleland 1995, p. 21)

This argument is an enthymeme; its hidden premise is that 'compute' is used univocally in the relevant theses, i.e., that 'compute' means the same thing on both the left and right sides of CT, CT_1, and CT_2. This premise is false. The locution 'f is effectively computable,' on the orthodox conception of Church's Thesis, does imply that there is an idealized agent capable of *following* an algorithm in order to compute f. But it hardly follows from this that when 'compute' is used in the locution 'f is Turing-computable' (or in the related locution 'TM M computes f'), the term 'compute' must have the same meaning as it does in connection with idealized agents. Certainly anyone interested in CT, and in defending it, would hasten to remind Cleland that the term 'compute' means one thing when embedded within CT's left side, and another thing when embedded within CT's right side.[32] Having said this, however, and having implicitly conceded the core

[31] This reasoning is certainly enthymematic (since it hides a premise to the effect that there are no other answers that can be given to question Q), but we charitably leave this issue aside.

[32] Unexceptionable parallels abound: We can say 'My friend told me that Burlington is a nice city,' and we can say 'My CD-ROM travel program told me that

mathematical point (viz., that at least some definitions of TMs and Turing-computability deploy 'compute' in the absence of the concept of "following"[33]), we should probably draw Cleland's attention to the formal approach we took in Chapter 1, where in order to characterize information-processing beyond the Turing Limit, we distinguished between a TM as a type of architecture, and a program which this architecture *follows* in order to compute.

Cleland never intended to literally refute CT_1 and CT_2. (As we have seen, she did intend to refute the heterodox CT_3, and for the sake of argument we agreed that here she succeeds.) But we conclude that she fails even in her attempt to cast doubt upon these theses, and that therefore CT is unscathed by her discussion. In contrast, our own case against CT targets this thesis with a deductive argument having no hidden premises and presupposing no convenient construal of CT. We have laid our cards on the table for all to see. We're pretty sure our hand is the best one hitherto revealed, but as to whether it wins, or merely marks another chapter in the interesting story of Church's Thesis, our readers must judge.

Burlington is a nice city,' but we needn't accept the view that 'told me' means the same in both utterances.

[33]Consider, e.g., one Bringsjord uses in teaching mathematical logic: A **Turing machine** is a quadruple (S, Σ, f, s) where

1. S is a finite set of **states**;
2. Σ is an alphabet containing the black symbol $-$, but not containing the symbols $\Leftarrow$ ("go left") and $\Rightarrow$ ("go right").
3. $s \in S$ is the **initial state**;
4. $f : S \times \Sigma \longrightarrow (\Sigma \cup \{\Leftarrow, \Rightarrow\}) \times S$ (the **transition function**).

Chapter 5

The Zombie Attack on Computationalism

5.1 Introduction

Consider PERI, the robot shown in Figure 5.1. This creature is located in Bringsjord's laboratory, and is designed (in accordance with a brand of AI — Psychometric AI — described in the final chapter) to perform at the level of a human on all sorts of tests for human intelligence and creativity. Do the researchers who work with this robot regard it to be the bearer of human consciousness? No.

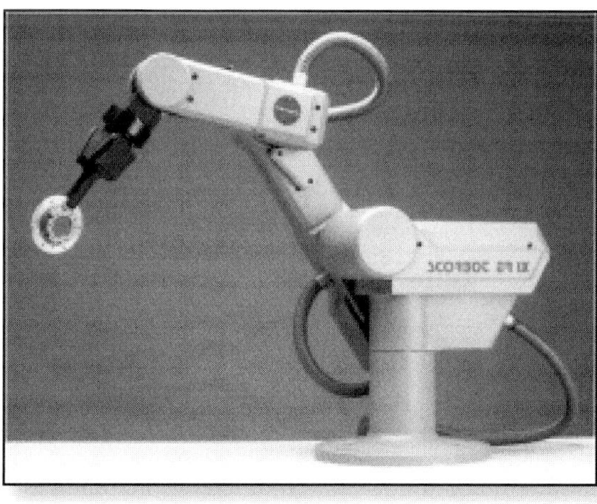

Figure 5.1: PERI, of the Rensselaer AI and Reasoning Laboratory

However, many proponents of computationalism are busy trying to build robots which *are* conscious. Two such proponents are the philosophers Daniel Dennett and John Pollock. For Dennett, the person-to-be is the robot COG, or a descendant thereof, a being "growing up" in Rodney Brooks' lab at MIT, with Dennett's help.[1] For Pollock, the creature in question is OSCAR, described in his two books, *How to Build a Person* (1989) and *Cognitive Carpentry: A Blueprint for How to Build a Person* (1995), Pollock argues that in the future his OSCAR system will be a full-fledged person.

Unfortunately, by our lights, Dennett and Pollock, and other like-minded thinkers, are in the grip of irrational anthropopathism. The reality is that the robots in RAIR, and also COG and OSCAR, are *zombies* (see Figure 5.2).

Is it really possible that what Pollock and Dennett and other computationalists are building is a creature whose overt behavior is as sophisticated as ours, but whose inner life is as empty as a rock's? We believe so. We also believe — for reasons to be specified in this chapter — that the mere *possibility* of zombies is enough to explode the computational conception of mind.[2]

Figure 5.2: Zombies Unmasked

[1]Dennett shares his vision in (Dennett 1994).

[2]Please note that nowhere in this chapter do we fallaciously argue that since it's possible that would-be person-builders are building zombies, they *are* building zombies. While we hold that would-be person-builders *are* building zombies (and that therefore it follows immediately that they are *possibly* building zombies), the argument of this chapter, in a nutshell, as we have just indicated, is that the possibility of zombies can be used to establish the falsity of computationalism.

A recent clash between Daniel Dennett and John Searle over zombies provides a tailor-made springboard to a sustained defense of the zombie attack against computationalism. Dennett, more than any other thinker, says that no philosopher of mind has anything to fear from zombies; in fact, he thinks that those philosophers who seriously ponder zombies (and Blockheads, Twin Earthlings, and Swampmen) have "lost their grip on reality" (Dennett 1996). Searle, on the other hand, believes that zombies threaten at least behavioral conceptions of mentality. In this chapter we try to show that Searle is right, and that he has laid the foundation for our new, rigorous attack on computationalism — the zombie attack. If this attack is sound, it will follow not only that aspiring person builders will fail, but that in failing they may indeed give us zombies.[3]

This chapter is structured as follows. In section 5.2 we focus the Dennett-Searle clash, and then argue that Searle seems to be the immediate victor. In section 5.3 we adapt the results of section 5.2 so as to specifically target computationalism. In section 5.4 we consider and destroy a number of rebuttals from Dennett, including one from his *Consciousness Explained* which seeks to exploit David Rosenthal's "higher order theory" of consciousness. In section 5.5 we consider and reject two final rejoinders, one of which presses the question, "Well then, why aren't *we* zombies?"

5.2 Dennett's Dilemma

Dennett is the arch-defender of the computational conception of mind that underlies the "person building project"; Searle, on the other hand, is the arch-attacker — and both relish their roles: Dennett, in a rather harsh review (1993) of Searle's recent *The Rediscovery of the Mind* (= RM, Searle 1992), affirms that, from the perspective of Searle and like-minded anti-computationalist thinkers, he is the "enemy," and the "target representative of [cognitive] orthodoxy." Searle, as is well known (from his Chinese Room Argument: 1980), and well-revealed repeatedly in RM, regards computationalism (and related positions on the mind, e.g., machine functionalism), to be a "stunning mistake."[4]

[3]When we refer to 'person builders' we refer to those who intend to replicate human persons in a *computational system*. Presumably there are more "biological" ways of striving to build persons — ways involving, e.g., cloning.

[4]The Dennett-Searle clash has reached a new level of ferocity: Dennett claims that Searle is at best an exceedingly forgetful man:

Dennett has claimed that it's *Searle* who has made a stunning mistake: his claim is specifically that Searle's inference from RM's central zombie thought-experiment is obviously flawed, and fatally so. But, as we'll soon see, the argument based upon this thought-experiment is not only competent: once formalized, it becomes transparently valid. Moreover, the Searlean zombie argument can easily withstand Dennett's recent computationalist *Consciousness Explained* (= CE, i.e., Dennett 1991), the achilles heel of which, interestingly, would appear to be precisely its vulnerability to zombie thought-experiments.

These thought-experiments arise from a situation lifted directly out of the toolbox most philosophers of mind today carry with them on the job: Your brain starts to deteriorate and the doctors replace it, piecemeal, with silicon chip workalikes, until there is only silicon inside your refurbished cranium.[5] Searle claims that at least three distinct possibilities arise from this *gedanken*-experiment:

V1 The Smooth-as-Silk Variation: The complete silicon replacement of your flesh-and-blood brain works like a charm: same mental life, same sensorimotor capacities, etc.

V2 The Zombie Variation: "As the silicon is progressively implanted into your dwindling brain, you find that the area of your conscious experience is shrinking, but that this shows no effect on your external behavior. You find, to your total amazement, that you are indeed losing control of your external behavior ... [You have become blind, but] you hear your voice saying in a way that is completely out of your control, 'I see a red object in front of me.' ... We imagine that your conscious experience slowly shrinks to nothing, while your externally observable behavior remains the same." (Searle 1992, pp. 66–7)

> Is it possible that although Searle has at one time or another read all the literature, and understood it at the time, he has actually forgotten the subtle details, and (given his supreme self-confidence) not bothered to check his memory? For instance, has he simply forgotten that what he calls his reductio ad absurdum of my position (81 [in (Searle, 1992)]) is a version of an argument I myself composed and rebutted a dozen years ago? There is evidence of extreme forgetfulness right within the book. For instance ... (Dennett 1993, p. 203)

In the next paragraph, speaking about another of Searle's supposed lapses, Dennett says, "But he forgets all this (apparently!) when forty pages later (107 [in (RM)]) he sets out to explain the evolutionary advantage of consciousness ..." (Dennett 1993).

[5]For example, the toolbox is opened and the silicon supplantation elegantly pulled out in (Cole & Foelber 1984).

V3 The Curare Variation: Your body becomes paralyzed and the doctors, to your horror, give you up for dead.[6]

Searle wants to draw a certain conclusion from V2, the zombie variation, and it's this inference which turns Dennett nearly apoplectic. Here's a summary of the moral Searle wants to draw from V2, in his own words:

> In [V2] we imagined that the mediating relationship between the mind and the behavior patterns was broken. In this case, the silicon chips did not duplicate the causal powers of the brain to produce conscious mental states, they only duplicated certain input-output functions of the brain. The underlying conscious mental life was left out. (Searle 1992, p. 68)

And here is Dennett's reaction:

> But that is only one of the logically possible interpretations of his second variation ... The other is the crucial one: while *you* ... are dying, *another* consciousness is taking over your body. The speech acts you faintly hear your body uttering are not yours, but they are also not nobody's! ... I cannot see how Searle could simply have overlooked this gaping loophole in his thought-experiment. But there it is ... I am baffled. (Dennett 1993, pp. 198–199)

But what *exactly* does Searle want from V2? He tells us explicitly on page 69 of *The Rediscovery of the Mind* that he wants to establish via V2 and V3 that a certain trio of propositions is inconsistent. The trio, reproduced verbatim (p. 69):

(1) Brains cause conscious mental phenomena.

(2) There is some sort of conceptual or logical connection between conscious mental phenomena and external behavior.

(3) The capacity of the brain to cause consciousness is conceptually distinct from its capacity to cause motor behavior. A system could have consciousness without behavior and behavior without consciousness.

We can put things a bit more perspicuously, and put ourselves in position to assess the Dennett-Searle clash, if we represent the three

[6]This scenario would seem to resemble a real-life phenomenon: the so-called "Locked-In" Syndrome. See (Plum & Posner 1972) (esp. the fascinating description on pages 24–25) for the medical details.

propositions using elementary logical machinery: Bx iff x is a brain; Mx iff x causes (a full range of) mental phenomena; and Ex iff x causes (a full range of) external behavior. Then the trio, with Searle's underlying modal notions brought to the surface, and a denoting the brain of the character in our thought-experiments, becomes

(1*) $\exists x(Bx \wedge Mx)$

(2*) $\Box\forall x((Bx \wedge Mx) \rightarrow Ex) \wedge \Box\forall x((Bx \wedge Ex) \rightarrow Mx)$

(3*) $\Diamond(Ba \wedge Ma \wedge \neg Ea) \wedge \Diamond(Ba \wedge Ea \wedge \neg Ma)$

The set {(1*), (2*), (3*)} is provably inconsistent, in garden variety contexts; the proof is trivial, for example, in quantificational S5 (which we happen to like) and the weaker T.[7] Dennett's objection, however, is that (3*) doesn't follow from V2. But this is hardly a gaping loophole; the situation is remedied merely by fine-tuning the zombie variation: Let $V2_1$ denote the one-(moribund)consciousness variation Searle describes, let $V2_2$ describe the two-consciousness variation Dennett describes (and, for that matter, let $V2_3$ denote the three-consciousness case, $V2_4$ the four, *ad infinitum*). Clearly, $\Diamond V2_1$ (as Dennett himself concedes in the quote above). And just as clearly this logical possibility implies the second conjunction of (3*) (and $\Diamond V3_1$ implies the *first* conjunction).

Now, Searle's ultimate aim is probably not to show {(1), (2), (3)} or its formal correlate inconsistent, for reaching this aim, as we have seen, is a matter of some pretty straightforward logic. Rather, Searle aims no doubt to refute the claim that there is a conceptual connection between

---

[7]Systems like T and S5 can be determined by specifying certain rules of inference (which in both cases include the rules of first-order logic) and axiom-schemata. The key axiom-schema in T is the one known by that name, viz., $\Box\phi \rightarrow \phi$; the key axiom-schema in S5 is 5: $\Diamond\phi \rightarrow \Box\Diamond\phi$. (S5 includes as a theorem the interesting $\Diamond\Box\phi \rightarrow \Box\phi$, which becomes relevant later in the chapter.) In both systems, moving a negation sign through a modal operator changes that operator (from diamond to box, and vice versa) in a manner perfectly analogous to the rule of quantifier negation in first-order logic. For a succinct presentation of the core ideas behind (propositional) S5 see Chapter 1 of (Chellas 1980), a book which includes discussion of T and other systems as well. Now, here is how the proof goes. Proposition (1*) is superfluous. Then, e.g., instantiate appropriately on axiom- schema T to get, with (2*), by *modus ponens*, $\forall x((Bx \wedge Mx) \rightarrow Ex)$; instantiate to $(Ba \wedge Ma) \rightarrow Ea)$, derive by propositional logic that $\neg((Ba \wedge Ma) \wedge \neg Ea)$, rewrite this by the rule known as necessitation to $\Box\neg((Ba \wedge Ma) \wedge \neg Ea)$, and in turn rewrite this as $\neg\Diamond\neg\neg((Ba \wedge Ma) \wedge \neg Ea)$, and then, by double negation, as $\neg\Diamond((Ba \wedge Ma) \wedge \neg Ea)$, which of course contradicts (3*)'s first conjunct.

conscious mentality and behavior, that is, he seeks to demonstrate the truth of (3*) and the falsity of (2*) — a result which follows when the inconsistency we have noted is combined with $\Diamond V2_1$, $\Diamond V3_1$ and $((\Diamond V2_1 \wedge \Diamond V3_1) \rightarrow (3^*))$.[8] Hereafter this argument is denoted by 'Arg_1'.

By this point the reader has doubtless realized that there is an opportunity for careful exegesis before us. In conceding the logical possibility of $V2_1$, Dennett does seem to grant all that Searle wants from the case. But must not Dennett somehow see Searle's aim differently? After all, why does he think it's crucial that the possibilities listed by Searle are *exhaustive*? Our objective herein is not to explain, or explain away, Dennett's apparent lapse; our aim is to overthrow computationalism. Accordingly, we are happy to have arrived at Arg_1, and in the next section we proceed without further ado to adapt this argument to one specifically targeting this theory of mind — after which we offer a sustained defense of both the adaptation and Arg_1. However, in the course of this defense we cite and develop *seven* possible responses from Dennett, including one supplied by him through direct communication. These responses provide ample material for attempting the exegesis in question, and though for each one we will offer suggestions for how it can anchor the exegesis, we will leave detailed attempts to readers more concerned with hermeneutics than with whether or not computationalism is misguided.

5.3 Targeting Computationalism

It's easy enough to refine and then adapt what we have called "Dennett's Dilemma" so that it targets computationalism.

The first refinement is to replace talk of 'mental phenomena' with something more specific that we discussed back in Chapter 1: **phenomenal consciousness**. As noted in that chapter, Ned Block, in an essay on consciousness in *Behavioral and Brain Sciences* (Block 1995), calls this brand of consciousness **P-consciousness**. Here, again, is part of his explication:[9]

---

[8] For textual evidence that this is indeed Searle's goal, see p. 69 of (Searle 1992).

[9] Block distinguishes between P-consciousness and **A-consciousness**; the latter concept is characterized as follows:

> A state is access-conscious (A-conscious) if, in virtue of one's having the state, a representation of its content is (1) inferentially promiscuous, i.e., poised to be used as a premise in reasoning, and (2) poised

of functionalism — brands including a modal conditional to the effect that if an organism o is in a certain compu-causal state s, then o is necessarily the bearer of a certain mental state s_m. In connection with this observation, it's important to note that the target of our adaptation of Searle is none other than a specification of such a modal conditional: (2_C). And that such a conditional be taken to capture the heart of computationalism is quite in keeping with the literature (to give just a few examples: Peters 1962, Barr 1983, Fetzer 1994, Johnson-Laird 1988, Bringsjord 1992, Searle 1980, Harnad 1991, Dietrich 1990, Newell 1980, Simon 1980), which takes computation to reflect the *essence* of thinking. The idea is that thinking *is* computing, not that computing can be so configured as to produce a thing that seems to think but really doesn't (as in a zombie). Here again is how Haugeland puts it:

> What are minds? What is thinking? What sets people apart, in all the known universe? Such questions have tantalized philosophers for millennia, but ... scant progress could be claimed ... until recently. For the current generation has seen a sudden and brilliant flowering in the philosophy/science of the mind; by now not only psychology but also a host of related disciplines are in the throes of a great intellectual revolution. And the epitome of the entire drama is *Artificial Intelligence*, the exciting new effort to make computers think. The fundamental goal of this research is not merely to mimic intelligence or produce some clever fake. Not at all. AI wants only the genuine article: *machines with minds*, in the full and literal sense. This is not science fiction, but real science, based on a theoretical conception as deep as it is daring: namely, we are, at root, *computers ourselves* (Haugeland 1981, p. 2).

As many readers will remember, functionalist views taking the form of modal conditionals like (2_C) have been the target of "arbitrary realization" arguments, which involve thought-experiments designed to show the logical possibility of an organism instantiating compu-causal state s but failing to have *any* mental states.[12] Searle's zombie scenarios could be understood as thought-experiments intended to play

[12] Bringsjord has devised such thought-experiments to refute new, ingenious versions of machine functionalism explicitly designed to resist older, more primitive thought-experiments of the same general type. The purportedly inoculated brands of functionalism are specified by John Pollock (1989); the *gedanken*-experiments

the same role as those at the core of arbitrary realization arguments: (2^*)'s second conjunct would be a formalization of the sort of modal conditional cited above, and $\Diamond V3_1$ would destroy this conjunct.[13]

This is not at all to say that V2 is, or can be distilled to, an arbitrary realization argument. In the tragic story of the poor soul in V2 who fades away when his brain is gradually replaced with silicon workalikes, there is nothing ridiculous: the workalikes are not tiny beer cans connected with string, for example. It's also important to distinguish the zombie in V2 from Swampman.[14] The Swampman has a more violent history than our zombie (and, as Bill Rapaport recently pointed out to Bringsjord, history is thought by many to be crucial in these cases; see, e.g., Millikan 1996), and it would be very difficult to flesh out the Swampman case with neuroscientific details (such fleshing out becomes important later in this chapter).

The second reason why the move under consideration on Dennett's behalf — reading (2) so that it makes no claim about an intimate connection between external behavior and internal mentality — is not available to him is that though many philosophers these days would resist the view that you can *define* the having of a type of mental state exhaustively in terms of external behavioral conditions, Dennett would not, given his well-known "intentional stance" theory. This is readily confirmed by looking to some of the relevant writings. For example, consider Dennett's (1978) *Brainstorms*. When at the start of that book Dennett introduces his notion of an *intentional system*,

---

can be found in "Chapter VI: Arbitrary Realization" of Bringsjord's (1992) *What Robots Can and Can't Be*. The older thought-experiments are due to Ned Block (1980) and, ironically enough, John Searle (1983).

[13]After all, to say, as we have said on Dennett's behalf, that (2) can be read as "... there is only *some* sort of conceptual or logical connection between conscious mental phenomena and external behavior ..." conveniently overlooks the fact that the modal (2^*) is designed specifically to capture the notion that the connection in question *is*, whatever its specifics, conceptual/logical.

[14]The Swampman was introduced by Donald Davidson:

> Suppose lightning strikes a dead tree in a swamp; I am standing nearby. My body is reduced to its elements, while entirely by coincidence (and out of different molecules) the tree is turned into my physical replica. My replica, The Swampman, moves exactly as I did; according to its nature it departs the swamp, encounters and seems to recognize my friends, and appears to return their greetings in English. It moves into my house and seems to write articles on radical interpretation. No one can tell the difference. (Davidson 1987, p. 441)

and the intentional stance built upon it, he makes clear that he rejects machine functionalism of the sort set out in our (2_C) (e.g., p. xvi of Dennett 1978). His version of this doctrine is laid out (on the same page) as follows.

(4) $\forall x$ (x believes that snow is white $\equiv x$ "realizes" some Turing machine k in logical state A)

In Chapter 2 of *Brainstorms*, Dennett gives his argument against (4), which is based on a thought-experiment featuring two different face-recognition systems, each designed by a different team of AI engineers. The crux is that while both systems are "well characterized as believing that p" (Dennett 1978, p. 26; p ranges over such things as "I've seen this face before"), by hypothesis they realize different Turing machines in different states. Proposition (4) is thus supposed to be overthrown by the time-honored scheme: there is some scenario wherein its antecedent is true while its consequent is false. Whether or not Dennett succeeds in refuting (4), given our purposes, is immaterial. What matters is that Dennett's attack has been mounted from his armchair; his blow against (4) comes from a *thought*-experiment. This implies that (4) is to be read as asserting a *principled* connection between belief and Turing machine computation. This in turn implies that (4) is a modal conditional of some sort, a soulmate for our (2_C). The equivalence in (4) is *logical* equivalence, not simply the material biconditional. For consider the following material biconditional.

(5) $\forall x$ (x is an American politician $\leftrightarrow x$ is corrupt)

This proposition cannot be overthrown by armchair reflection. The cynic who affirms it will not be obliged to change his attitude upon hearing that a philosopher has cooked up a coherent story about a virtuous Senator.

Dennett encapsulates his intentional stance via a conditional having precisely the form of (4), viz.,

(6) $\forall x$ (x believes that snow is white $\equiv x$ can be predictively attributed the belief that snow is white) (Dennett 1978, p. xvii)

And (6) cannot be affirmed while at the same time the connection between mentality and behavior is said — *via* the rebuttal we are considering on Dennett's behalf — to be a non-conceptual/non-logical one.

How else might Dennett try to dodge his dilemma? By carefully analyzing CE, and by ultimately asking Dennett himself, we can isolate and reject the remaining candidate responses:

5.4.1 Dennett's Objection from Method

Dennett might remind us that his complaint is about Searle's *method*. Well, true, Dennett strenuously objects in CE to Searle's emphasis on first-person introspection, which he regards to be a benighted vestige of Cartesian folk-psychology. Dennett would supplant Searle's method with his own, a method called "heterophenomenology" — but the problem is, heterophenomenology is neutral on the possibility of zombies! Dennett is quite explicit about this:

> [W]hat about the zombie problem? Very simply, heterophenomenology by itself cannot distinguish between zombies and real, conscious people, and hence does not claim to solve the zombie problem or dismiss it. (Dennett 1991, p. 95)

It may be worth pointing out that Dennett's complaint about method, when applied to RM's central thought-experiment, seems aimed at a straw man: Propositions (1)–(3), as well as their formal counterparts (1*)–(3*), are apparently third-personish.[15] And, surely Dennett can't be saying that Searle's view is that we can establish $\Diamond\phi$, for any and all ϕ, only if ϕ can be conceived and "experienced" *via* the interior, "What-does-it-feel-like-to-me-on-the-inside-?" first-person point of view (since, e.g., Searle would hold, with us all, that it's logically possible that the Brooklyn Bridge turn instantly to jello, and would hold this in the absence of any phenomenological gymnastics).

5.4.2 Dennett's "Oops" Objection: Zombies vs. Zimboes

Faced with what our analysis has uncovered, Dennett might say that (oops) in his quoted reaction above he meant by the phrase "logically possible interpretation" not "account which describes a logical possibility," which is our reading, and surely the natural one, but something weaker like "candidate interpretation." This would be a rather

[15]Kieron O'Hara has pointed out to one of us that Mx is "available" only to the first-person, hence our insertion of the qualifier 'apparently' in the preceding sentence. Nothing substantive hinges on the claim that (1)-(3) are third-personish.

desperate move. The problem with it, of course, is that not only is it (obviously!) logically possible that someone *offer* $V2_1$, but $V2_1$ *is* itself logically possible.[16] After all, Searle could, at the drop of a hat, provide a luxurious novel-length account of the scenario in question (or he could hire someone with the talents of a Kafka to do the job for him).[17] Besides, if Dennett seriously maintains $\neg\Diamond V2_1$, where is the contradiction that the zombie scenario must then entail?

It might be said that we have moved hastily in the preceding paragraph, especially in light of another zombie-relevant section in Dennett's CE. More specifically, it might be said that Dennett not only *might* say that $\neg\Diamond V2_1$, he *does* say this in CE, and not only that: he produces an *argument* in CE for this position. It does appear that there is such an argument (on which Dennett has recently placed his chips[18]); it spans pages 304–314 of CE, and begins with

> In a series of recent papers, the philosopher David Rosenthal (forthcoming, 1986, 1989, 1990*b*, 1990*a*) has analyzed [the] everyday concept of consciousness and its relation to our concepts of reporting and expressing. He uncovers some structural features we can put to good use. First, we can use his analysis to ... show how it discredits the idea of zombies ... (Dennett 1991, p. 304)

What is the relevant part of Rosenthal's position? The answer, courtesy of his (Rosenthal forthcoming), can be put in declarative form:

Def 1 s is a conscious mental state at time t for agent $a =_{df}$ s is accompanied at t by a higher-order, noninferential, occurrent, assertoric thought s'

[16]As nearly all those who write on the zombie topic agree. See, for example, (Dretske 1996, Block 1995, Flanagan & Polger 1995, Harnad 1995).

[17]Despite having no such talents, Bringsjord usually spends twenty minutes or so telling a relevant short story to students when he presents zombies via V2. In this story, the doomed patient in V2 — Robert — first experiences an unintended movement of his hand, which is interpreted by an onlooker as perfectly natural. After more bodily movements of this sort, an unwanted sentence comes out of Robert's mouth — and is interpreted by an interlocutor as communication from Robert. The story describes how this weird phenomenon intensifies ... and finally approaches Searle's "late stage" description.

[18]In his recent "The Unimagined Preposterousness of Zombies" (1995*a*) Dennett says the argument from CE which we are about to examine shows that zombies are not really conceivable.

for a that a is in s, where s' is conscious or unconscious.[19]

Def 1 is said to be the **higher-order theory** (HOT) of consciousness. What sorts of examples conform to HOT? Dennett focuses on the state *wanting to be fed.* On Rosenthal's view, this state *is* a conscious state — and the reason it is is that it's the target of a higher-order thought, viz., the thought that I want to be fed. Rosenthal's Def 1, of course, leaves open the possibility that the higher-order thought can be itself unconscious.

How can Dennett be read as using Def 1 as a defense against the zombie attack (i.e., against Arg_1 and Arg_1^C)? We can construct a counter-argument on Dennett's behalf which involves not only zombies, but also a variant that Dennett introduces to ward off zombies: zimboes. "A zimboe," Dennett says, "is a zombie that, as a result of self-monitoring, has internal (but unconscious) higher-order informational states that are about its other, lower-order informational states" (Dennett 1993, p. 310). The corresponding argument is expressed by Dennett in prose, but it can be charitably reconstructed from pages 310–311 of CE as a *reductio* aimed at Searle:

[19]Def 1's time index (which ought, by the way, to be a *double* time index — but that's something that needn't detain us here) is necessary; this is so in light of thought-experiments like the following. Suppose (here, as we ask you to suppose again below) that while reading Tolstoy's *Anna Karenina* you experience the state *feeling for Levin's ambivalence toward Kitty.* Denote this state by s^*; and suppose that you have s^* at 3:05 pm sharp; and suppose also that you continue reading without interruption until 3:30 pm, at which time you put down the novel; and assume, further, that from 3:05:01 — the moment at which Levin and Kitty temporarily recede from the narrative — to 3:30 you're completely absorbed in the tragic romance between Anna and Count Vronsky. Now, if you report at 3:30:35 to a friend, as you sigh and think back now for the first time over the literary terrain you have passed, that you feel for Levin, are we to then say that at 3:30:35 s^*, by virtue of this report and the associated higher-order state targeting s^*, becomes a conscious state? If so, then we give you the power to change the past, something you cannot be given.

Arg$_2$

	(7)	$\Diamond V2_1$	supp.
	(8)	If $\Diamond V2_1$, then zimboes are logically possible.	
	(9)	If zimboes are logically possible, then Turing Test-passing zimboes are logically possible.	
∴	(10)	Turing Test-passing zimboes are logically possible.	7, 8, 9
	(11)	Turing Test-passing zimboes are *not* logically possible.	
∴	(12)	$\neg\Diamond V2_1$	4, 7, 8

This argument is obviously formally valid (as can be seen by symbolizing it in the propositional calculus); premises (8) and (9) seem to us to be above reproach; (7) is an assumption for indirect proof; and (10) is an intermediate conclusion. This leaves (11); why is this proposition supposed to be true? This proposition follows from a supporting sub-argument, viz.,

Arg$_3$

	(13)	If Turing Test-passing zimboes are logically possible, then it's logically possible that, in response to queries from the judge in the Turing Test, a zimboe's lower-level states become the target of higher-level, noninferential, occurrent, assertoric states.	
	(14)	Def 1: s is a conscious mental state at time t for agent $a =_{df} s$ is accompanied at t by a higher-order, noninferential, occurrent, assertoric thought s' for a that a is in s, where s' is conscious or unconscious.	
∴	(15)	If Turing Test-passing zimboes are logically possible, it's logically possible that a Turing Test-passing zimboe is conscious.	13, 14
	(16)	It's necessarily the case that a zimboe is *un*conscious.	
∴	(11)	Turing Test-passing zimboes are *not* logically possible.	15, 16

Here, again, the reasoning can be effortlessly formalized (in propositional modal logic) and thereby shown to be valid. We grant (13), because Dennett's view that the judge in the Turing Test will unavoidably catalyze (in some sense of the term) self-monitoring via questions to contestants like "Why did you say, a few minutes back, that ..." is quite plausible. This leaves proposition (16) and Def 1 itself as the only potentially vulnerable spots. But (16) is true because *by definition* zimboes are not conscious. So we are left having to evaluate, in earnest, Def 1 — no small task, since this definition is an account David Rosenthal has assembled, refined, and defended over many years in a number of intelligent papers.

Readers inclined to affirm our zombie attack, and to sustain it in the face of Dennett's HOT-based objection, may think there is a shortcut that obviates having to grapple with HOT. Specifically, the idea might be that Dennett begs the question: that $\text{Arg}_2+\text{Arg}_3$, and indeed any argument against $\Diamond V2_1$ which has Def 1 for a premise, is a case of *petitio principii.*

This move, we concede, is at least initially promising. For it would seem that Def 1 automatically implies that zombies are impossible, because according to this definition an unconscious state targeted at a lower-level state immediately implies (by *modus ponens* right-to-left across the biconditional) that the lower-level state is conscious. Isn't this circular? After all, the zombie thought-experiment is designed to reveal that it's perfectly conceivable that a behaviorally complex zombie (one that can pass itself off in human discourse as conscious, and therefore *a fortiori* one which can excel in the Turing Test[20]) exists! Rosenthal's views are undeniably convenient for Dennett, but perhaps they are *too* convenient. One way to focus this complaint is perhaps to ask: Why should we accept (13)? The intuition behind this premise, after all, was the same as that behind

(8) If $\Diamond V2_1$, then zimboes are logically possible;

and this proposition was to incarnate the intuition that we have only to make minor changes in the original zombie thought-experiment in order to have it portray zimboes. But now suddenly we find that that which the original zombie thought-experiment is specifically designed to exclude — namely, intention-bearing states — is surreptitiously stipulated to be inherent the zimboe thought-experiment! Is Dennett's move like rejecting the claim that time travel is logically possible by taking as *premise* a view of space-time according to which time travel is impossible?

No. The charge of *petitio* fails, for two reasons. First, it conflates 'intentional state' with 'P-conscious state.' The original zombie

[20]We are now talking about Harnad (1991)'s *Total* Turing Test, the passing of which requires not only human-level linguistic behavior but sensorimotor behavior as well.

gedanken-experiment (and the amended version designed to overthrow computationalism) insists that P-consciousness is absent; it *doesn't* insist that *intentional* states (which at least on HOT may or may not be P-conscious) are excluded. So the move against Dennett under consideration is like rejecting the claim that time travel is logically *im*possible by taking as *premise* a view of space-time according to which time travel is *possible*. While most will agree that Dennett does not prove much by simply affirming Rosenthal's HOT, we must concede that neither does one prove anything by denying it. What is needed, then, *is* a direct attack on HOT itself — something we now provide.

5.4.3 A Direct Attack on Rosenthal's HOT

At first glance, it seems that HOT is quickly killed off by taking note of everyday experiences in which one is in a conscious state that is *not* the target of any higher-order state. For example, consider the state, s^*, *feeling for Levin's ambivalence toward Kitty*, experienced while reading about these two characters as they move through Tolstoy's immortal *Anna Karenina*. Suppose that this state is experienced by you, as reader, while you are completely "transported" by the narrative. In such a case — so the story goes — there are no higher-order states directed at s^*.

But such thought-experiments are hardly conclusive. Rosenthal, in response, simply bites the bullet and insists that, unbeknownst to us, there just *is* a conscious-making higher-order state directed at s^*. He would say: "Look, *of course* you're not aware of any conscious states directed at s^*. This is just what my HOT predicts! After all, the state directed at s^* doesn't have a state directed at it, so it's unconscious; and whatever is unconscious is by definition something you're not aware of."

We think this response only gives Rosenthal a temporary reprieve, for there are other, sharper thought-experiments: Let s'' be a paradigmatic P-conscious state, say *savoring a hefty spoonful of deep, rich chocolate ice cream*. Since s'' is P-conscious, "there is something it's like" to be in this state. As Rosenthal admits about states like this one:

> When [such a state as s''] is conscious, there is something it's like for us to be in that state. When it's not conscious, we do not consciously experience any of its qualitative properties; so

> then there is nothing it's like for us to be in that state. How can we explain this difference? ... How can being in an intentional state, of whatever sort, result in there being something it's like for one to be in a conscious sensory state? (Rosenthal forthcoming, pp. 24–25)

Our question exactly. And Rosenthal's answer? He tells us that there are "factors that help establish the correlation between having HOTs and there being something it's like for one to be in conscious sensory states" (Rosenthal forthcoming, p. 26). These factors, Rosenthal tells us, can be seen in the case of wine tasting:

> Learning new concepts for our experiences of the gustatory and olfactory properties of wines typically leads to our being conscious of more fine-grained differences among the qualities of our sensory states ... Somehow, the new concepts appear to generate new conscious sensory qualities. (Rosenthal forthcoming, p. 27)

We confess we cannot help but regard Rosenthal's choice of wine tasting as tendentious. In wine tasting there is indeed a connection between HOTs and P-conscious states (the nature of which we don't pretend to grasp). But wine-tasting, as a source of P-consciousness, is unusually "intellectual," and Def 1 must cover all cases — including ones based on less cerebral activities. For example, consider fast downhill skiing. Someone who makes a rapid, "on-the-edge" run from peak to base will have enjoyed an explosion of P-consciousness; such an explosion, after all, will probably be the main reason such an athlete buys expensive equipment and expensive tickets, and braves the cold. But expert downhill skiers (as well as, for that matter, Bringsjord and Zenzen), while hurtling down the mountain, surely don't analyze the ins and outs of pole plants on hardpack versus packed powder surfaces, and the fine distinctions between carving a turn at 20 mph versus 27 mph. Fast skiers ski; they plunge down, turn, jump, soar, all at incredible speeds. Now is it really the case, as Def 1 implies, that the myriad P-conscious states $s_1, \ldots, s_n$ generated in a screaming top-to-bottom run are the result of higher-level, noninferential, assertoric, *occurrent* beliefs on the part of a skier k that k is in s_1, that k is in s_2, k is in s_4, ..., k is in s_n? Wine tasters do indeed sit around and say such things as that, "Hmm, I believe this 87 Barolo has a bit of a grassy taste, no?" But what racer, streaking over near-ice at 50 mph, ponders thus: "Hmm, with these new parabolic skis, 3 millimeters thinner at the wait, the sensation of this turn is like turning a corner in a fine vintage Porsche." And who would claim that such thinking *results in* that which it's like to plummet downhill?

C	F	P	Y
J	M	B	X
S	G	R	L

Figure 5.3: 3 × 4 Array of Random Letters

HOT is threatened by phenomena generated not only at ski areas, but in the laboratory as well. We have in mind an argument arising from the phenomenon known as **backward masking**. Using a tachistoscope, psychologists are able to present subjects with a visual stimulus for periods of time on the order of milliseconds (one millisecond is $\frac{1}{1000}^{th}$ of a second). If a subject is shown a 3×4 array of random letters (see Figure 5.3) for, say, 50 milliseconds (msecs), and is then asked to report the letters seen, accuracy of about 37% is the norm. In a set of very famous experiments conducted by Sperling (1960), it was discovered that recall could be dramatically increased if a tone sounded after the visual stimulus. Subjects were told that a high tone indicated they should report the top row, a middle tone the middle row, and a low tone the bottom row. After Figure 5.3 was shown for 50 msec, to be followed by the high tone, recall was 76% for the top row; the same result was obtained for the other two rows. It follows that a remarkable full 76% of the array is available to subjects after it appears. However, if the original visual stimulus is followed immediately thereafter by another *visual* stimulus in the same location (e.g., circles where the letters in the array appeared; see Figure 5.4), recall is abysmal; the second visual stimulus is said to **backward mask** the first. (For the seminal study see Averbach & Coriell 1961). Suppose, then, that a subject is flashed a series of visual patterns p_i, each of which appears for only 5 msec. In such a case, while "there is something it's like" for the subject to see p_i, it is very doubtful that this is because the subject thinks that she is in p_i. In fact, most models of human cognition on the table today hold that information about p_i never travels "far enough" to become even a potential object of any assertoric thought (Ashcraft 1994).[21]

[21]It would seem to be easy enough to turn the phenomenon of backward masking into a working computer simulation using the ACT-R system (Anderson & Lebiere

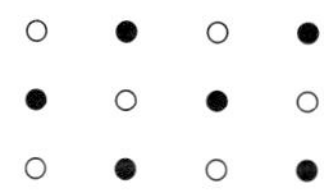

Figure 5.4: Visual Stimulus in the Same Location

So, for these reasons, Def 1 looks to be massively implausible, and therefore the zombie attack has yet to be disarmed.

Interestingly enough, Dennett himself doesn't take Rosenthal's definition seriously. Nor, for that matter, does he feel, at this point in CE, any pressure to disarm the zombie attack:

> If ... Rosenthal's analysis of consciousness in terms of higher-order thoughts is rejected, then zombies can live on for another day's thought experiments. I offer this parable of the zimboes tongue in cheek, since I don't think either the concept of a zombie or the folk-psychological categories of higher-order thoughts can survive except as relics of a creed outworn. (Dennett 1991, p. 313–14)

Since we do indeed reject Rosenthal's HOT, zombies live on, *by Dennett's own admission*, in the thought-experiments with which we began our investigation.

5.4.4 Dennett's Objection from Racism

Dennett expresses this objection as follows:

> Notice, by the way, that this equivocation between two senses of "epiphenomenal" also infects the discussion of zombies. ['Zombie'] can be given a strong or weak interpretation, depending on how we treat ... indistinguishability to observers. If we were to declare that in principle, a zombie is indistinguishable from a conscious person, then we would be saying that genuine consciousness is epiphenomenal in the ridiculous sense. That is just silly. So we could say instead that consciousness might be epiphenomenal [in the sense that] although there was some way of distinguishing zombies from real people (who knows, maybe zombies have green brains), the difference doesn't show up as

1998). This simulation would not involve doxastic intricacies, but rather just rapid firing of production rules.

> a functional difference to observers ... On this hypothesis, we would be able in principle to distinguish the inhabited bodies from the uninhabited bodies by checking for brain color. This is also silly, of course, and dangerously silly, for it echoes the sort of utterly unmotivated prejudices that have denied full personhood to people on the basis of the color of their skin. It is time to recognize the idea of the possibility of zombies for what it is: not a serious philosophical idea but a preposterous and ignoble relic of ancient prejudices. (Dennett 1991, pp. 405–406)

Bringsjord (1995*b*) has elsewhere discussed the **penetrability** of zombies, where, put roughly, the idea is that zombies are penetrated if they are unmasked as empty-headed, if, to use Harnad's (1995) phrase, there is found to be "nobody home." Penetrability seems to essentially correspond to what Dennett here calls "distinguishability;" so let's unite his property and Bringsjord's under the predicate 'D:' for a zombie z, Dz iff z is distinguished. In addition, let us invoke another predicate, O, which holds of a property G (we thus allow second-order constructs) just in case G is observed, and another property, F, which holds of a property if and only if that property "makes a functional difference." Finally, let's take explicit note of the fact that when Dennett refers to green brains, he must have in mind *any* property that could serve to distinguish zombies from "real people" — so that we can refer to any distinguishing property Δ. (No moderately intelligent defender of $\Diamond$V2_1 thinks that in the key thought-experiments the color of the "brains" involved, or any other simple property like this, is in any way relevant.) With this simple machinery in hand, we can charitably set out Dennett's argument in the previous quote, and we can connect it to the zombie attack we're promoting in this chapter:

Arg$_4$

	(17)	If the Bringsjord-Zenzen zombie attack on computationalism (Arg_1^{C}) is sound, then either $\Diamond\Box\forall z\neg Dz$ or $\forall z\exists\Delta\Diamond(O\Delta\wedge\neg F\Delta\wedge(O\Delta\rightarrow Dz))$.	
	(18)	It's not the case that $\Diamond\Box\forall z\neg Dz$.	
	(19)	If $\forall z\exists\Delta\Diamond(O\Delta\wedge\neg F\Delta\wedge(O\Delta\rightarrow Dz))$, then the prejudices behind racism are acceptable.	
	(20)	The prejudices behind racism are unacceptable.	
∴	(21)	It's not the case that $\forall z\exists\Delta\Diamond(O\Delta\wedge\neg F\Delta\wedge(O\Delta\rightarrow Dz))$.	19, 20
∴	(22)	B-Z's zombie attack on computationalism fails, i.e., Arg_1^{C} is unsound.	17, 18, 21

Arg_4 is of course formally valid. As to soundness, what of premises (18), (19), and (20)? The first of these is quite plausible; in fact, if the modal operators are interpreted in line with the modal system S5 visited above, we may even be able to give a little sub-proof of (18): $\Diamond\Box\phi \rightarrow \Box\phi$ is a (distinctive) validity in S5, so the first disjunct in (17)'s consequent, if we assume S5, becomes simply

$$\Box\forall z \neg Dz.$$

But this seems just plain wrong. Couldn't an observer always in principle be enlightened about "nobody homeness" by an oracle, or by God, or by other exotic but nonetheless coherent means? If so, then (18) is true.

But now what about (19)? Well, (19), we submit, is irreproachable, and who will contradict us? This leaves (20) — and here we think Dennett faces some rather rougher sledding. Why does he think this premise is true? Why does he think that affirming

(23) $\forall z \exists \Delta \Diamond (O\Delta \wedge \neg F\Delta \wedge (O\Delta \rightarrow Dz))$

entails an affirmation of racism? After all, isn't it somewhat implausible (not to mention harsh) to claim that those who affirm this formula are racists, or, in virtue of this affirmation, are at least willing to accept such prejudice? It's worth remembering, perhaps, that many of those who reject compu-causal functionalism will affirm (23). Are they thereby racist? We doubt it. It seems to us that everyone, Dennett included, must concede that racism, though perhaps aptly deemed a "cognitive sin," is usually understood to be a phenomenon a good deal less recherché than embracing the formula in question! Affirming (23) entails a certain attitude toward abstract properties in a thought-experiment; it does not entail a certain attitude toward *actual* beings. The quick way to grasp this is to simply change (23)'s quantification over zombies to (23′) quantification over a victimized race or group, and to then inquire if (23′) captures a racist attitude toward this group (it doesn't). The more rigorous route proceeds as follows.

First, note that (23)'s predicate O is ambiguous between "observed by a human" and "observed by an oracle (or a god ...)." As we noted when affirming (18), the second of these construals for O is the sort of thing we need to take seriously in our context. Let's use the subscripts $_h$ and $_o$ to disambiguate (23) into

(23_h) $\forall z \exists \Delta \Diamond (O_h\Delta \wedge \neg F\Delta \wedge (O_h\Delta \rightarrow Dz))$

(23_o) $\forall z \exists \Delta \Diamond (O_o \Delta \wedge \neg F \Delta \wedge (O_o \Delta \rightarrow Dz))$

Now, Dennett's premise (17) must allow both of these in the disjunction forming its consequent. To quickly make this change, let us stipulate that O as it occurs in Arg_4 is a *disjunctive* property allowing for observation in either the O_h *or* O_o sense. It then becomes possible, we think, to show that (19) is false by devising a thought-experiment in which its antecedent is affirmed by those who are clearly not racists.[22]

Suppose that there are people — Oraks — living on the Earth-like planet Orak who believe that rocks, plants, individual molecules, grains of sand, electrons (and other sub-atomic particles) are not P-conscious. Suppose, furthermore, that Oraks have all thought long and hard about whether the coordinated movement of such objects changes anything: about whether such moving objects are P-conscious. Their conclusion, to a person, is that P-consciousness is still nowhere to

[22]Though all this talk of oracles and gods may seem to be carrying us toward mysticism, there is actually an analogue for propositions of the form of (23_h) and (23_o) to be found in logic, specifically in the area of undecidable problems: It is well-known (and we covered this in Chapter 1) that there is no Turing machine which can decide whether or not a fixed Turing machine (or computer program, etc.) is ever going to halt. On the other hand, for every Turing machine m, there is a fact of the matter as to whether or not m halts; m is either a halter or a non-halter (for elegant formal coverage of these matters see Boolos & Jeffrey 1989). Now whereas it is generally thought to be impossible for a human to decide whether or not a Turing machine halts (because, by computationalism itself, people don't have powers beyond Turing machines), such is not the case for "oracles." Here, for example, to revisit a quote from Chapter 1, is what we read in one of today's standard logico-mathematical textbooks:

> Once one gets used to the fact that there are explicit problems, such as the halting problem, that have no algorithmic solution, one is led to consider questions such as the following. Suppose we were given a "black box" or, as one says, an *oracle*, which somehow can tell us whether a given Turing machine with given input eventually halts. Then it is natural to consider a kind of program that is allowed to ask questions of our oracle and to use the answers in its further computation. (Davis & Weyuker 1994, p. 197; emphasis theirs)

More specifically, if we reinterpret the predicate Dx as "decides whether or not x halts," and take the variable m to range over Turing machines, the following two propositions are coherent and not implausible.

- $\forall m \exists \Delta \Diamond (O_o \Delta \wedge \neg F \Delta \wedge (O_o \Delta \rightarrow Dm))$
- $\neg \forall m \exists \Delta \Diamond (O_h \Delta \wedge \neg F \Delta \wedge (O_h \Delta \rightarrow Dm))$

be found. Let us imagine, in fact, that their conclusion stems from an affirmation of the arbitrary realization argument: they have considered scenarios in which, say, beer cans and string implement certain information-processing functions the Oraks have detected in their own brains, and have concluded that such contraptions would ("Obviously!", they declare) lack P-consciousness. As to why Oraks *themselves* have P-consciousness, Oraks have a ready answer: the Oracle bestows a non-functional property upon their bodies, and *this* produces (or just is) P-consciousness. In fact, Oraks sometimes hear their Oracle speak: it thunderously says such things as, "Live!" — whereupon inert bodies glow for an instant and then move like those of healthy Oraks. Oraks, each and every one, affirm (23_o). Each and every Orak also lives a saintly life: they are loving, caring, altruistic, even-tempered, self-sacrificial, and so on. Finally, every now and then the Oraks come upon a creature that looks like them, but which never bore the evanescent tell-tale glow, and cannot point to a time when the Oracle performed (what the Oraks call) ensoulment. These beings (known as "zombaks") the Oraks lovingly treat as they treat all other Oraks, but they nonetheless never reject (23_o).

This case seems perfectly coherent.[23] But then (19) is false, and Arg_4 fails.

5.5 Two Final Moves

We see two remaining moves available to the computationalist in the face of Arg_1 and Arg_1^C. The first is to retreat to the claim that Arg_1's conclusion — (1), (2), (3) inconsistent because of $V2_1$ and $V3_1$ — is trivial, because no one maintains (2), but rather something like $(2'_C)$, where this proposition is the result of changing the necessity operator in (2_C) to one of *physical* necessity.[24] This retreat is blocked by the brute fact that, as explained above, computationalists have

[23]Note that this *gedanken*-experiment is *not* offered to establish that it's logically possible that it's logically possible that there are zombies. To sell the case as such would of course be to beg the question against certain opponents of our zombie attack, because $\Diamond\Diamond\phi \rightarrow \Diamond\phi$ is a validity in (e.g.) S5. What the case *does* show, it seems to us, is that in order to have racism one needs to have, in addition to an affirmation of something like (23), a corresponding prescription for how to think about and treat zombies or zombaks.

[24]Proposition $(2')$ would say that it's *physically necessary* that a brain's causing mental phenomena implies corresponding external behavior.

traditionally advanced the likes of the stronger (2).[25]

Even those who suddenly decide to champion an idiosyncratic version of computationalism (according to which there is only a nomological connection between cognition and computation) are not out of the woods. The reason is that there would appear to be no reason why $V2_1$ ought not to be regarded *physically* possible: Why couldn't a neuroscience-schooled Kafka write us a detailed, compelling account of $V2_1$, replete with wonderfully fine-grained revelations about brain surgery and "neurochips"? Then, to generate the physics counterpart to Arg_1/Arg_1^C, we have only to change the modal operators to their physics correlates — $\Box$ to $\Box_p$ and $\Diamond$ to $\Diamond_p$, perhaps — and then invoke, say, some very plausible semantic account of this formalism suitably parasitic on the standard semantic account of logical modes.[26]

---

[25] By the way, the retreat would have for Dennett the welcome effect of trivializing Searle's piéce de résistance, for it implies that Searle's famous Chinese Room thought-experiment, designed to show that there is no *logical/conceptual* connection between symbol manipulation and mental phenomena, is trivial. Unfortunately, though Dennett isn't alone in wanting to dodge the Chinese Room, the view that the argument is trivial is an exceedingly solitary one. (For reasons apparently noted only by Bringsjord, it's nonetheless true that Dennett has given the most powerful objection against CR. See Bringsjord's essay in (Bishop & Preston 2002).) Note also that some have given versions of the Chinese Room designed from scratch to be physically possible. See, for example, "Chapter V: Searle," in (Bringsjord 1992).

[26] For a number of such accounts, see (Earman 1986). Note that the thought-experiment we have in mind, combined with such an account, does not merely establish that $\Diamond\Diamond_p V2_1$. Such a proposition says that there is a possible world at which $V2_1$ is physically possible — which is verified by imagining a possible world w in a cluster of worlds $w_1, \ldots, w_n$ comprising those which preserve the laws of nature in w, where $V2_1$ is true not only at w, but at at least one w_i. Let α be the actual world; let W_α^P denote the set of worlds preserving the laws of nature in α. Our thought-experiment scrupulously stays within W_α^P. Each and every inch of the thought-experiment is to be devised to preserve consistency with neuroscience and neurosurgery specifically, and biology and physics generally. Our approach here is no different than the approach taken to establish that more mundane states of affairs are physically possible. For example, consider a story designed to establish that brain transplantation is physically possible (and not merely that it's logically possible that it's physically possible). Such a story might fix a protagonist whose spinal cord is deteriorating, and would proceed to include a step-by-step description of the surgery involved, each step described to avoid any inconsistency with neuroscience, neurosurgery, etc. It should be easy enough to convince someone, by such a story, that brain transplantation, at α, is physically possible. It is of course much easier to convince someone that it's *logically* possible that it's physically possible that Jones' brain is transplanted: one could start by imagining (say) a world whose physical laws allow for body parts to be removed, isolated, and then made

Let us make it clear that we can easily do more than express our confidence in Kafka: We can provide an *argument* for $\Diamond V2_1^{TM}$ given that Kafka is suitably armed. There are two main components to this argument. The first is a slight modification of a point made by David Chalmers (1996), namely, when some state of affairs ψ seems, by all accounts, to be perfectly coherent, the burden of proof is on those who would resist the claim that ψ is logically possible.[27] Specifically, those who would resist need to expose some contradiction or incoherence in ψ. We think most philosophers are inclined to agree with Chalmers here. But then the same principle would presumably hold with respect to *physical* possibility: that is, if by all accounts ψ seems physically possible, then the burden of proof is on those who would resist affirming $\Diamond_p \psi$ to indicate where physical laws are contravened.

The second component in our argument comes courtesy of the fact that $V2_1^{TM}$ can be modified to yield $V2_1^{NN}$, where the superscript 'NN' indicates that the new situation appeals not to Turing machines, but to **a**rtificial **n**eural **n**etworks, which are said to correspond to actual flesh-and-blood brains.[28] So what we have in mind for $V2_1^{NN}$ is this: Kafka really knows his stuff: he knows not only about natural neural nets, but also about artificial ones, and he tells us the sad story of Smith — who has his neurons and dendrites gradually replaced with artificial correlates in flawless, painstaking fashion, so that information flow in the biological substrate is perfectly preserved in the artificial substrate ... and yet, as in $V2_1^{TM}$, Smith's P-consciousness withers away to zero while observable behavior runs smoothly on. Now it certainly seems that $\Diamond_p V2_1^{NN}$; and hence by the principle we isolated above with Chalmers' help, the onus is on those who would resist $\Diamond_p V2_1^{NN}$. This would seem to be a *very* heavy burden. What physical laws are violated in the new story of Smith?

---

contiguous, whereupon the healing and reconstitution happens automatically, in a matter of minutes.

[27]Chalmers gives the case of a mile-high unicycle, which certainly seems logically possible. The burden of proof would surely fall on the person claiming that such a thing is logically impossible. This may be the place to note that Chalmers considers it *obvious* that zombies are both logically and physically possible — though he doesn't think zombies are *naturally* possible. Though we disagree with this position, it would take us too far afield to consider our objections. By the way, Chalmers refutes (1996, p. 193–200) the only serious argument for the logical impossibility of zombies not covered in this chapter, one due to Sydney Shoemaker (1975).

[28]A quick encapsulated account of ANNs is provided in Chapter 6. ANNs are described in the context of AI in (Russell & Norvig 1994).

Some may retort that if the "physics version" of the zombie attack is sound, then beings with our behavioral repertoire, but without P-consciousness, could *in fact* have evolved in the actual world, on this very planet, under the constraints imposed by our laws of nature. Why then, they may go on to say, aren't *we* zombies? This question has already been eloquently raised in a slightly different form by those who merrily endorse $\Diamond V2_1$, for example, by Flanagan & Polger (1995). We think the question can be cashed out in an explicit argument against either the core notion of P-consciousness or the claim that $\Diamond V2_1$. The gist of the argument is this:

> "Look, evolution implies that every significant mental property corresponds to some concrete behavioral 'payoff,' something that has survival value. But P-consciousness, in light of the arguments you two promote, corresponds to no such payoff. (The purported payoffs from P-consciousness can all be explained through information-processing mechanisms involved in zombiehood; e.g., see Flanagan & Polger 1995.) Since evolution is of course true, it follows either that P-consciousness is a mirage, or $\neg\Diamond V2_1$."

There are at least two general ways to counter this argument.[29] The first is to remember that evolution does allow for outright accidents, and to then point out that P-consciousness could be adventitious. (As Ned Block has pointed out to Bringsjord, since at least all mammals are probably P-conscious, the accident would had to have happened quite a while ago.) The second response, and the one we favor, is to step up to the challenge and show that certain behaviors *do* correspond to P-consciousness. As you may recall, we provided the heart of this response in Chapter 1, and pointed out there that we have elsewhere (e.g., Bringsjord 1995*c*, Bringsjord 1997*b*) offered sustained arguments for the position that creativity, for example the creativity shown by a great dramatist, requires P-consciousness.[30]

---

[29] Actually, there are *three*. We leave aside for later the third one: the response that evolution isn't true. By 'later' we refer here specifically to Chapter 7, where we present two attacks (one mature, one to be specified in publications subsequent to the present book) on the view that all earthly creatures are the products of evolution.

[30] Henrik Ibsen wrote:

> I have to have the character in mind through and through, I must penetrate into the last wrinkle of his soul. I always proceed from the individual; the stage setting, the dramatic ensemble, all that comes naturally and does not cause me any worry, as soon as I am certain

The second and final move open to the computationalist still bent on resisting our zombie attack is one Dennett has made in personal communication:[31] concede $\Diamond V2_1$, and concede our corresponding arguments, but issue a reminder that zombies are not a "serious possibility."[32] In this regard zombies are said to be like gremlins in the internal combustion engine: it's logically possible that those little sedulous creatures are what make your car run, but no one takes this possibility seriously.[33] Unfortunately, this is just to change the subject. It's true that no one needing to fix an ailing Ford pops open the hood and reaches for bread crumbs to feed the gremlins; and it's also true that few indeed are those among us who wonder whether their friends are zombies. But such facts are perfectly consistent with each and every premise used above to refute the computational conception of mind. To make his point another way, Dennett asks if we would be impressed if somebody argued that he could imagine an entity that acted and looked just like a cat (and had the same DNA etc.) but was not alive? He points out that we certainly would not take the person to have shown that vitalism should be taken seriously. Why, then, so the line here goes, should we be any more impressed when somebody says that he can imagine a zombie? The fatal problem with this is that

---

> of the individual in every aspect of his humanity. (reported in (Fjelde 1965), p. xiv)

Ibsen's *modus operandi* is impossible for an agent incapable of P-consciousness. And without something like this *modus operandi* how is one to produce creative literature? This is discussed in (Bringsjord & Ferrucci 2000).

[31]To Bringsjord after he presented an ancestor of this chapter at the 1994 Eastern APA Meeting.

[32]There is a passage in CE consistent with this move. It is the last time Dennett discusses the zombie attack in his book:

> This book [argues] that if we are not urbane verificationists, we will end up tolerating all sorts of nonsense: epiphenomenalism, zombies, indistinguishable inverted spectra, conscious teddy bears, self-conscious spiders. (Dennett 1991, p. 459)

The odd thing is that, as we have seen, Dennett nowhere in CE explains, let alone proves, that zombie thought-experiments, and the associated arguments, e.g., Arg_1, are nonsensical. Moreover, these experiments and arguments are part of the canonical arsenal used against verificationism!

[33]Dennett claims that in light of this, zombies have become a public relations nightmare for philosophers, for when scientists hear that a major philosophical controversy boils down to zombies, they wear silly grins. But shouldn't we be concerned with constructing sound arguments and discarding unsound ones, regardless of how people *feel* about these arguments?

thought-experiments featuring zombies aren't intended to substantiate any "ism." We have not claimed that there is some spooky theory of mind that zombie thought-experiments show to be true. The vitalism thought-experiment is irrelevant. We have claimed that zombie thought-experiments have *negative* force: they allow for refutations of a particular "ism," viz., computationalism. Finally, while the gremlin and cat scenarios promise to remain but esoteric examples dreamed up by philosophers, it is not exactly difficult to imagine a future in which the question of whether behaviorally sophisticated robots are or are not zombies is a pressing one. The movie *Blade Runner* shows such a future.

5.6 Conclusion

Where does this leave us? Well, if computationalism is true, then (2_{C}), essentially the claim that appropriate computation suffices for P-consciousness, is as well. If argument $\mathrm{Arg}_1^{\mathrm{C}}$ is sound, (2_{C}) falls, as does, by *modus tollens*, computationalism itself. $\mathrm{Arg}_1^{\mathrm{C}}$, as we have seen, can be formalized as a modal disproof patterned on Arg_1, the argument fashioned from the Dennett-Searle clash on zombie thought-experiments. (Arg_1 is based on *gedanken*-experiments designed to establish $\Diamond V2_1$ and $\Diamond V3_1$; $\mathrm{Arg}_1^{\mathrm{C}}$ is based on one designed to establish $\Diamond V2_1^{\mathrm{TM}}$.) $\mathrm{Arg}_1^{\mathrm{C}}$ is formally valid. As to this argument's premises, they have emerged intact from a dialectical crucible fueled by the best objections computationalists can muster. Finally, attempts to tweak (2_{C}) so as to produce a view not targeted by $\mathrm{Arg}_1^{\mathrm{C}}$ fail. (Here we considered zombie stories written (hypothetically) by Kafka to establish $\Diamond_p V2_1^{\mathrm{TM}}$ and $\Diamond_p V2_1^{\mathrm{NN}}$.) In the end, then, the zombie attack proves lethal: computationalism is dead.

We turn now to the argument against computationalism from the irreversibility of phenomenal consciousness.

Chapter 6

The Argument from Irreversibility

6.1 Introduction

In this chapter we refute computationalism *via* specification and defense of the following argument: Computation is reversible; cognition isn't; ergo, cognition isn't computation. The specification of the argument involves a quartet: (i) certain elementary theorems from computability theory, according to which computation is reversible; (ii) the doctrine of agent materialism, according to which, contrary to any sort of dualistic view, human agents (= human persons) are physical things whose psychological histories are physical processes; (iii) the introspection- and physics-supported fact that human cognition is *not* reversible; and (iv) the by-now-familiar claim — put roughly for now — that cognition is computation. The basic structure of the argument is straightforward: the conjunction of (i), (ii) and (iii) entails the falsity of (iv).

Our plan is as follows. In section 6.2 we unpack the "cognition is computation" slogan. In section 6.3 we provide a rudimentary but sufficient-for-present-purposes account of recursion-theoretic reversibility. In section 6.4 we present the Argument from Irreversibility. The final section, 6.5, is a sustained dialectic arising from objections to the argument — a dialectic which generates a triumphant version of the argument presented in section 6.4.

6.2 The Computational Conception of Mind

As you know, we characterized computationalism in Chapter 1. Using the predicate calculus, we now set out a version of computationalism that will prove convenient for us in the present chapter. With the predicates Mx iff x is a Turing machine and Px iff x is a person, let's say, in a sharpening of proposition P=aTM from Chapter 1:[1]

Proposition 1. $\forall x\,(Px \Rightarrow \exists y\,(My \,\wedge\, x = y))$.

Later on we'll propose an even more "fine-grained" version of P=aTM, one which makes explicit reference to both cognition and computation. But before moving toward this version, one remark: There have of course been prior attacks on computationalism. To our knowledge, however, none of these attacks, with the possible exception of (Maudlin 1989) and Chapter IX of (Bringsjord 1992), have attempted to show that the falsity of this doctrine *follows from the very foundations upon which it's built.* In the case of Searle's (1980) Chinese Room Argument, everything hinges on an ingenious but undeniably fanciful thought-experiment. The Argument from Irreversibility is based, for the most part, on provable facets (e.g., reversibility) of computation which appear by the lights of both common-sense and elementary physics to be at odds with what we can come to know about cognition.

6.3 Rudiments of Reversibility

Computationalism relies upon automata like Turing machines (TMs) to fix the concept of computation. But what's a Turing machine? We answered this question in Chapter 3, but we return now to the elementary aspects of computability theory upon which the Argument from Irreversibility is built, and in doing so we place special emphasis on those elements that are central to reversibility.

As we pointed out in Chapter 3, TMs include a **two-way infinite tape** divided into squares, a **read/write head** for writing and erasing **symbols** (from some finite, fixed **alphabet**) on and off this tape, a **finite control unit** which at any step in a computation is in one particular state from among a finite number of possible states, and a set of **instructions** (= the program) telling the machine what to do,

[1]Please note that as in Chapter 2, we use '$\Rightarrow$' for the material conditional, and reserve '$\longrightarrow$' for computation.

depending upon what state it's in and what (if anything) is written on the square currently scanned by its head. Formally speaking, a TM is a set, specifically a quintuple (S, Σ, s, h, f), where: S is a finite set (of states); Σ is an alphabet containing the special "blank" symbol b, but not the symbols L (for "left") and R (for "right"); $s \in S$ is the **initial state**; $h \notin S$ is the **halt state**; and f is a **transition function** from $S \times \Sigma$ to $(S \cup \{h\} \times (\Sigma \cup \{L, R\})$.[2]

This formalism makes it easy to render the notion of a **configuration** of a TM precise,[3] but an informal account will suffice for our purposes. We'll say that a configuration is composed of four pieces of information: first, the state the TM is in; second, information representing what's to the left of the square currently scanned by the read/write head; third, the single symbol being scanned; and fourth, the contents of the tape to the right of the head. In a moment, we'll concretize the notion of a configuration by explaining and analyzing the particular TM specified in Figure 6.1. (The particular TM here was one we visited earlier in Chapter 3. You may recall that we also visited there relatives as well.) But first some notation for computations: Let $C_1, C_2, ...$ denote configurations. We write $C_i \vdash_M C_k$ to indicate that TM M has moved "in one step" from configuration C_i to C_k. For every TM M, $\vdash_M^*$ is the reflexive, transitive closure of $\vdash_M$. A **computation** by some TM M is a sequence of configurations $C_1, C_2, C_3, ..., C_n$, where $n \geq 1$ such that

$$C_1 \vdash_M C_2 \vdash_M C_3 \vdash_M \cdots \vdash_M C_n.$$

There are many ways to steamline the full set-theoretic description of TMs. One such method is the state diagram approach used in Figure 6.1. This TM, "Gordon's 19 in 186," is designed to start on a 0–filled infinite tape and produce, after 186 steps, 19 1's. Though we discussed this particular machine (and its relatives in the 6-state quadruple Busy Beaver Competition) in Chapter 3, we do so again here in order to make the present chapter self-contained.

Let's "hand simulate" an initial segment of the computation of Gordon's TM — let's label the machine G — so that we completely

[2]As we pointed out earlier, a TM needn't have dedicated halt states. A state from which nothing further can happen is tantamount to a halt state; e.g., examine the last configuration in Figure 6.3.

[3]Formally, keeping in mind that after a while, in both directions, the tape is populated by infinitely many blanks, a configuration is a member of the set $S \cup \{h\} \times \ldots b \circ \Sigma^* \times \Sigma \times \Sigma^* \circ b \ldots$

argument. Non-useful ways of reversing TM computation are easy to come by; here's one: First, go get some paper, a pencil, and an eraser. Next, watch a given TM M in action, recording each configuration through which it passes. Finally, using your list of configurations, build a new TM M_R which goes from the last entry on your list to the first in such a way that it visits your entries in reverse sequence.

Though we are informal here, it should be obvious that (in keeping with the inaugural writings of Turing and Post, in which, as we pointed out in the Preface, the notion of a human computist is primitive) we have here a straightforward *algorithm*, call it $\mathcal{A}$, for reversing computation. You are encouraged to try your hand at simulating $\mathcal{A}$ on the TM shown in Figure 6.1. All you need is some patience, and, as we say, a pencil, an eraser, and sufficient paper.

If you're short of time, we suggest a shortcut constructive "proof" as a substitute: Purchase the software "Turing's World"^{TM},[6] run one of the TMs that come bundled with this package, and then use the REVERSE command to step back through the configurations your selected machine has run. You can be sure that the existence of some such algorithm as $\mathcal{A}$ is precisely what enables you to carry out this reversal using "Turing's World"^{TM}. Though we can't run the software in question for you now, we can communicate the relevant ideas through snapshots taken of *our* use of the software in question. What you see in Figures 6.2 and 6.3 are six snapshots of Gordon's machine as it operates in Turing's World^{TM}, the first and last three steps of its operation, respectively. What the REVERSE command allows you to do is simply move the TM in question from the last state to the penultimate state, to the antepenultimate state, ..., to the third state, to the second state, to the first state. The fact that a computer program exists for doing this in the case of the states shown in Figures 6.2 and 6.3 immediately proves by construction that the algorithm $\mathcal{A}$ of Theorem 1 does is fact exist. Here is the theorem:[7]

Theorem 1. For every computation $C_1 \vdash_M C_2 \vdash_M C_3 \vdash_M \cdots \vdash_M C_n, n \in \mathbf{Z}^+$, there exists a computation $C_n \vdash_{M'} C_{n-1} \vdash_{M'} C_{n-2} \vdash_{M'} \cdots \vdash_{M'} C_1$ which can be obtained from the original computation via some algorithm $\mathcal{A}$.

---

[6]The "Turing's World"^{TM} software comes with (Barwise & Etchemendy 1993).

[7]Note that by Church's Thesis Theorem 1 is interchangeable with this theorem (which is easy to prove *without* CT):

Theorem 1′. For every computation $C_1 \vdash_M C_2 \vdash_M C_3 \vdash_M \cdots \vdash_M C_n, n \in \mathbf{Z}^+$, there exists a computation $C_n \vdash_{M'} C_{n-1} \vdash_{M'} C_{n-2} \vdash_{M'} \cdots \vdash_{M'} C_1$ which can be obtained from the original computation via some TM M^*.

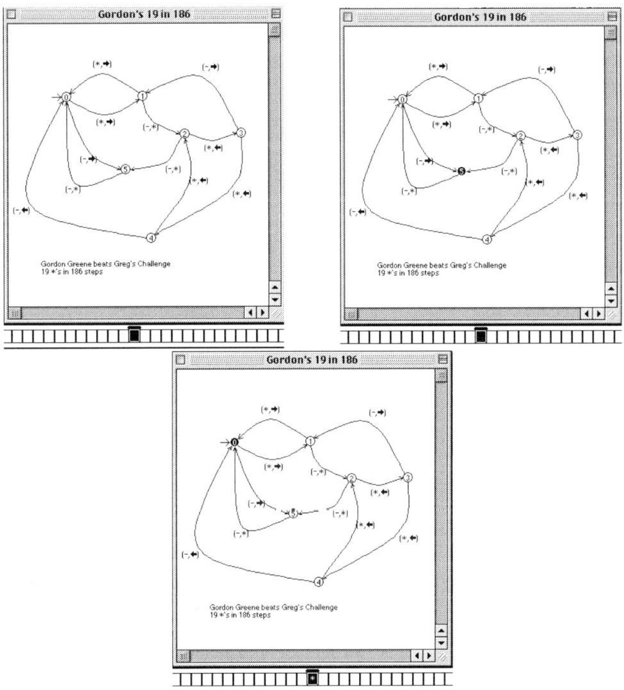

Figure 6.2: First Three Configurations of Gordon's 6–19 Turing Machine. *Notice the blackened circles. They indicate what state the machine is in.*

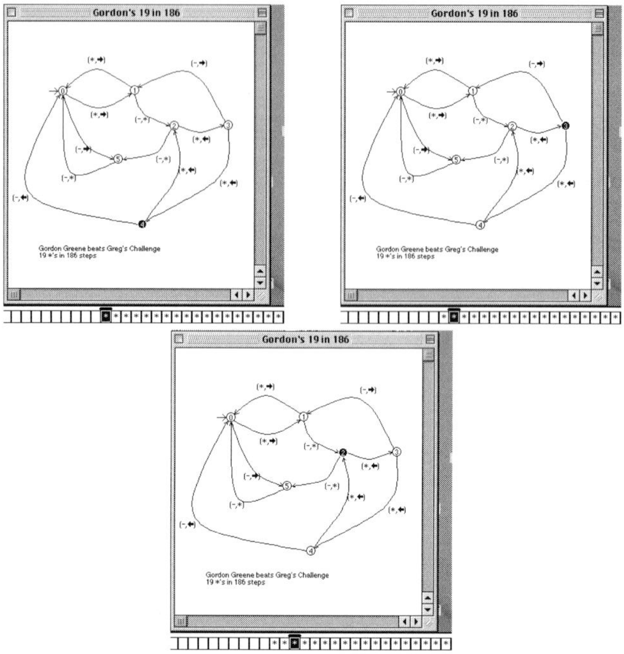

Figure 6.3: Last Three Configurations of Gordon's 6–19 Turing Machine

6.4 The Argument from Irreversibility

6.4.1 The Starting Point: Proposition 1

We begin by refining Proposition 1 so that it clarifies and narrows the "cognition is computation" slogan:

Proposition 1$'$. $\forall x\,(Px \wedge x$ is conscious from t_i to $t_{i+k} \Rightarrow \exists y\,(My\ \wedge\ x = y \wedge C_j \vdash_y C_{j+1} \vdash_y \cdots \vdash_y C_{j+p}))$, where this computation is identical to the consciousness x enjoys through $[t_i, t_{i+k}]$.

6.4.2 Clarifying 'Consciousness'

Note that Proposition 1$'$ marks a shift from talk of cognition to talk of consciousness. 'Consciousness' is a narrower term than 'cognition.'

Cognition, that is, includes consciousness.[8] It follows that if consciousness isn't reversible, then the broader phenomenon of cognition isn't either. (Of course, many elements of cognition *are* computable.)

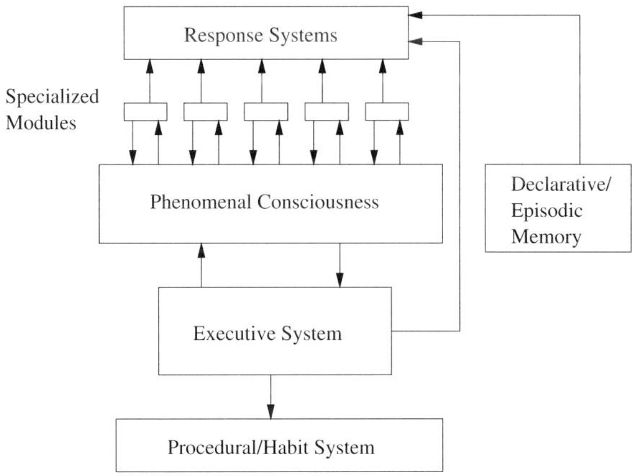

Figure 6.4: Schacter's Model

It's important to realize that we have something rather specific in mind when we use the term 'consciousness' in this chapter: we have in mind what is sometimes called **phenomenal consciousness**. Ned Block (1995), in a recent essay on consciousness in *Behavioral and Brain Sciences*, calls this brand of consciousness **P-consciousness**. We have exploited Block's characterization at previous points in this book; here again is the relevant explication:

> So how should we point to P-consciousness? Well, one way is *via* rough synonyms. As I said, P-consciousness is experience. P-conscious properties are experiential properties. P-conscious states are experiential states, that is, a state is P-conscious if it has experiential properties. The totality of the experiential

[8]This is borne out by, among other things, looking at proposed comprehensive models of cognition in cognitive science: these models now often include some component claimed to account for consciousness. (Recall that we pointed out in Chapter 1 that these days consciousness, even *phenomenal* consciousness, is regarded to be cognitive.) For example, Schacter (1989) gives us the picture of cognition shown in Figure 6.4. For a penetrating analysis of such models (including Schacter's) see (Block 1995). For those interested in charting the first-order formalization of our argument, the relevant sentence in first-order logic would be a full symbolization of $\forall x(x$ is cognizing $\Rightarrow x$ is conscious), where 'cognizing' is understood to indicate the full scope of human cognition as purportedly captured in models like Schacter's.

> properties of a state are "what it is like" to have it. Moving from synonyms to examples, we have P-conscious states when we see, hear, smell, taste and have pains. P-conscious properties include the experiential properties of sensations, feelings and perceptions, but I would also include thoughts, wants and emotions. (Block 1995, p. 230)

Block distinguishes between P-consciousness and access consciousness, or just **A-consciousness**; the latter concept is characterized as follows:

> A state is access-conscious (A-conscious) if, in virtue of one's having the state, a representation of its content is (1) inferentially promiscuous, i.e., poised to be used as a premise in reasoning, and (2) poised for [rational] control of action and (3) poised for rational control of speech. (Block 1995, p. 231)

Note that it's plausible to regard certain computational artifacts to be bearers of A-consciousness (e.g., theorem provers with natural language generation capability), whereas we shall now attempt to establish that P-consciousness (hereafter referred to by simply 'consciousness') is beyond computation.

6.4.3 The Argument from Irreversibility

Fix some arbitrary human person; call him 'Bob.' Suppose that Bob is conscious from t_1 to t_{44}. From Proposition $1'$ it follows directly (by elementary first-order logic) that there is a TM M_{Bob}, identical to Bob, such that $C_1 \vdash_{M_{Bob}} C_2 \vdash_{M_{Bob}} \cdots \vdash_{M_{Bob}} C_j$, where this computation is identical to Bob's consciousness from t_1 to t_{44}. But Theorem 1 implies that some algorithm $\mathcal{A}$ reverses the computation in question. Hence, by Leibniz' Law, $\mathcal{A}$ reverses Bob's consciousness from t_1 to t_{44}. But consciousness, whether Bob's or yours or mine, *can't* be reversed. By indirect proof it follows that Proposition $1'$ is false (since Theorem 1, the other possible culprit, is just that: a theorem). And since this proposition is "Strong" AI (or computationalism) incarnate, it follows that this view is in turn false.

6.5 Dialectic

6.5.1 Objection 1

The first objection is rather obvious: "You baldly assert that a stretch of consciousness cannot be reversed. But the irreversibility of consciousness is far from self-evident. I can certainly imagine a person coming into or passing out of consciousness — which seems to make it a reversible process. Do we not also routinely say that people can 'change their minds'?"

The first problem with this objection is that it appeals to senses of 'consciousness' other than the sense we are employing. As we said above, we are specifically considering the question of whether subjective awareness (P-consciousness) over some stretch of time can be reversed. When Jones admits that he has "changed his mind," he isn't saying anything remotely like, "I just experienced the beauty of a sunset backwards." And when we say that Jones received a blow that caused him to lose consciousness, and that presently he is once again alert, we are saying at most that Jones was at one point in such and such a state of P-consciousness, at a subsequent point *not* in a state of P-consciousness, and then thereafter once again in some state of P-consciousness.

As to whether a stretch of P-consciousness is irreversible, we concede *this* much: the proposition in question isn't self-evident *for those who haven't contemplated its truth value.* In order to begin to gain an appreciation for the plausibility of the claim that consciousness is irreversible, one has but to try to pull off the reversal in one's own case. We ask, accordingly, that you indulge in a bit of introspection; we ask that you tackle the challenge of reversing a stretch of your own mental life.[9] In order to fix things a bit, let's suppose that you are quite powerfully affected by hearing Beethoven's Ninth; suppose, more specifically, that you feel energized, deep down, when listening

[9]It's important to realize that we're talking about your *mental* life: we're not talking about reversing some sequence of physical actions which can be described in such as way as to imply that it is a discrete chain. For example, suppose that you move block a from on top of block b to a position next to block c, and then move block c to on top of block b. You might say that this sequential action is easily reversed by first moving block c to its original position, and then moving block a on top of block b. Though as a matter of fact — as we shall see below — you haven't truly reversed the sequence *qua* physical process, the present point is that we are concerned with reversing a stretch of conscious experience, not a sequence of ostensibly discrete bodily actions.

to an initial segment of the choral part of the Ninth during the interval of time t_1 to t_{10}. (If you've never had this experience, merely substitute something similar.) Now, remember your experience during this stretch, fix it in your mind, and proceed to reverse it — live it backwards, so to speak.

What happened? Not much, we wager. Part of the problem seems to be that a stretch of conscious experience can be what might be called *indivisible.* Perhaps if you could divide your Beethoven experience into sub-chunks corresponding to t_1 to t_3, t_4 to t_6, and so on, you could — thought-experimentally speaking — reassemble them in reverse. But for many, if not most, stretches of P-consciousness such division isn't possible.[10]

We should not be read as proposing divisibility as a quodlibet around which the entire issue revolves. In fact, we're inclined to think that reversing a stretch of P-consciousness is simply *incoherent.* The challenge under consideration seems to us to be analogous to the challenge of imagining that $2 + 2 = 7$, or that *modus ponens* is false. Such things can't be imagined — because they are incoherent (to use a philosophical concept, they are *logically impossible*).

Now to say that a reversed experience of the Ninth is logically impossible is of course to make an exceedingly strong claim. For our main argument it suffices that such a reversed experience be *physically* impossible — a more circumspect claim, and one we as a matter of fact retreat to below. However, after meditating on challenges like the one just posed, and on some states of affairs which are agreed to be logically impossible, it really does seem to us (and to those "subjects" who, at our request, have genuinely attempted to conceive of some reversed stretches of P-consciousness) that the kind of reversal

[10] Compare this sort of indivisibility with the type Descartes famously ascribed (perhaps incorrectly) to the mind when he said:

> In order to begin this examination, then, I here say, in the first place, that there is a great difference between mind and body, inasmuch as body is by nature always divisible, and the mind is entirely indivisible. For, as a matter of fact, when I consider the mind, that is to say, myself inasmuch as I am only a thinking thing, I cannot distinguish in myself any parts, but apprehend myself to be clearly one and entire; and although the whole mind seems to be united to the whole body, yet if a foot, or an arm, or some other part, is separated from my body, I am aware that nothing has been taken away from my mind. (Descartes 1911, p. 196)

being called for in our challenge *is* logically impossible. We have no outright *proof* that P-consciousness is irreversible in this strong modal sense. Our rationale involves the observation that there are states of affairs deemed by all to be logically impossible (e.g., there being a book both 150 and 200 pages in length at the same time) not because they entail some violation of a truth in logic or mathematics — though the book just alluded to presumably entails some violation of the Peano axioms —, but rather because when one ponders them, nothing can be envisaged. The same "nothingness" seems to attach to experiencing the Ninth in reverse. And Beethoven's Ninth isn't unique: other more mundane stretches of P-consciousness seem to be just as resistant to reversal. Experiences of a dish of fresh strawberries and cream, a vigorous ski run, the reading of a short story, etc. — the reveral of all these scenarios seems to share the "nothingness" of the 150/200 page book. It is easy enough to imagine swimming the Atlantic unaided, being twenty feet tall, enjoying immortality, finding a counter-example to Boyle's Law, and so on. But it's a tad harder to imagine that the circle has been squared — about as hard, it seems to us, as imagining phenomenal awareness in reverse.

Nonetheless, we imagine that some opponents will say some such thing as:

> "Look, if the stream of P-consciousness is like a movie (or a *Brave New World*-style "feelie") — a standard comparison — clearly experience *is* reversible. What could be simpler to imagine than this? No need to imagine anything like "2+2=7" (that's ridiculous); just run the film backwards!"

Unfortunately, this objection is legerdemain. It substitutes a *representation* of a stretch of P-consciousness for the stretch itself. A movie, of course, represents a series of events. Think about the issue in terms of a mental image: Suppose that your friend, Amanda, experiences some sort of protracted agony — and that you witness her misfortune. Perhaps for the rest of your life you remember the black events in question; perhaps you are haunted by the "mental movie" that "plays" when you recall Amanda's experience. This mental movie can certainly be reversed: conceivably, you could suddenly watch (in the "mind's eye" sense of 'watch') Amanda writhing (with "backwards" body movements), and then passing back to the painless state she was in before her misery began. But this back-to-front mental movie is obviously not the same thing as Amanda's experience literally happening in reverse. What our opponent must imagine is Amanda's (or the subjects' in the movie she has in mind) reversed P-consciousness. To reiterate, *this* is something the two of us can't imagine, and we wager that, if you're honest, you'll report the same inability.

6.5.2 Objection 2 (From Physics)

The next reply to our rebuttal is likely to be that such phenomenological meanderings are unreliable, and that we seem to be flirting with dualism. People are physical things; consciousness is hence ultimately a physical brain process; ergo, since the Beethoven experience is a stretch of conscious experience (and Amanda's, etc.), it corresponds to some brain process stretching from t_1 to t_{10} — and surely (so the reply here continues) such a process can be reversed.

Unfortunately, this reply fails. Recall that we explicitly ruled out dualism from the outset by affirming agent materialism. (This doesn't imply that *as a matter of fact* supermentalism is no friend to dualism: after all, see Chapter 7. The point is that *in this argument* we are assuming agent materialism.) In keeping with this affirmation [an affirmation, specifically, of the proposition we labeled (ii)], and with the spirit of the present objection in mind, let's view the brain in overtly materialistic terms: Suppose that it's composed, not of neurons and dendrites and the like, but of tiny colliding billiard balls. A brain state is thus a "snapshot" of these billiard balls which "freezes" them in a certain configuration. Suppose, further, that we have on hand a microscopic but prodigious helper named Max, a (distant) relative of Maxwell's demon. Max can manipulate neurological billiard balls *par excellence.* It should be possible for Max to reverse billiard ball mentation, if genuine mentation, bound as it is to be infinitely more complex than its homey pool-hall analogue, is reversible. But Max, no matter how clever, no matter how fast, no matter how conscientious, can't pull it off.

In order to see this let's return to Bob's mentation from t_1 to t_{44}. In keeping with our billiards setup, we are now free to view Bob's mentation as the temporally extended interaction of tiny billiard balls inside his cranium. Suppose that Max has kept his eyes on this t_1–t_{44} progression, and that he remembers, impeccably, the passage of these balls through time — their velocities, spatial and temporal positions, and so on. And suppose that we give Max the opportunity to reverse the mentation. Can he manage? No. And the reason he can't is quite straightforward: Suppose he begins by noting that the 8 ball, which moved from place p_4 to p_6, must be reversed; so he moves the 8 ball from p_6 to p_4 (making sure that the temporal factors match up perfectly with the "forward" recording of the 8 ball). In order to perform his task, the demon must remember what he has done. However, at some point Max's memory will be filled and he will have to erase some stored information. Since erasure is a setting process (i.e., an operation which brings all members of an ensemble to the same observational state) and this operation increases the entropy of the system (billiard balls plus demon) by an amount equal to or greater than the entropy decrease made possible by the newly available memory capacity, the demon cannot execute the reversed mentation.

"Slow down!" exclaims our critic. "If Max's memory is large enough to store the entire sequence of states, then it's large enough to do the reversing; and the obvious way to reverse a sequence of billiard ball collisions is to simply reverse the velocities of the balls in the final state and let the whole system cycle backwards. There is no theorem of statistical mechanics which rules out reversing physical processes."

The central claim here is that since the dynamics of the billiard ball system is completely described by classical statistical mechanics, all we need do to reverse the system is to change the algebraic signs of the motion variables. Any state of the system, so the story goes, is sufficient to allow us to compute any previous and subsequent state.

Unfortunately, this objection conflates the formal system called 'classical' or 'rational mechanics' with a particular instantiation of it by point particles colliding elastically. Reversing is achieved for our critic by manipulating the *equations* of classical statistical mechanics, that is, it is a formal operation, not a physical one. We introduced Max to see what it would take to *actually* reverse a physical process, not simply reverse the formalism used to represent a physical process. Max is supposed to bring into focus the conditions and consequences of executing such a reversal. So it doesn't suffice to say "reverse the velocities of the balls in the final state and let the whole system cycle backwards." We want to know *how* this is done — and this leads us to consider fanciful creatures like Max.

And what Max reminds us of is that while the idealized (point particles, elastic collisions) billiard ball model is logically reversible, the non-ideal (non-point particles, inelastic collisions) is not. The former is logically equivalent to the formalism it re-represents, and there is a one-to-one correspondence between the discrete states of the idealized model and the continuous state of the mathematical description.

The non-ideal model is non-ideal precisely because it lacks those characteristics: there is no simple one-to-one correspondence between an "inelastic collision" and the formalism which tries to capture it. And this is why Max cannot win when working in the real world of friction, dissipation, inelastic collisions, uncertain initial conditions, etc. His memory is soon exhausted; he needs to erase, but the erasure only causes him further difficulties and increases his task.

Our metaphorical Max can be supplanted with a technical discussion making the same point, but such a discussion would quickly render this chapter inaccessible to most. This discussion would begin by charting the research devoted to instantiating purely conceptual, reversible models in terms of physical models. This research has lead to a family of devices with intriguing properties: one such device is is the **ballistic computer**, which shows, *in principle*, how a computation can be performed without dissipating the kinetic energy of its signals. The device employs a set of hard spheres and a number of fixed barriers with which the "balls" collide and which cause the "balls" to collide with each other. The collisions are elastic, and between collisions the balls travel in straight lines with constant velocity according to Newton's second law. But what happens when one departs slightly from the idealizations upon which the ballistic computer is based? We find what Max already showed us: When we move from the realm of the ideal and ask how ballistic computers might be physically embodied, we immediately confront two problems: (i) the sensitivity of the ballistic trajectories to small perturbations and (ii) how to make the collisions truly elastic. Initial random errors in position and or velocity of one part in 10^{15} are successively magnified with each generation of collisions, so that after a few dozen generations, the trajectories cannot sustain a computation. Even if we had perfect accuracy and a perfect apparatus, the ballistic computer would still be subject to fluctuating gravitational forces in the environment, so that after a few hundred collisions, the "perfect" trajectories would be spoiled. This dynamical instability of the ballistic computer could be countered by corrections after every few collisions, but then we would no longer have a thermodynamically reversible device: the computer becomes dissipative and logically irreversible. (For a detailed survey of the "Max phenomenon," in the form not only of ballistic computers, but also the enzymatic Turing machine and the Brownian computer, see (Bennett 1982). For a detailed, book-length treatment of irreversibility in general, see the book one of us (Zenzen — with Hollinger) has written on the subject: (Zenzen & Hollinger 1985).)

Some skeptics might persist in objecting that our argument founders in its use of physics — and the dialectic could go on, at the cost of producing a chapter suitable only for consumption by students and practitioners of physics. Someone might say, for example, that with the possible exception of some rare weak interactions, all of fundamental physics is thought to be reversible in the most straightforward sense. One of the central puzzles of statistical thermodynamics, so the objection goes, is exactly how to explain why the world seems to contain irreversible processes when at base it doesn't.

The puzzle alluded to here gives rise to a fundamental rift in physics between those who hold that the world, despite appearances, can't contain irreversible processes (because the formalisms designed to capture the microlevel imply that all physical processes are in principle reversible), and those who hold that no one ought to deny what appears to be clear as day (for reasons of the sort canvassed above in our discussion above of ballistic computers), viz., that *real* physical processes are irreversible. And because there is such a rift, the present objection is uncompelling. In order to see this, first consider someone who, taking P-consciousness seriously, makes a careful case C for property dualism (the best example is probably Jacquette 1987). Now consider an eliminative materialist who literally denies the existence of P-consciousness, for example Dennett (1991). And next consider someone who, apprised of this clash, objects to C as follows. "Look, one of the great puzzles in philosophy is that there appears to be this phenomenon called 'P-consciousness' which resists reduction to purely physical terms. But by the tenets of physicalism, the world contains nothing that cannot be captured in physical terms." This objection leaves C intact for all those who, unlike the eliminative materialist, haven't unalterably placed their bets. Similarly, our argument is intact for all those who haven't permanently affirmed the view that certain formalisms in use in physics imply that nothing is irreversible.[11]

[11]Additional objections are possible, but not powerful. For example, someone might claim that appeals to thermodynamics have no place here since neither Max nor the system being manipulated need be closed. The short answer to this is that only equilibrium thermodynamics requires closed systems. Indeed, our arguments depend on recognizing important differences between equilibrium and non-equilibrium situations. Consciousness necessarily involves departure from equilibrium; computation doesn't.

6.5.3 Objection 3

At this point those who would resist our argument might turn away from attacks that spring from physics, and claim instead that we are attacking a straw man. They might say: "Look, there are plenty of AI and CogSci researchers who don't affirm anything like Propositions 1 and $1'$. In fact, one can be a darn good AI researcher and not take a stand on the relation between computation and cognition."

This objection is of course easily cast aside — because it fails to take account of the target we have set for ourselves. We are not targeting a brand of AI concerned (say) exclusively with engineering a computational correlate to the olfactory component of rat brains.[12] We are concerned with what, following tradition, we've called "Strong" AI (and CogSci): a discipline which aims at replicating human cognition in part by identifying, scientifically, cognition with computation.[13] Propositions 1 and $1'$ simply reflect our focus on this brand of AI/Cog Sci.

6.5.4 Objection 4

Against Proposition $1'$ (and, for that matter, Proposition 1) it might be said, "This proposition leaves out of the picture something we know to be essential for mentation of the sort we humans enjoy, viz., interchange with the environment. An agent's mentation over some stretch of time doesn't consist solely in computation divorced from the surrounding environment; it consists of computation which reflects a symbiosis with the environment — because human persons have sensors (for taking in information from the environment) and effectors (for manipulating the environment)."

---

[12] Bringsjord does personally conduct a lot of "Weak" AI, in the form of an attempt to engineer systems capable of passing determinate tests. (This brand of "Weak" AI is discussed in the final chapter of this volume.) See, for example, the programs featured in (Bringsjord & Ferrucci 2000), which are designed to produce an agent capable of passing S^3G (Bringsjord 1998*a*).

[13] We make veiled reference here to a distinction now mentioned a number of times earlier in the book, viz., the distinction between *simulation* and *replication*, e.g., see the discussion of in section 2.3. Put roughly here, we say that to simulate some human chess player *via* AI techniques would be to produce a computer program whose overt behavior (i.e., actual chess moves) resembles that of some human, whereas to replicate Kasparov would be to construct an artifact that literally has the same inner life he has, the same emotions, plans, experiences, mental images, memories, etc.

Though we have argued elsewhere that transaction with the "outside" environment is, at least in principle, entirely superfluous when it comes to consciousness,[14] we grant here for the sake of argument the claim that sentience requires such interchange. This concession does imply that Propositions 1 and 1′ are indeed unacceptable. But (as we indicated in a slightly different context: Bringsjord 1991), it should be easy enough to remedy the situation. We have only to build in the environmental interchange to Proposition 1′. In order to do so, let M^*x hold iff x is a TM with sensors and effectors; then we have both a corresponding picture (Figure 6.5) and a new proposition, viz.,

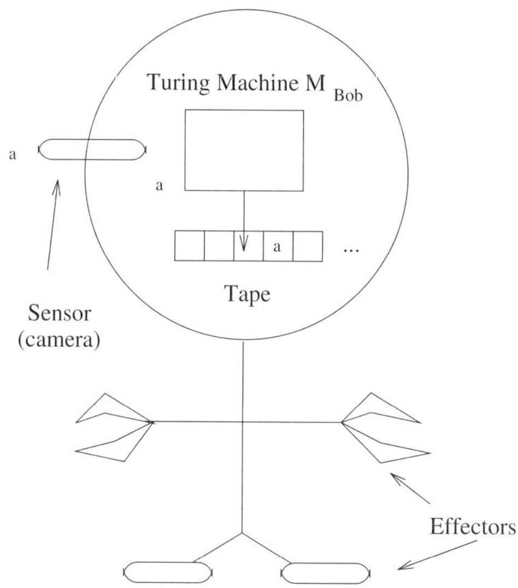

Figure 6.5: Bob as Transactional Turing Machine

Proposition 1″. $\forall x\,(Px \wedge x$ is conscious from t_i to $t_{i+k} \Rightarrow \exists y\,(M^*y \wedge x = y \wedge C_j \vdash_y C_{j+1} \vdash_y \cdots \vdash_y C_{j+k}))$, where this computation — partly determined by causal interaction with the environment — is identical to the consciousness x enjoys through $[t_i, t_{i+k}]$.

Now, in order to rebut Objection 4, let's simply expand Max's powers: give him control not only over Bob's "neuro" billiard balls,

[14] In (Bringsjord & Zenzen 1991) we devise and exploit a variation on the classic brain-in-a-vat gedanken-experiment in order to make the case for this view. Bringsjord (1995*a*) argues, *contra* Harnad (1991), that Turing Testing, in order to test for consciousness, needn't include a test for the ability of a would-be AI to interact with its environment.

but over the homey analogue for the molecular motion involved in the use of sensors and effectors. (We leave the details of the more elaborate analogue to your imagination, but a good start is had by simply assuming that the neuro billiard balls behave the way they do in part because billiard balls from the outside interact with them.) Does this change the picture in any way? No. In fact, Max's handiwork is now all the *more* irreversible, for reasons already covered, and so reasoning which parallels our argument in the obvious way arises.

6.5.5 Objection 5

This objection, more subtle and powerful than its predecessors, begins by bringing before us the argument as it stands given the dialectic to this point:

> Fix some arbitrary human person; call him 'Bob.' And suppose that Bob is conscious from t_1 to t_{44}. From Proposition 1$''$ it follows directly (by elementary first-order logic) that there is a sensor-and-effector-outfitted (physical) TM M_{Bob}, identical to Bob, such that $C_1 \vdash_{M_{Bob}} C_2 \vdash_{M_{Bob}} \cdots \vdash_{M_{Bob}} C_m$, where this computation — partly determined by causal interaction with the environment — is identical to Bob's consciousness from t_1 to t_{44}. But Theorem 1 implies that some algorithm $\mathcal{A}$ reverses the computation in question. Hence, by Leibniz' Law, $\mathcal{A}$ reverses Bob's consciousness from t_1 to t_{44}. But consciousness, whether Bob's or yours or mine, *can't* be reversed (as can be seen both *via* introspection and a look at the elementary physics of irreversibility). By indirect proof it follows that Proposition 1$''$ is false (since Theorem 1, the only other possible culprit, is just that: a theorem). And since this proposition is "Strong" AI (or computationalism) incarnate, it follows that this view is in turn itself false.

Now, the new objection runs as follows: "The inference at ...*But Theorem 1 implies that...* is invalid, because Theorem 1 applies to abstract mathematical objects (of a sort studied in *theoretical* computer science), whereas M_{Bob} is now, given the rebuttal to Objection 4, a *physical* object. To link the two as the argument now does, to ascribe to M_{Bob} properties possessed by *mathematical* machines, is to commit a fallacy."

The first problem plaguing this objection is that "Strong" AI and CogSci are founded upon the notion that there is a level of analysis, in the study of the mind, which corresponds to computation. As Dennett (1993) puts it:

> The central doctrine of cognitive science is that there is a level of analysis, the information-processing level, intermediate between the phenomenological level (the personal level, or the level of consciousness) and the neurophysiological level. (Dennett 1993 p. 195).

It may well be, for example, that at the neuromolecular level, computation is a dispensable notion. It certainly seems that at the phenomenological level computation is unwieldy, to say the least. (We routinely give exclusively "folk-psychological" (= phenomenological) explanations of human behavior, as when, e.g., we say such things as that "Bob left the room because he wanted to.") But the information-processing level, the level at which AI and CogSci is canonically done, is a different story: it's a level surely governed by results in computability theory, and one such result is none other than our Theorem 1.

There will doubtless be those inclined to sustain the fight for the position that an unbridgeable chasm separates computability theory from M_{Bob} (and the like), so let's explore, in a bit more detail, the relation between such results and the "real" world.

Some well-known theorems in computability theory, in particular those concerning *un*computability, clearly enjoy direct connections with (or at least have immediate implications concerning) the physical world. Consider, for example, the halting problem, introduced in this book back in Chapter 1.

In order to briefly review this problem, we begin by writing

$$M : u \longrightarrow \infty$$

to indicate that TM M goes from input u through a computation that never halts. We write

$$M : u \longrightarrow \textbf{halt}$$

when machine M goes from input u through a computation that *does* halt. And we write

$$M : u \longrightarrow v$$

when machine M goes from input u through a computation that prints out v and then halts.

Let the traditional property of decidability be handled by way of the symbols **Y** ("yes") and **N** ("no"). For example, the problem of whether a given object a is a member of a set A is decidable iff some machine M' exists which is such that

$$M' : a \longrightarrow \mathbf{Y} \text{ iff } a \in A$$

$$M' : a \longrightarrow \mathbf{N} \text{ iff } a \notin A.$$

Finally, let n^M represent an encoding, in the form of a natural number, of machine M.[15]

The halting problem now amounts to the theorem (call it 'HP') that it's not the case that there is a TM M such that for every input u, and every M^*:

$$M : n^{M^*} \longrightarrow \mathbf{Y} \text{ iff } M^* : u \longrightarrow \textbf{halt}$$

$$M : n^{M^*} \longrightarrow \mathbf{N} \text{ iff } M^* : u \longrightarrow \infty$$

Now that we have HP firmly on the table, let's return to our dialectic. The question before us is whether there is an unbridgeable chasm separating HP from the physical world. Put another way: Does HP apply to the physical world, as well as the mathematical?

The answer is an unwavering "Yes." In fact, HP can be "physicalized," in the sense that it can be reworded so as to make a most concrete assertion. One possibility is what we might call HP^P, which says that it's not *physically* possible that one *build* a TM M (or, if you like, a *standard computer*) such that for every input u (in the "real world" sense of 'input' attached to the word in, say, business and engineering applications), and every *physical* M^*:

$$M : n^{M^*} \longrightarrow \mathbf{Y} \text{ iff } M^* : u \longrightarrow \textbf{halt}$$

$$M : n^{M^*} \longrightarrow \mathbf{N} \text{ iff } M^* : u \longrightarrow \infty$$

Since HP applies to the corporeal world in the form of HP^P the objection in question looks to be evaporating.

---

[15]One method for such encoding comes via Gödel numbers; we resorted to this method earlier in the book. For an example of such coding explained with help from convenient tables, see (Boolos & Jeffrey 1989). Some of the philosophical issues related to this particular form of coding are discussed in (Bringsjord 1998*b*).

But perhaps it will be said, "You've just been lucky with HP. This theorem does link the 'Platonic' realm with the corporeal one; with this I agree. But this is a coincidence. Other theorems, and in fact perhaps the *bulk* of computability theory, stands inseparably apart from the physical realm. If I had to guess, I would say that HP turned out to apply because it's a *negative* result. Notice that you yourself turned first to *un*computability. Things will turn out differently for theorems that are not negative. And when things do turn out this way, we can return to scrutinize Theorem 1 and its relatives, which are themselves not negative."

Well, let's see if this is right; let's pick a theorem which isn't negative. Specifically, let's consider a fact mentioned above, viz.,

"Theorem" 2.[16] Turing machines with multiple tapes are no more powerful than standard one-tape TMs.

For any fixed natural number k, a k-**tape Turing machine** has k two-way infinite tapes, each of which has its own read/write head. We assume that a k-tape TM can sense in one step the symbols scanned by all heads and, depending upon what those heads find, can proceed with standard actions (erase, write, move right or left).

Though a fully worked out proof of Theorem 2 is perhaps a bit daunting in its detail, the key idea behind the proof is actually rather simple (specifying the proof, once one grasps this key idea, is downright tedious). It is that k tapes can be converted into one tape with k "tracks." An example will make the idea clear: Suppose that we have a 3-tape TM whose configuration, at some moment, is captured by

...	0	0	1	0	0	0	...
			↑				

...	0	0	1	1	1	0	...
					↑		

...	1	0	1	0	0	0	...
	↑						

Then the algorithm at the heart of Theorem 2 would simply convert this into one tape with three tracks (where the single-tape TM's alphabet is suitably composed), as in

[16]We place scare quotes around 'Theorem' because we describe the result very informally. For a more precise statement of the theorem, as well as a more precise account of the proof than what we provide below, see (Lewis & Papadimitriou 1981).

...	0	0	1 ↑	0	0	0	...
...	0	0	1	1	1 ↑	0	...
...	1 ↑	0	1	0	0	0	...

Now, the question we face is whether

"Theorem" 2^P. Physical Turing machines with multiple tapes are no more powerful than standard one-tape physical TMs.

is true.[17] And the correct answer would seem to be an affirmative one — because it seems physically possible to follow the algorithm at the heart of the proof of Theorem 2 in order to convert a physical k-tape TM into a single tape physical TM.[18]

Objection 5, then, is beginning to look like a dead end. There does appear to be a link between computability theory and the physical world sufficiently strong to undergird the Argument from Irreversibility. And this is just what we would expect, given that computability theory provides much of the mathematical framework for computationalism.

Nonetheless, someone might say: "If no physical process is reversible, and all computational processes are reversible, then so-called 'computers' really aren't. Yet you want to argue that provable limits on (standard) computation translate into actual limitations on the abilities of these same so-called 'computers.' If you're right, you would not just show that *we* aren't Turing machines; it would equally show that *robots* (of the sort that you describe, in your final chapter, yourselves building) aren't Turing machines!"

[17]We leave aside a complexity-based construal of 'power.' Obviously, a physical TM with multiple tapes could sometimes solve problems *faster* than an ordinary one-tape TM. However, there are no problems which are unsolvable by a standard TM yet solvable by a multi-tape machine. Likewise, a quantum computer can solve certain problems very fast, but can't solve any Turing-unsolvable problems (Deutsch 1985). See, however, (Stannett forthcoming).

[18]It's easy enough to imagine the details of this conversion in some cases. For example, suppose that we present you with a 4-tape TM roughly in the form of a model railroad set. That is, suppose that: the tapes are divided into squares, the read/write head for each tape is a lone boxcar, there is some way to give simple instructions to this railroad TM, etc. Now imagine converting this physical TM to a one-tape TM via the trick at the heart of Theorem 2. You would have to find some way to link the tracks together into one unit. If you have any experience with model railroading, you probably can visualize yourself tackling the process.

If the upshot of our case from irreversibility is indeed that we aren't TMs, and that, moreover, robots aren't either, we wouldn't be unhappy in the least. Why would we be threatened by such a result? After all, it would be a result that puts us in rather good company, for Searle may be understood as mounting a sustained case for the two-fold view that (courtesy of the Chinese Room Argument: Searle 1980) we aren't mere (standard) computing machines, *and* (courtesy of Searle 1992) 'computer' is a term so empty and confused that it applies to either everything or nothing. If the present chapter demonstrates that 'computation,' as a term ascribed to persons and their laptops, is nonsensical, we would certainly have dealt a fatal blow to computationalism.

At this point some may be inclined to sustain Objection 5 in the following manner: "The two of you say that computation is reversible, and with this I heartily agree. But you also go on to point out that no physical process is reversible; and you then capitalize on this fact. But what you fail to appreciate is that at the information-processing level to which Dennett has drawn our attention, consciousness *is* reversible. The problem is that you have surreptitiously moved at the same time to a level *beneath* this level, the level of Max and entropy and thermodynamical equilibrium, which is indeed a level where reversibility fails to apply. Your argument is nothing more than sleight-of-hand."

The problem with this objection is that it conveniently ignores the fact that computationalism is wed not only to information processing, but also to agent materialism, the view that cognizers are physical things, and that therefore cognition is a physical process. In light of this, introducing at least elementary considerations from physics, as we have done, is not only natural, it's unavoidable. And, as we have shown, once these considerations are introduced, the Argument from Irreversibility is off and running.

The situation can be specified by returning to the sketch of the argument we offered at the outset of the chapter. There, as you may recall, we said that the argument would involve a quartet of propositions numbered (i) through (iv), now unpacked as

(i) Theorem 1;

(ii) Agent Materialism;

(iii) P-consciousness is irreversible, as can be seen by both introspection and an analysis informed by elementary physics; and

(iv) Proposition 1$''$.

The problem for computationalism is that the denial of (iv) is entailed by {(i), (ii), (iii)}.

Please note that our argument is of necessity a good deal trickier than this one: "TMs are abstract entities that do not exist in space/time and between which only logical relations obtain, while minds are causal systems in space/time. Logical processes are reversible while causal processes are not. Therefore minds are not TMs." Our argument is trickier because computationalism doesn't hold that minds (or persons) are abstract entities. On the contrary, computationalists as a rule hold that minds are physical things. They must also hold, however, that minds, as physical things, abide by the principles of computation — and this is what gets them into trouble, as we are in the process of showing. The simplest distillation of our argument is that it is a proof of inconsistency holding between four propositions, as we indicated at the outset, and as we explain in more detail in the present section.

It's important to note that our conclusion needn't worry those who seek to pursue only "Weak" AI, essentially the program satisfied with engineering intelligent systems — systems not intended to replicate those properties (e.g., self-consciousness, command of a language, autonomy, etc.) traditionally thought to be constitutive of personhood. "Weak" AIniks are free to reject Proposition $1''$.[19] We return to this point at the end of the chapter.

6.5.6 Objection 6

The next objection marks a return to physics: "You assume that all physical processes are irreversible. But they are not. In fact, under

[19]They are *not* free to reject Theorem 1, however. Two different and determinate levels may be addressed by Theorem 1 (and the like): perhaps there is the purely mathematical level of the theorem itself, and and then perhaps also what might be called the "logic gate level" of computer engineering — a level indispensable for "Weak" AI. Computer engineers have on hand physical instantiations of Turing machines which they can combine in order to generate increasingly sophisticated computational artifacts, but there is no guarantee, say, that such engineers will be familiar with the purely formal set theoretic definition of a TM and a TM configuration. Nonetheless, both theoretical computer science *and* computer engineering are constrained and informed by Theorem 1 and its relatives. Work in AI and CogSci, whether of the "Weak" or "Strong" variety, is carried out at both of these levels. To abandon these levels and carry out a research program which in its entirety is "below" them, say research exclusively at the neuromolecular level, is to abandon AI for a variant on bio-engineering.

certain conditions, namely those where the system is allowed to move from state A to state B in such fashion as to always be arbitrarily close to thermodynamic equilibrium, it's possible to execute a reversed process. All we have to do is avoid generating or allowing the system to generate an entropy increase which is transferred to the environment and is thus not retrievable. One way to do this is to proceed very slowly so as to not generate heat, in other words, we execute an **adiabatic process**."

This objection flies in the face of what we said above in Objection 2 regarding Max and ballistic computers, which were designed by our opponent to capitalize on such tricks as adiabatic processes.[20] But let's suppose for the sake of argument that the barriers to real, physical irreversibility described in connection with Max and ballistic computers can somehow be surmounted. Under this charitable supposition, how does the present objection fare? Not well; here's why.

Again, let's suppose for the sake of argument that an automaton can be made reversible at every step *in the real physical world*, and that this allows not only an *in-principle* thermodynamically reversible automaton, but a "real world" one: a concrete one that saves all intermediate results, avoids irreversible erasure, prints the desired output and reversibly disposes of all undesired intermediate results as it retraces the machine's configurations in reverse order. To accomplish this, in keeping with the ideas behind ballistic computers, the automaton must be kept or keep itself arbitrarily close to thermodynamic equilibrium. For erasure to be thermodynamically reversible, the initial ensemble of memories must be distributed equally among possible states of phase space. Let's call such a process **r.w.i**, for "real-world irreversible."

Now, given that consciousness isn't currently associated with r.w.i. processes, why isn't the objection a complete non-starter? The answer must be that somehow — at least in the realm of thought and thought-experiment — the processes associated with consciousness can, at least in principle, become r.w.i. But notice that since the objective of our opponent is to show that reversing *consciousness* is doable, if the present objection is to have a chance of succeeding, it must assert not only that the neurological processes in question can become r.w.i, but also that they could become r.w.i. *without thereby preventing the rise of the consciousness with which they are connected.* And so a

[20] Again, for a detailed discussion of Max, ballistic computers, and other, similar devices, see Bennett (1973, 1984).

putationalism in disguise.[23] Here, it will suffice to give an intuitive recapitulation of the chain in question — a chain which establishes the interchangeability of neural nets and Turing machines.

Before we sketch this chain, let's pause to make clear that it underlies a certain proposition which can be conjoined with Proposition 2 in order to adapt the Argument from Irreversibility. This proposition is

Proposition 3. $\forall x\,((N^*x \wedge x$ passes through some process through $[t_i, t_{i+k}]) \Rightarrow \exists y\,(My \wedge C_j \vdash_y C_{j+1} \vdash_y \cdots \vdash_y C_{j+p})$, where this computation is identical to the process x enjoys through $[t_i, t_{i+k}]$).

It's easy to prove in elementary logic (using such rules as universal elimination and *modus ponens*) that Proposition 2, conjoined with Proposition 3, reenergizes the Argument from Irreversibility. But why is Proposition 3 true? In order to answer this question, we need to look first at (standard = ultimately-equivalent-to-TMs) artificial neural nets. After that, even a casual look at "two-dimensional" TMs should make it plain that Proposition 3 is true.

Neural nets are composed of **units** or **nodes**, which are connected by **links**, each of which has a numeric **weight**. It is usually assumed that some of the units work in symbiosis with the external environment; these units form the sets of **input** and **output** units. Each unit has a current **activation level**, which is its output, and can compute, based on its inputs and weights on those inputs, its activation level at the next moment in time. This computation is entirely local: a unit takes account of but its neighbors in the net. This local computation is calculated in two stages. First, the **input function**, in_i, gives the weighted sum of the unit's input values, that is, the sum of the input activations multiplied by their weights:

$$in_i = \sum_j W_{ji} a_j.$$

In the second stage, the **activation function**, g, takes the input from the first stage as argument and generates the output, or activation level, a_i:

$$a_i = g(in_i) = g\left(\sum_j W_{ji} aj\right).$$

[23]This paper is based on the classic statement of connectionism given by Paul Smolensky (1988*a*, 1988*b*).

One common (and confessedly elementary) choice for the activation function (which usually governs all units in a given net) is the step function, which usually has a threshold t that sees to it that a 1 is output when the input is greater than t, and that 0 is output otherwise.[24] (This is supposed to look "brain-like" to some degree, given the metaphor that 1 represents the firing of a pulse from a neuron through an axon, and 0 represents no firing.)

As you might imagine, there are many different kinds of (standard) neural nets. The main distinction is between **feed-forward** and **recurrent** nets. In feed-forward nets, as their name suggests, links move information in one direction, and there are no cycles; recurrent nets allow for cycling back, and can become rather complicated. But no matter what neural net you care to talk about, Proposition 3's deployment of its universal quantifier remains justified. Proving this would require a chapter rather more substantial than the present one, but there is a way to make the point in short order. The first step is to note that neural nets can be viewed as a series of snapshots capturing the state of its nodes. For example, if we assume for simplicity that we have a 3-layer net (one input layer, one "hidden" layer, and one output layer) whose nodes, at any given time, or either "on" (filled circle) or "off" (blank circle), then here is such a snapshot:

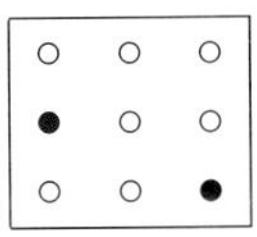

As the units in this net compute and the net moves through time, snapshots will capture different patterns. But Turing machines can accomplish the very same thing. (Indeed, recall our use of the term **snapshot** when characterizing computation through $\mathcal{L}$ in Chapter 1.) In order to show this, we ask that you briefly consider **two-dimensional** TMs. We saw k-tape TMs above; and we noted the equivalence between these machines and standard TMs. One-head two-dimensional TMs are simpler than k-tape machines, but (in the present multidisciplinary context) more appropriate than their standard k-tape cousins. Two-dimensional TMs have an infinite two-dimensional grid instead of a one-dimensional tape. As an example, consider a two-dimensional

[24] McCulloch and Pitts (1943) showed long ago that such a simple activation function allows for the representation of the basic Boolean functions of AND, OR and NOT.

TM which produces an infinite "swirling" pattern. We present a series of snapshots (starting with the initial configuration, in which all squares are blank, and moving through the next eight configurations produced by the "swirl" program) of this machine in action.[25] Here's the series:

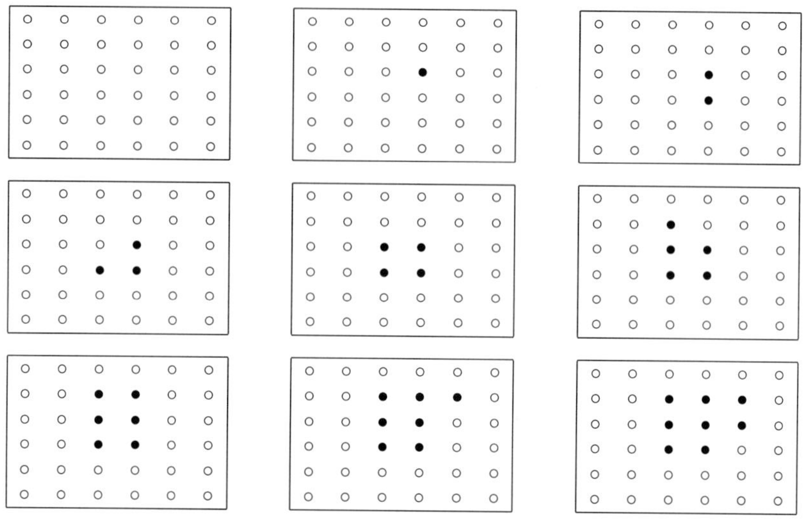

The point of this series of snapshots is to convey that snapshots of a neural net in action can be captured by a one-head two-dimensional TM (and, more easily, by a k-tape, k-head machine). Hopefully you can see why this is so, even in the absence of the proof itself. The trick, of course, is to first view the neural net in question as an $n \times n$ array (we ignore the part of the infinite grid beyond this finite array), as we did above. Of course, it's necessary to provide a suitably enlarged alphabet for our neural-net-copying TM: it will need to have an alphabet which

[25]The machine itself, as a series of quadruples (with 0 for blank, 1 for filled, R and L for "right" and "left," resp., A for "any," etc.) is (read left to right, top to bottom):

1011	1102	2012	21L3	3013
31U4	4014	41U5	5015	51R6
6016	61D7	70U9	71U8	8AR6
9AD10	100110	101L11	111R9	110R12
12AL13	130113	131U14	141D12	140D15
15A16	160116	161R17	171L15	170L8

The implementation of this machine — in Turing's World^{TM} — sent upon request.

contains a character corresponding to all the states a node can be in. For this reason, our swirl TM is a bit limited, since it has but a binary alphabet. But it's easy enough to boost the alphabet (and thereby produce some entrancing pyrotechnics).[26]

The proponent of Objection 7 might say that a sequence of configurations of a Turing machine M proposed to capture a neural net N as it passes through time isn't complete, because such a sequence includes, at best, only *discrete* "snapshots" of the *continuous* entity N. But suppose that the sequence includes snapshots of N at t ($= N_t$) and $t+1$ ($= N_{t+1}$), and that our opponent is concerned with what is left out here, i.e., with the states of N between t and $t+1$. The concern is easily handled: one can make the interval of time during which the state of N is ignored arbitrarily small. Doing so makes it the case that the states of N which had formerly been left out are now captured.

6.5.8 Objection 8

The next objection is based on an analogy: "Look, a chess game is nothing but a sequence of moves of the pieces in accord with the rules of chess. No one would claim that the ontological status of chess games is particularly mysterious. The sequence of moves in any chess game is reversible: there exists a sequence which consists in exactly the same positions on the board but in opposite order. The backward sequence is not, of course, a chess game. The 'initial' position is not the legal initial position for chess, and some pieces, e.g., pawns, will move in a way not legally permissible. So the chess game cannot be reversed, in the following sense: the reverse sequence of positions is not a legal chess game. Does one want to use Leibniz's Law to conclude that a particular chess game is not identical to the sequence of moves in it, because the one is reversible and the other isn't? I wouldn't want to conclude this, but in any case it doesn't matter. In whatever sense one cares about it, the chess game is nothing over and above the sequence of moves: if similarly consciousness were nothing over and above a computation, then the defenders of 'Strong' AI win."

The analogy offered here is clever, and if we were to accept it and its implied purely syntactical view of chess, then our master argument would probably be threatened. But if we remember that chess is a se-

[26]Readers wanting to see some of them are encouraged to consult Poundstone's (1985) classic discussion of Conway's (1982) *Game of Life*.

quence of moves made amid rules, intentions, beliefs, plans, goals, and so on, there is no reason to accept an analogy which *ab initio* commits us to a reductionist view of consciousness. In fact, the objection, upon reflection, can be exposed as fundamentally flawed — because it's easy enough to *show*, courtesy of a thought-experiment, that what the discussant takes to be absurd (a chess game is something over and above a sequence of board configurations) is *true*: Begin by fixing some legal sequence S of chess moves from the required initial situation to one in which the white king is checkmated. Suppose that this sequence is one which has never been actualized in the past. Now imagine a chess board lying on the surface of an uninhabited planet, with chess pieces beside it.[27] Next imagine that, as a wild coincidence would have it, the lifeless winds blowing over the surface of this planet happen to jumble the chess pieces over an interval of time in such a way that S is actualized. Has a chess game taken place on this planet? It would seem not (if a game took place, who won and who lost?). But then the key premise in Objection 8 (viz., a chess game is nothing over and above a sequence of moves) is in grave doubt, and hence Objection 8 evaporates.[28]

6.5.9 Objection 9

Here's the next objection: "It's not clear that the sense in which you claim computation is reversible is relevant. You identify a computation as a sequence of total states of a machine, where the total state specifies the machine state, read/write head location, and tape register contents. You then assert that there is *some other* program M' which will cause the computer to run through those states in reverse order. But the *identity* of machine states is not a fact separable from the program actually running on a machine. Machine state "1," for example, is not that very state because it has a "1" pasted to it, but because of the transitions between it and other states which are caused by the program the machine is running. So in switching from the original program M to M', we no longer have the same machine states available, so the sequence induced by M' isn't the reverse of the original. And

[27] Don't worry about how the chess board and pieces got there in the first place. Perhaps our props were launched into space centuries ago, when the human race still existed. Perhaps the board and the pieces were formed from mud by the random winds we have imagined ...

[28] It is at any rate undeniable, given this gedanken-experiment, that the ontological status of chess games *is* a bit mysterious.

one can't reverse a computation by reversing the program in any interesting way. There is, for example, a simple two-state program which will erase a sequence of 1s of any length, but one cannot 'reverse' it to get a two-state program that will write out such a sequence."

This objection saddles computationalism with **machine state functionalism** (MSF), according to which our mental states are to be identified with the *machine states* of a TM rather than the *configurations* of such a machine. Unfortunately, while it's true that the output of the algorithm $\mathcal{A}$ of Theorem 1 is never a TM which in any sense "reverses" the machine states of its input, machine state functionalism has long ago been buried; no contemporary computationalist advances this view. (The *locus classicus* of MSF is due to Putnam (1975), who has himself rejected the doctrine.) The reasons MSF is a carcass are myriad; they are nicely catalogued in Chapter 8 of (Stillings et al. 1995). One problem with MSF is the apparent unboundedness of human mental states. It has seemed to many that humans can enter any of an infinite number of mental states. (One could believe that 1 is the successor of 0, that 2 is the successor of 1, that 3 is the successor of 2, ..., and so on *ad infinitum*. And of course we would need to consider states involving not only beliefs about arithmetic, but also hopes, fears, dreams, mental images, and so on.) But every TM has a fixed and finite set of machine states (while on the other hand even tiny TMs are capable of entering an infinite number of configurations.) Another agreed-upon defect plaguing MSF is that according to it, two TMs which compute the same function f but which differ in their machine states and the arcs connecting them (to use the critic's scheme) are classified as giving rise to different cognition. But this implies that if you share with us (say) a love of climbing roses of the "Blaze" color, underlying our attitude must be one TM with the same exact states — which hasn't seemed too plausible to most.

Finally, the present objection is problematic for another reason having nothing to do with the history of computationalism: Computationalism is the view that cognition (including consciousness) is *computation*, but computation is *not* a machine state (or a collection of such states, or a collection of such states linked by arcs). Computation, in the terms our critic prefers, isn't a program; rather, computation is a program *in progress*. That is, computation is a sequence of configurations (Boolos & Jeffrey 1989, Bringsjord 1994), as we have explained above.

6.5.10 Objection 10

The final objection comes from those (e.g., Bill Rapaport) who are unusually observant: "Look closely at your Theorem 1 (or, for that matter, at Theorem 1′). It has the following abstract structure, does it not?

$$\forall x \exists y (\exists z \ldots R(z, x, y) \ldots)$$

Now, one way to read this, informally, but, as you must grant, accurately, is as

For every x, there exists a y which is the result of z's reversing x.

Your first move is to instantiate x to some stretch of P-consciousness had by some arbitrary human person (recall your boxed 'proof' from above: Bob was your arbitrary person). You then proceed to consider the reversal of that stretch *in that very same person*. But how then are you abiding by the quantifier structure in question? According to that structure, the reversed stretch of P-consciousness might take place in *another* person, no?"

This objection is easily disposed of. The problem with it is that if it is unimaginable that *you* experience a stretch of your P-consciousness in reverse, it would certainly seem to be unimaginable that, say, a *friend of yours* experience a stretch of your P-consciousness in reverse! At any rate, our opponent gives us no reason whatsoever to think that what we have discovered in the first-person case doesn't apply to others. Finally, the physics-based considerations which tell against the prospects of reversing phenomenal consciousness, catalogued above, aren't touched by noting the niceties of quantifier structure in Theorem 1.

Chapter 7

What are We? Where'd We Come From?

This chapter is divided into two parts, corresponding to the two questions that together compose the title of the chapter.

7.1 What, at Bottom, Are We?

"What, at bottom, are we?" is short for a query regarding which **ontology of mind** the supermind doctrine is to be aligned with. Following Jacquette (1994), we can say that the ontologies in question fall into three basic categories that will be familiar to most philosophers, but perhaps not to those coming to AI and/or CogSci from other disciplines: **dualism**, **eliminativism**, and **reductivism**. It will turn out that supermentalism is to be associated with relevant ontolo*gies*. More specifically, to preview the situation, it will turn out that supermentalism is to be associated with: computationalism, a standard form of reductivism that is at once **materialist** and **functionalist**, and much discussed earlier (Chapter 1); "hypercomputationalism," which is functionalist in nature but may or may not also be materialist in nature, depending upon whether or not hypercomputation is physically possible and/or physically real; and either **property dualism** or **substance** (= **agent**) **dualism**. We now proceed to explain all this talk of "isms."

First, a brutally rough review of what the highest-level "isms" mean: Dualism is the view that mind and body are distinct. Eliminativism is the view that there is no such thing as mind, and that talk of cognitive phenomena can and should be completely replaced with talk of observable behavior, brain processes, algorithms, computations, and so on. Reductivism doesn't deny the existence of mental phenomena, but claims that all true propositions making use of mentalistic terms can be expressed equivalently in a vocabulary having no such terms. Each of these three main ontologies of mind can be specialized in various ways, some of which we'll momentarily discuss.

Now for our second step. Since superminds (at the level of persons in the hierarchy of Table 1.3) are by our lights real people enjoying real "Cartesian" properties like phenomenal awareness, eliminativism can be discarded. (Our list of "axioms" regarding persons presented — and affirmed — in Chapter 1 is patently inconsistent with eliminativism.) Next, what about reductivism? Since attributes like phenomenal awareness are not only said in supermentalism to be real, but inexpressible in any third-person format, one might think that reductivism can likewise be jettisoned. However, things are a bit more tricky than they seem, for reasons pertaining to the supermind doctrine as pictured in Figure 7.1 (which, as you'll recall, we first presented in Chapter 1).

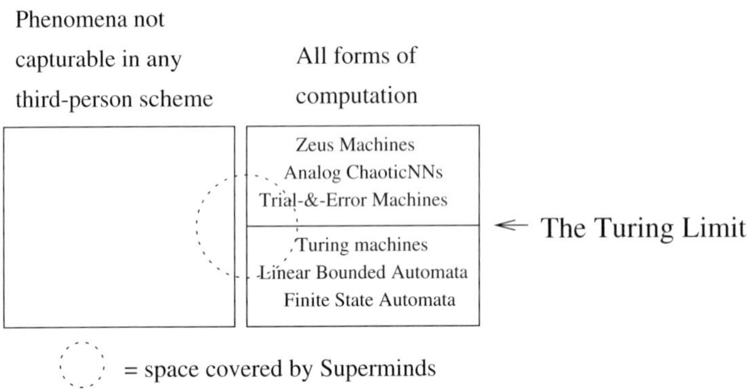

Figure 7.1: Superminds Include Parts of Three Spaces

According to this picture, as reflected in the propositional presentation of the supermind view in Chapter 1, persons as superminds make use of information processing that is presumably embodied, at least in part, in brains. This means that some supermind *cognition* conforms to reductivism. Following Jacquette (1994) again, we know that reductivism comes in three main subtypes: **behaviorism**, **materialism**, and **functionalism**. Computationalism, set out in Chapter

1, is a particular form of functionalism, and without question is the most popular ontology of mind today. Behaviorism is wholly inconsistent with looking "inside" people and cognition via introspection, armchair deduction, neuroscience, computational modeling, and so on. Since such looking is part and parcel of supermentalism, we can confidently cast behaviorism aside.[1] So what about the other two types of reductivism? These two types apply directly to two of the subspaces shown in Figure 7.1; the application is shown pictorially in a modification of this figure (see Figure 7.2). The basic idea here is as follows. Standard computationalism will apply to the cognition that is information processing below the Turing Limit. And this cognition is almost without question embodied in brain processes; the brain, in general, certainly looks to be a digital computer, or an equivalent. But we need a new form of computationalism to attach to the cognition that is hypercomputation: 'hypercomputationalism.'

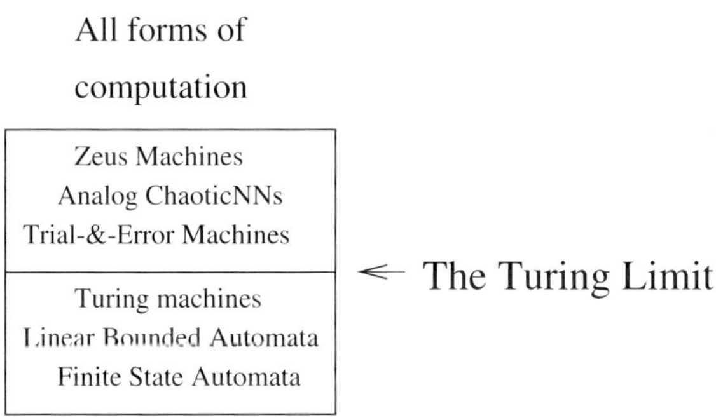

Figure 7.2: Two Spaces To Which Reductivist Schemes and Methods Can Be Applied

Unfortunately, so far we haven't really come to grips with the "What, at bottom, are we?" question. What we have at this point is a general scheme describing what some of our *cognition*, at bottom is; that is what Figures 7.1 and 7.2 show. What about *us*? What about *persons*?

In order to classify persons, let's continue to follow Jacquette (1994), who partitions dualism into three varieties: **substance** dualism, **prop-**

[1] Behaviorism, as a theory of mind, is at any rate simply dead. A relatively small number of scientists continue to use behaviorist methodologies in their empirical work (Bringsjord received a number of personal letters from practicing behaviorists expressing objections to Bringsjord & Bringsjord 1993), but as a fundamental account of mind, behaviorism is without question a carcass.

erty dualism, and **idealist** dualism (or just *idealism*). The most well-known version of idealism is due to George Berkeley (1949*b*, 1949*a*), whose famous slogan "*Esse est percipi aut posse percepere*" ("To be is to be perceived or a perceiver") encapsulates his view that only two kinds of things exist: minds that perceive, and the percepts these minds perceive. Since Berkeley denies that there are physical objects in any real sense, and this is a denial at the heart of idealism, unsurprisingly, there are few idealists around today. The two of us, along with you, certainly believe that there are rocks, trees, electrical circuits, human brains, etc., all of which underlie objects as they are experienced by persons (and other creatures). This leaves substance dualism and property dualism as the two remaining options.

Now, recall our position on personhood, as set out in Chapter 1, specifically through the PER_i theses and the list L_p. This position requires more, ontologically speaking, than what mere property dualism can offer. In this connection, note that Jacquette's (1994) own view is that though some mental properties (clustered around intentionality) are irreducibly mental (this part of his view qualifies him as a property dualist), persons are illusions — or, more felicitously put, the *self* is an illusion. For us, as superminds, human persons are as real as can be — an attitude revealed, again, by our affirming the likes of PER_i and L_p. We are either physical things that have non-physical properties that cannot be reduced to behavioral, material, or functional properties (perhaps an incoherent view), or we are non-physical objects (with such properties).[2] Bringsjord is inclined to affirm substance dualism of a type that has been articulated and defended by his teacher Roderick Chisholm. In fact, no small part of the reason Bringsjord inclines toward substance dualism is the last defense of substance dualism given by Chisholm before his death, which seems to fare very well when matched against views of the self as an extended thing. One new and fascinating argument in favor of the view that persons or selves are physical things is due to Andy Clark and David Chalmers (1998), who argue that persons are not only embodied, but *extensibly* embodied. We now proceed to investigate how Chisholm's simple souls fare against the extended (and decidedly non-super) selves propounded by

---

[2] Zenzen (along with a colleague we both share: Jim Fahey) is inclined to affirm emergentism; see (Fahey & Zenzen n.d.), wherein it is shown that emergentism, *contra* Kim (1992), in light of quantum mechanics, is immune from the objection that the intrinsic properties of macro-things must always derive from the intrinsic properties of micro-constituents.

Clark and Chalmers.

7.2 Perhaps Superminds are Simple Souls

As the bibliography for his volume (XXV) of *The Library of Living Philosophers* attests, and as his many students confirm, Roderick Chisholm grappled tirelessly for decades with problems related to personhood. In recent years Chisholm directed his attention toward the view of the self encapsulated under the phrase "simple souls." Proponents of simple souls include Augustine, Descartes, and Bolzano, but the clearest expression of the doctrine, and the most formidable defense of it, has been given by Chisholm (1991, 1989). Following the other just-mentioned famous friends of simple souls, Chisholm uses the term 'soul' or 'self' to mean the same thing as 'person'; he is thus talking about *us*: beings who persist through time and exemplify diverse mental properties — judging, wondering, desiring, believing (in the occurrent sense), wishing, hoping, enjoying — which themselves have certain key (to-be-discussed-below) properties. According to the doctrine of the simplicity of the soul, we are unextended and incorporeal substances.[3] Here is Chisholm just before defending the simplicity of the soul:

> I will defend the thesis according to which there is something that is metaphysically unique about persons: we have a nature wholly unlike anything that is known to be true of things that are known to be compound physical things. I will attempt to show how this thesis coheres with the traditional doctrine of "the simplicity of the soul." And I will argue that the doctrine of the simplicity of the soul is, in William James' terms, very much a live option. (Chisholm 1991, p. 167)

After briefly discussing the defense to which Chisholm here alludes, we present an attack on it that Chisholm never anticipated: one built from the view, recently defended by Andy Clark and David Chalmers, that selves extend beyond the human bodies in which they are usually

[3]The simplicity of the soul ought not to be confused with the doctrine of divine simplicity, a radically different notion. Note also that the use of 'soul' in Chisholm's simplicity account contrasts with its Aristotelian usage, according to which we could only be said to *have* souls, not *be* them. These issues, and others, are discussed in a recent paper by Philip Quinn (1997) on simple souls, to which Chisholm (1997) replies.

taken to be "enclosed." We then show that Chisholm and other friends of simple souls have nothing to fear from this attack.

7.2.1 Chisholm's Argument

In the space we have here, Chisholm's argument for the simplicity of the soul can't be treated in the detail it deserves, but a level of detail sufficient for the purposes at hand can probably be reached.

Chisholm starts by encouraging us to examine our mental properties — *fearing ghosts*, *intending to work another two hours*, *believing that Frege's Axiom V has been successfully replaced by the axiom schema of separation*, etc. Such examination, according to Chisholm, reveals that our mental properties have five structural properties, viz.,

(S1) Mental properties are **open**, in the sense that a property is open iff it is possibly such that, for any number n, there are n substances that have it, and n substances that don't.

(S2) Mental properties are **repeatable**, where a property is repeatable iff it is possibly such that there is something that doesn't have it, but did have it and will have it.

(S3) Mental properties are not **compositive**, where a property is compositive iff it is necessarily such that whatever is composed of things that have it is itself a thing that has it.

(S4) Mental properties aren't **divisive**, where a property is divisive iff it is necessarily such that any compound thing that has it has a proper part that has it.

(S5) Mental properties are **internal** properties of substances, where a property is an internal property of substances iff it is necessarily such that whatever has it is a substance, and either it is necessary to whatever has it or it is necessarily such that whatever has it has every open and repeatable property it implies.

Ignoring certain technical details that needn't detain us,[4] Chisholm then defines a **qualitative property** as one that is open and repeatable, neither compositive nor divisive, and internal. Now here is his argument (Chisholm 1991, p. 177) for simple souls:

[4]There are certain disjunctive properties that have all five properties on this list, and they need to be excluded from the class of qualitative properties if qualitative properties are to attach to things that are capable of thinking. Certain conjunctive properties need to be excluded for similar reasons. Like Quinn (1997), we ignore these details, which are outside the focus of this chapter.

$\mathbf{Arg}_1$

(1) We have qualitative properties.
(2) Every qualitative property that we are acquainted with is known to be possibly such that it is exemplified by simple substances.
(3) No qualitative property is known to be such that it may be exemplified by compound substances.
∴ (4) Some of our properties are known to be such that simple substances can have them and are not known to be such that compound substances can have them.
∴ (5) We have a nature which is wholly unlike the nature that anything known to be a compound physical thing is known to have.

Those who, like Chisholm himself, find the argument to be a powerful defense of simple souls, must apply disjunctive syllogism in a manner that falls beyond Arg_1 — a manner that relates directly to the two possibilities we alluded to above: persons are non-physical objects versus persons are physical objects having non-physical properties. As Chisholm writes:

> The conclusion of this argument leaves us with two possibilities: either (a) the soul is an unextended substance or (b) souls have a type of property that extended physical substances are not known to have. The latter option is defended by those who have argued that the fact of thinking indicates the presence of a peculiar type of "emergent property" in nature. (Chisholm 1991, p. 177)

The fact of the matter is that the soundness of Arg_1 is perfectly consistent with some such view as that we are extended souls (brains, perhaps, or maybe proper physical parts of brains) that have properties not known to be possessed by extended things. So for Chisholm, and for other contemporary friends of simple souls (e.g., Philip Quinn 1997), the power of Arg_1 hinges on whether or not some form of emergentism (according to which our mental properties somehow arise from the "ordinary" physical realm) is plausible. Chisholm, as you might guess, rules out option (b) from the previous quote; he then arrives at a complete argument for the doctrine of simple souls, viz., Arg_1 sound $\rightarrow$ (a) $\vee$ (b); Arg_1 sound $\wedge \neg$(b).

What rationale does Chisholm give for $\neg$(b)? Note, first, that, as Quinn (1997) seems to acknowledge, perhaps the only sort of formidable emergentism is one aligned with supervenience. Quinn appeals to a brand of supervenience due to Jaegwon Kim, defined as follows.

(D1) A-properties supervene on B-properties $=_{df}$ Necessarily, for any object x and A-property a, if x has a, then there is a B-property b such that (i) x has b, and (ii) necessarily, if anything has b, it also has a. (Kim 1984, p. 165)

The idea here would be that mental properties supervene on certain physical properties; such a scheme would lend credence to option (b). Chisholm's response is vintage Chisholm:

> If we take [this bolstering of (b)] seriously, then we should consider how we might go about applying it to some property that is quite obviously mental and not merely sensory. Let it be the property of concluding that it is about to rain. We should have at least *one* plausible version of how the explication might go. We would like to know, for example, how to complete the following thesis:
>
> (T) The mental property of concluding that it is about to rain supervenes upon the physical property of —.
>
> How are we to fill in the blank? At present, the only examples we have are make-believe, such as "The mental event supervenes on the Z excitation of the C-fibers," where one offers no real terms with which "Z excitations" and "C-fibers" might be replaced. Have there been any real attempts to answer this question? Or have there been any conflicting theories about what the answer might be? When we try to consider *all* the things that can influence one's mental life, doesn't it seem clear that *the* physical upon which *any* mental event depends would need to be put as an extraordinarily complex disjunctive property? One cannot help but wonder whether the finding of such a property would have to await the completion of the natural sciences. (Chisholm 1997, p. 69)

Perhaps you find this line of defense to be at least somewhat successful.[5] But there is *another* sort of attack on Arg_1, one Chisholm

[5] Recall that Zenzen looks favorably upon emergentism (see note 2), for reasons specified in (Fahey & Zenzen n.d.). But note that these reasons don't seem to protect against Chisholm's line of defense. This is the case because though these

never anticipated. The attack appeals to the view, recently defended by Andy Clark and David Chalmers (1998), that selves are compound physical entities extending beyond the human bodies to which they are usually taken to be limited. The counter-argument we have in mind is this one:

Arg_2

(1) We have qualitative properties.
(6) We are composed (at least in part) of compound physical objects that extend beyond our bodies.
∴ (7) $\neg(3)$, i.e., Arg_1 is unsound.

This argument is formally valid.[6] Premise (1) is of course one Chisholm himself champions in Arg_1. So if premise (6) is true, or even if it's judged to be plausible, the doctrine of simple souls would be threatened in a way that Chisholm didn't anticipate.[7] We turn now to an evaluation of the case made for (6) by Clark and Chalmers.

7.2.2 The Unexpected Attack from Extended Selves

In "The Extended Mind," Andy Clark and David Chalmers (1998) advance three main theses about the reach of mentation beyond skull and skin:

T_1 Cognition extends into the external world to include physical objects other than our bodies and body parts (e.g., pencil and paper, computers, and perhaps — in the future — neural implants and the like).

T_2 Dispositional beliefs can reside in, or at least be partly constituted by, external physical objects.

---

reasons include a principled rejection of Kim's views on the matter (views Chisholm implicitly affirms here), the fact remains that emergentism is committed to *some* kind of correspondence between the ordinary physical and the emergent. Whatever correspondence this might be, Chisholm's point about it being unspecifiable seems (to Bringsjord, anyway) to be telling.

[6] We don't consider arguments for

(6′) We are composed (at least in part) of compound physical objects.

Clearly, though, such arguments would be relevant. Such arguments are rather hard to find, however.

[7] In fact, Chisholm said: "Surely, whatever else I may be, I am not identical with any bodily thing having parts that are not shared by *this* body." (1991, p. 168)

T_3 = (6) We are composed (at least in part) of compound physical objects that extend beyond our bodies.

No doubt we aren't alone in finding this trio provocative, and the case presented on its behalf imaginative, but, as will be seen, C&C fail to deliver compelling argumentation, at least when it comes to T_3, which, for purposes of articulating and defending supermentalism, is the heart of the matter.[8]

Because the argument for T_3 has T_2 as a premise, we need to review the case for T_2. We will try to show that T_2 should probably be supplanted with a more circumspect relative, but even if we're wrong about this, C&C's argument for T_3 will still fail in light of what we uncover in our investigation of their case for T_2.

At the heart of the case for T_2 stand thought-experiments involving C&C's duo of Inga and Otto.[9] Inga hears that there is an exhibition at the Museum of Modern Art, and decides to see it. She remembers that the Museum is on 53rd Street, walks to that location, and enters. Otto, on the other hand, is plagued by Alzheimer's disease; hence he carries a notebook around in which he writes down new information as he learns it. When he needs some old information, he looks it up in his notebook. As C&C put it: "For Otto, his notebook plays the role usually played by biological memory" (Clark & Chalmers 1998, p. 12). When Otto hears about the exhibition at the Museum, he too decides to see it. So he consults his notebook, finds the address, walks to 53rd Street, and enters.

C&C claim that Otto believed the Museum was on 53rd even before he consulted his notebook, just as Inga believed the Museum was so located before accessing her memory. Since Otto is arbitrary, the idea is that we can generalize from Otto's particular dispositional belief to T_2; we readily grant this inference. But what argument do C&C give for the proposition that Otto dispositionally believes, along with Inga, that the Museum is on 53rd?

[8] T_1 strikes us as unexceptionable and routinely affirmed. The term 'cognition' is an abstract noun whose denotation, by the lights of the cognitive science literature, includes composite systems having as parts agents (conventionally delineated) *and* external objects. A mundane but convincing example is to be found in an account of arithmetical cognition — multiplication, say — in children, an account that includes not only Johnny, but Johnny's use pencil and paper to follow familiar algorithms. For textual evidence that cognition is routinely understood to be extended, consult any popular textbook in cognitive psychology; e.g., see (Ashcraft 1994).

[9] A recent movie would perhaps supply a tailor-made character to Clark and Chalmers: we refer to Lenny in *Momento*. Lenny's mind is extended if anyone's is.

This is not an easy question to answer; after much reading and reflection, we find at best two rapid enthymemes. The first is designed to counter an analysis of Otto that explains his museum attendance in terms of his occurrent desire to go to the museum, his standing belief that the museum is at the location inscribed in his notebook, and the accessible fact that the notebook says the Museum is on 53rd. This analysis — call it A_o — is said by C&C to be

> pointlessly complex, in the same way that it would be pointlessly complex to explain Inga's actions in terms of beliefs about her memory. The notebook is a constant for Otto, in the same way that memory is a constant for Inga; to point to it in every belief/desire explanation would be redundant. In an explanation, simplicity is power. (Clark & Chalmers 1998, p. 14)

Before we try to excavate the argument here, note that though they regard it to be "pointlessly complex," clearly C&C see an analysis of Inga's actions that parallels A_o. This second analysis, A_i, would presumably explain Inga's museum attendance in terms of her occurrent desire to go to the museum, her standing belief that the museum is at the location "recorded" in her memory, and the accessible fact that her memory "says" the museum is on 53rd. Let us refer to the general schema of analysis and explanation at work here as $A^\star$.

Now, what exactly is the argument for T_2? Clearly, it revolves around some such general principle as

P_1 If there are two competing explanations E_1 and E_2 for S's action a, and E_1 invokes fewer terms than E_2, then E_1 is to be preferred.

The idea would then be to instantiate E_2 to an explanation appealing to A_o, and E_1 to an explanation that treats Otto's dispositional belief as being partly constituted by his notebook and features thereof. We concede for the sake of argument that E_1 does invoke fewer terms. (We envisage a first-order formalization of the explanations, with 'terms' having the customary meaning in such systems.) But there are many serious problems. For starters, P_1 is false. To see this, suppose Jones explains the movement of your arms and hands as you play a game of bridge by appeal to a lone, super-strong gremlin pedaling inside your skull, whereas Smith explains the movement by appeal to all the customary ins and outs of anatomy and intention and so on. We would not want to say that the gremlin story is preferred, despite its simplicity. The upshot would seem to be that the competing explanations in

P_1 must both be at least genuine contenders; both must be — to use one covering term in the interests of space — plausible.

P_1 is plagued by other problems. For example, in what *context* are E_1 and E_2 to be judged? Consider again the movement of your hands and arms as cards are played and tricks are taken. Suppose again that Smith gives an explanation that mixes folk psychology with facts about neuroscience, whereas Brown accurately explains things by appeal to technical biomechanics. Which explanation is to be preferred? I should think that the answer depends on the background metric at play. For a crowd of psychologists, Smith is on the right track; for a group of biomechanical engineers, Brown is. This time the upshot seems to be that some metric needs to be included in P_1. So what we have at this point is

P_2 If there are two competing, plausible explanations E_1 and E_2 for S's action a, both to be judged under metric M, and E_1 invokes fewer terms than E_2, then E_1 is to be preferred.

Despite these and other niceties, in the end, it would appear that all that will ever follow from P_2 in C&C's argument is its instantiated consequent, viz., that from the standpoint of M we should *prefer* an explanation in which Otto's notebook partly constitutes his dispositional belief. How will it ever follow that *in fact* dispositional beliefs can reside in, or at least be partly constituted by, external objects? All that C&C will obtain via their present tack is something like

T_2' From the standpoint of M, dispositional beliefs ought to be regarded to possibly reside in, or be at least partly constituted by, external physical objects.

With M set to something like 'seeking economical folk psychological explanations,' we have in T_2' a thesis we gladly affirm. As we scour C&C's paper, we find no arguments for anything stronger.[10]

With the stage set, let's turn now to the argument for T_3 = (6), presented in its entirety in the penultimate paragraph of C&C's paper:

[10]C&C's other main argument for T_2 comes in the form of a challenge to their opponents to show that "Otto's and Inga's cases differ in some important and relevant respect" (Clark & Chalmers 1998, p. 14–15). We grant for the sake of argument that this challenge expresses the kernel of a careful and compelling argument. As will become clear, we treat Otto's and Inga's cases the same on behalf of friends of simple souls.

> What, finally, of the self? Does the extended mind imply an extended self? It seems so. Most of us already accept that the self outstrips the boundaries of consciousness; my dispositional beliefs, for example, constitute in some deep sense part of who I am. If so, then these boundaries may also fall beyond the skin. The information in Otto's notebook, for example, is a central part of his identity as a cognitive agent. What this comes to is that Otto *himself* is best regarded as an extended system, a coupling of biological organism and external resources. To consistently resist this conclusion, we would have to shrink the self into a mere bundle of occurrent states, severely threatening its deep psychological continuity. Far better to take the broader view, and see agents themselves as spread into the world. (Clark & Chalmers 1998, p. 18)

This too is painfully enthymematic, but easy enough to unpack. First, to ease exposition, reword T_2 so that (i) it drops mention of the stronger notion that dispositional beliefs *reside in* physical objects in favor of the more palatable relation of partial constitution, and (ii) in place of the modal 'can' appearing in the original version, reference is made to the type $\mathcal{T}$ of dispositional beliefs pointed to by Otto's "notebook-based" one. (Membership conditions for $\mathcal{T}$ could be extracted from the "features of of [C&C's] central cases" (1998: 17).[11]) Then the argument can be reconstructed as follows.

[11] We could say something like:

> A dispositional belief b that p for agent s at t is in $\mathcal{T}$ iff (i) the physical object in which p is represented by r is (a) a constant in the life of s and (b) directly available to s without difficulty; (ii) upon retrieving r s automatically endorses p; and (iii) at some $t' < t$ s consciously endorsed p via r.

7.3.1 Pinker's Position

Pinker unambiguously expresses his main position a number of times at the beginning of his book. For example, he tells us in the Preface that the "the mind is a system of organs of computation designed by natural selection to solve problems faced by our evolutionary ancestors in their foraging way of life" (Pinker 1997, x). And again soon thereafter in Chapter 1 we read:

> The mind is a system of organs of computation, designed by natural selection to solve the kinds of problems our ancestors faced in their foraging way of life, in particular, understanding and outmaneuvering objects, animals, plants, and other people. (Pinker 1997, p. 21)

Pinker's main thesis is thus computationalism with an evolutionary twist. Let's label this thesis $\mathcal{C}^{\mathcal{E}}$, and (following labelling established at the start of the present book) let's use $\mathcal{C}$ to denote "straight" computationalism, the view that minds are computing machines, with $\mathcal{E}$ abbreviating the proposition that minds arose from evolution by mutation and natural selection. So, $\mathcal{C}^{\mathcal{E}}$ $(= \mathcal{C} \wedge \mathcal{E})$ entails $\mathcal{C}$ $(\mathcal{E})$, and hence by *modus tollens* any argument that overthrows $\mathcal{C}$ $(\mathcal{E})$ also overthrows $\mathcal{C}^{\mathcal{E}}$. Now, as all our readers know, many do reject $\mathcal{C}$ $(\mathcal{E})$ — and on the basis of *reasoned argument*, no less. We gave a list of our arguments against $\mathcal{C}$ in the Preface. Bringsjord also rejects $\mathcal{E}$, and he is far from alone. Soulmates include Kurt Gödel, Noam Chomsky, Roger Penrose, Alvin Plantinga, Paul Davies, Edgar Allen Poe, and many others — all of whom base their rejection not on some wild-eyed emotionalism or religious fanaticism, but on cool ratiocination.[12] So, what reasoning impels Bringsjord to reject $\mathcal{E}$? This reasoning begins with "Wallace's Paradox," which stands at the heart of Pinker's book.

[12]We don't have sufficient space to discuss the ratiocination in question. Let us say only that both Pinker's *HTMW* and Dennett's like-minded *Darwin's Dangerous Idea* (*DDI*) mark in some real sense the synthesis of Darwin and Turing, and perhaps the combined genius of this august pair is so great as to bestow upon these books a *prima facie* plausibility. But such an appeal to authority, over and above the fact that it constitutes a fallacy, would be dangerous for proponents of *HTMW* and *DDI* (and, thereby, $\mathcal{C}^{\mathcal{E}}$): Gödel was as great a mind as any that has graced our planet, and certainly stands as the terrestrial god of demonstration, yet he regarded both $\mathcal{C}$ and $\mathcal{E}$ to be demonstrably false for mathematical reasons: see (Wang 1974, p. 326).

7.3.2 Wallace's Paradox

The foundation for Pinker's *HTMW* is Chapter 5, in which he presents and attempts to solve Wallace's Paradox (WP), which Pinker (rightly) regards to be a direct and singular threat to $\mathcal{C}^{\mathcal{E}}$. This paradox takes its name from the co-inventor, with Darwin, of the theory of evolution: Alfred Russel Wallace.[13] WP arises from the existence of mental "overkill:" mental powers which seem to have no explanation from the standpoint of evolution. Specifically, these powers seem to be massively superfluous for problems faced by our evolutionary ancestors in their foraging way of life (such foragers were called "savages" in Wallace's time). Wallace pointed toward his paradox when he wrote:

> Our law, our government, and our science continually require us to reason through a variety of complicated phenomena to the expected result. Even our games, such as chess, compel us to exercise all these faculties in a remarkable degree. Compare this with the savage languages, which contain no words for abstract conceptions; the utter want of foresight of the savage man beyond his simplest necessities; his inability to combine, or to compare, or to reason on any general subject that does not immediately appeal to his senses. ...
>
> ... A Brain one-half larger than that of the gorilla would ... fully have sufficed for the limited mental development of the savage; and we must therefore admit that the large brain he actually possesses could never have been solely developed by any of those laws of evolution, whose essence is, that they lead to a degree of organization exactly proportionate to the wants of each species, never beyond those wants. ... Natural selection could only have endowed savage man with a brain a few degrees superior to that of an ape, whereas he actually possesses one very little inferior to that of a philosopher.[14] (Wallace, in Pinker 1997, p. 300)

---

[13]Darwin received a rather shocking letter from Wallace in which was set out the essentials of evolution through mutation and natural selection — a letter which prompted the delaying Darwin to quickly hammer out *Origin* (1859) — and now one and a half centuries later, few and far between are the schoolchildren (or, for that matter, members of the intelligentsia) who know of Wallace. The Darwin-Wallace relationship was a rather complicated one; for two accounts, see (Desmond & Moore 1991) and (Richards 1987).

[14]That Wallace links the full power of the brains of *Homo sapiens* to the cognitive activities of philosophy is, we dare say, deliciously ironic given the the exceedingly minor roles Pinker prescribes for psychology and philosophy in *HTMW*. See his "Chapter 8: The Meaning of Life."

To his considerable credit, Pinker admits that WP is a fundamental threat to his enterprise. In fact, he calls WP a "central problem of psychology, biology, and the scientific worldview" (1997, p. 300), and a "foundation-shaking mystery" (1997) — but a mystery he promises to solve. As we'll see shortly, because we're superminds, the promise can't be kept. The problem for Pinker will be two of the three kinds of cognition comprising (see again Figure 7.1): phenomenal consciousness and free will, etc.; and hypercomputation. We move now to showing that despite Pinker's toil, WP is remains stubbornly intact.

7.3.3 Wallace's Paradox Still Stands

7.3.3.1 Clarifying Wallace's Paradox

Wallace's Paradox, as it stands, is painfully obscure, and to our knowledge it hasn't been analyzed rigorously elsewhere.[15] We therefore begin this section with an attempt to get clearer about Wallace's Paradox by way of a particular idealization. This idealization will appeal to **problems** faced by, and **powers** possessed by, **intelligent agents**, information-processing creatures that form the central subject matter of the field of AI. Powers will be intimately connected to the notion of a **logical system**. We take the time to set out this idealization in order to make possible a substantive treatment of Wallace's Paradox, and Pinker's response to it.

The "Bible" of agent technology in AI is a wonderfully written, comprehensive textbook intended to cover literally *all* of AI: Russell and Norvig's (1994) *Artificial Intelligence: A Modern Approach* (*AIMA*), the cover of which tellingly displays the phrase "The Intelligent Agent Book." The overall, informal architecture for an intelligent agent is shown in Figure 7.3. According to this architecture, agents take percepts from the environment, process them in some way that prescribes actions, perform these actions, take in new percepts, and continue in the cycle.[16]

---

[15]In addition, the version intimated in the quote above from Wallace seems to contain a premise (that the "laws of evolution ... lead to a degree of organization exactly proportionate to the wants of each species, *never beyond those wants* ..."; emphasis ours to highlight the doubtful part of this premise) that is false. This premise isn't needed in the more sophisticated version of WP that we articulate below.

[16]The cycle here is strikingly similar to the overall architecture of cognition described by Pollock (1995).

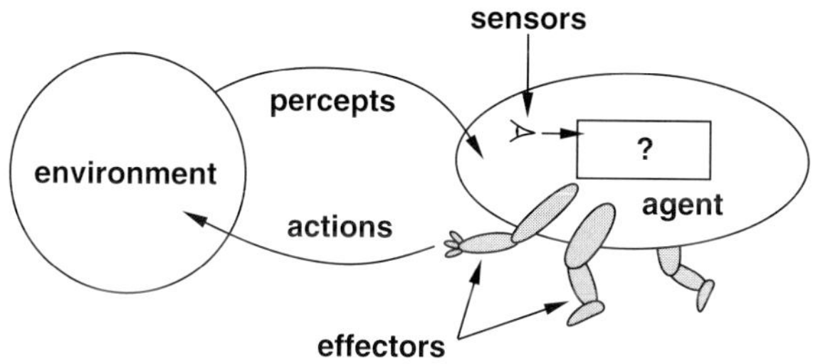

Figure 7.3: The Architecture of an Intelligent Agent (From *AIMA*.)

In our idealization, powers will be procedures that take percepts as input and generate actions, in order to secure goals. In general, problems will be conceived as decision problems: Given percepts, which actions should be taken? Examples will follow shortly, but first we need to have on hand the notion that agents, powers, and problems fall along continua. We begin with a brief discussion of the range of agents.

In *AIMA*, intelligent agents fall on a continuum from least intelligent to more intelligent to most intelligent. The least intelligent artificial agent is a "TABLE-DRIVEN-AGENT," the program (in pseudocode) for which is shown in Figure 7.4. Suppose that we have a set of actions each one of which is the utterance (through a suitable effector) of a color name ("Green," "Red," etc.), and suppose that percepts are digital expressions of the color of an object taken in by the sensor of a table-driven agent. Then given Table 7.1 our simple intelligent agent, running the program in Figure 7.4, has (e.g.) the power to utter (through a voice synthesizer, assume) "Blue" if its sensor detects 100. The problem is just to announce the right color for the right percept. Of course, this is a stunningly dim agent. What about smarter agents?

```
function TABLE-DRIVEN-AGENT(percept) returns action
  static: percepts, a sequence, initially empty
          table, a table, indexed by percept sequences, initially fully specified

  append percept to the end of percepts
  action ← LOOKUP(percepts, table)
  return action
```

Figure 7.4: The Least Intelligent Artificial Agent (From *AIMA*.)

Table 7.1: Lookup Table for TABLE-DRIVEN-AGENT

Percept	Action
001	"Red"
010	"Green"
100	"Blue"
011	"Yellow"
111	"Black"

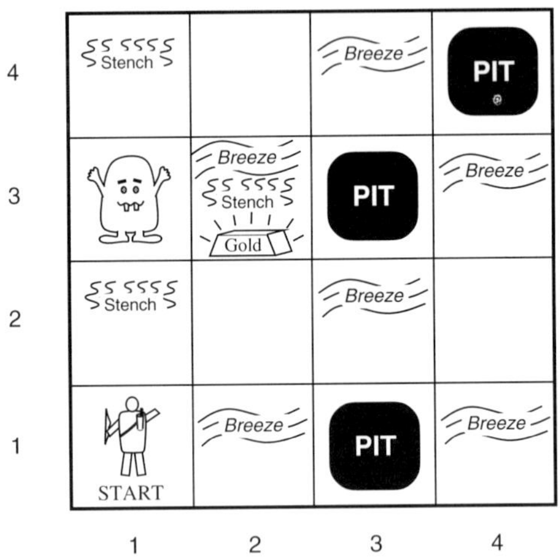

Figure 7.5: A Typical Wumpus World (From *AIMA*.)

A smarter agent would, for example, be one clever enough to leave home, find food and kill prey, and return home better off; such an agent would be doing things at least in principle relevant to our foraging ancestors. There is a fairly well-known testbed in AI for presenting intelligent agents with challenges at least somewhat like these: the so-called "wumpus world." An example of such a world is shown in Figure 7.5. The objective of the agent that finds itself in this world is to find the gold and bring it back without getting killed. As Figure 7.5 indicates, pits are always surrounded on three sides by breezes, the wumpus is always surrounded on three sides by a stench, and the gold glitters in the square in which it's positioned. The agent dies if it enters a square with a pit in it (interpreted as falling into a pit) or a wumpus in it (interpreted as succumbing to an attack by the wumpus). The percepts for the agent can be given in the form of quadruples. For example,

(Stench,Breeze,Glitter,None)

means that the agent, in the square in which it's located, perceives a stench, a breeze, a glitter, and no scream. A scream occurs when the agent shoots an arrow that kills the wumpus. There are a number of other details involved, but this is enough to demonstrate how command over the logical system known as the **propositional calculus**[17] can give an agent a level of power that will allow it to succeed in the wumpus world. For the demonstration, let $S_{i,j}$ represent the fact that there is a stench in column i row j, let $B_{i,j}$ denote that there is a breeze in column i row j, and let $W_{i,j}$ denote that there is a wumpus in column i row j. Suppose now that an agent has the following five facts in its *KB*.

1. $\neg S_{1,1} \wedge \neg S_{2,1} \wedge S_{1,2} \wedge \neg B_{1,1} \wedge B_{2,1} \wedge \neg B_{1,2}$
   - In English: "There is no stench at location 1 1, no stench at location 2 1, no stench at location 1 2, no breeze at location 1 1, no breeze at location 2 1, and no breeze at location 1 2."
2. $\neg S_{1,1} \rightarrow (\neg W_{1,1} \wedge \neg W_{1,2} \wedge \neg W_{2,1})$
   - In English: "If there is no stench at location 1 1, then the wumpus is not at location 1 1, not at 1 2, and not at 2 1."
3. $\neg S_{2,1} \rightarrow (\neg W_{1,1} \wedge \neg W_{2,1} \wedge \neg W_{2,2} \wedge \neg W_{3,1})$
4. $\neg S_{1,2} \rightarrow (\neg W_{1,1} \wedge \neg W_{1,2} \wedge \neg W_{2,2} \wedge \neg W_{1,3})$
5. $S_{1,2} \rightarrow (W_{1,3} \vee W_{1,2} \wedge W_{2,2} \wedge W_{1,3})$

Then in light of the fact that $W_{1,3}$ can be deduced from $\{1, \ldots, 5\}$ in the propositional calculus,[18] the agent can come to know that the wumpus is at location column 1 row 3 — and this sort of knowledge directly enables the agent to succeed. So, fundamentally, robots able to

[17] As a reminder, the propositional calculus allows for simple declarative statements to be represented by propositional variables $p_1, p_2, \ldots$ which can be put together with help from the familiar truth-functional connectives $\neg$ (not), $\vee$ (or), $\wedge$ (and), $\rightarrow$ (if then), $\leftrightarrow$ (if and only if). (Of course, the form of the propositional variables can vary in accordance with desired expressivity. For example, variables can be set up to form as mnemonics, as in the wumpus world situation reasoned about momentarily.)

[18] The proof is left to sedulous readers.

successfully negotiate the wumpus world have a power corresponding to reasoning in a particular logical system: the propositional calculus.

Figure 7.6: A Real-Life Wumpus-World-Winning Robot in the Rensselaer AI & Reasoning Laboratory. *(Observant readers may note that the wumpus here is represented by a figurine upon which appears the (Vulcanized) face of the Director of the RAIR Lab: Bringsjord.)*

Perhaps we should confess that though the wumpus world can be negotiated by an agent operating merely at the level of the propositional calculus, full first-order logic does make things much easier. There is a way you can see this for yourself, because a former student of Bringsjord's, John Escobar, has written a program that allows a user to play the wumpus world game over the web. This program works with the assistance of an automated theorem-prover which can help the player deduce the positions of dangerous objects. The particular theorem-prover the program uses is the popular (among AI enthusiasts) OTTER, and the representational system used is the situation calculus. An excerpt of OTTER's output appears in Figure 7.8, in which it was asked to deduce the location of the wumpus in the scenario pictured in Figure 7.7. The player, represented by the happy face, smells stench, which is a sign that the Wumpus is near. Knowing that it cannot be North or West of this location (because these spaces have already been explored), and that it cannot be South because of the edge of the dungeon, the program deduces that the Wumpus is located to the East, in the lowest and rightmost square. This sort of deduction can be used for any number of pits as well, and is easily extensible to other objects in similar situations. Escobar's program is accessible over the web: see `http://kryten.mm.rpi.edu/otter`. (Simply email Bringsjord if you have a problem.)

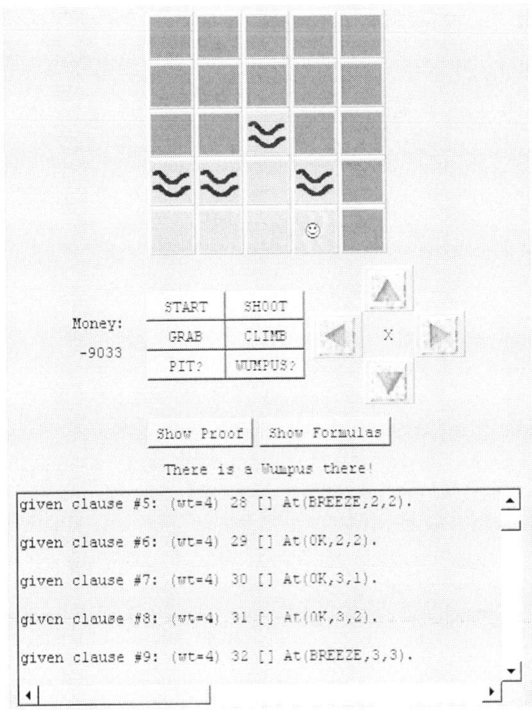

Figure 7.7: Deducing the location of the Wumpus

At this point we have enough context to set out the idealization for clarifying Wallace's Paradox. Suppose that problems start with the utterly easy and pass gradually on up to the tremendously hard in a continuum:

$$\mathcal{P} = p_1, p_2, p_3, \ldots$$

At the easy end of the spectrum we might have a straightforward task of getting from point A to point B, or of detecting light and moving toward it, or perhaps the problem our table-driven agent solved. Somewhere after problems like these we would find problems like those involved in the wumpus world, and those involved in the real-world analogue of a foraging way of life. Moving toward the part of $\mathcal{P}$ populated by truly difficult problems, we would have problems requiring more powerful logical systems than the propositional calculus (e.g., first-order logic), and then we would head on up toward problems solved by novelists, professional logicians, mathematicians, and so on.

```
Length of proof is 1.  Level of proof is 1.

---------------- PROOF ----------------
22 [] -At(STENCH,x,y)| -At(OK,INC(x),y)|
-At(OK,x,INC(y))| -At(OK,DEC(x),y)| -At(OK,x,DEC(y)).
30 [] At(OK,3,1).
35 [] At(OK,4,2).
36 [] At(STENCH,4,1).
40 [] INC(1)=2.
46 [] INC(4)=5.
59,58 [] DEC(4)=3.
65,64 [] DEC(1)=0.
79 [] At(OK,x,0).
82 [] At(OK,5,1).
87 [para_from,40.1.1,22.3.3,demod,65,unit_del,79]
-At(STENCH,x,1)| -At(OK,INC(x),1)| -At(OK,x,2)| -At(OK,DEC(x),1).
304 [para_into,87.2.2,46.1.1,demod,59,unit_del,36,82,35,30] $F.
------------ end of proof -------------
```

Figure 7.8: Output from OTTER

For an example of a problem requiring a logical system as powerful as first-order logic,[19] We offer the "Dreadsbury Mansion Mystery," which would seem to at least to a slight degree resemble problems solved by "real-world" detectives.[20] Here is the mystery.

> Someone who lives in Dreadsbury Mansion killed Aunt Agatha. Agatha, the butler, and Charles live in Dreadsbury Mansion,

[19] As was the case with respect to the propositional calculus, we assume our readers to be familiar with first-order, or ordinary quantifier, logic, which adds to the machinery of the propositional calculus the quantifiers $\forall x$ (for all x) and $\exists x$ (there exists an x), and also predicate letters to represent individual properties. With these additions, statements in natural language impossible to capture in the propositional calculus (e.g., 'All students are bored by at least two professors") can be captured in first-order logic (e.g., the example in the previous parenthetical: $\forall x(Sx \rightarrow (\exists y \exists z(y \neq z \wedge Py \wedge Pz \wedge Byx \wedge Bzx)))$).

[20] We've "Norwegianized" the problem as it appears in (Pelletier 1986). For a fuller discussion of such real-world reasoning in connection with AI and logical systems see (Bringsjord & Ferrucci 2000).

> and are the only people who live therein. A killer always hates his victim, and is never richer than his victim. Charles hates no one that Aunt Agatha hates. Agatha hates everyone except the butler. The butler hates everyone not richer than Aunt Agatha. The butler hates everyone Agatha hates. No one hates everyone. Agatha is not the butler.
>
> Now, given the above clues, there is a bit of a disagreement between three Norwegian detectives: Inspector Bjorn is sure that Charles didn't do it. Is he right? Inspector Reidar is sure that it was a suicide. Is he right? Inspector Olaf is sure that the butler, despite conventional wisdom, is innocent. Is he right?

If you would like to try to solve this problem on your own, give it a shot now, and check your reasoning by looking at the footnote referenced at the end of this sentence.[21]

Of course, in Chapter 3, we gave a series of problems (illusions, as they were called) that might be worth your while to revisit now in connection with cognition at level of FOL. For the sake of completeness, here is an even more difficult problem that one of us (Bringsjord) is in the process of gathering subject data on. In this case we don't provide the answer; be forewarned that this is a rather tricky problem. (If you get desperate, email Bringsjord for a solution.) Here it is:

> Problem #1 Suppose that
>
> (1) Charles loves Darla.
>
> (2) No one is such that, if he or she loves Alvin, everyone loves Alvin.

[21]Here is a solution in the form of an informal proof that appeals to rules of inference in first-order logic. Suppose Charles is the killer. Then, given that a killer always hates his victim, Charles hates Agatha. We know that Agatha hates herself (because she hates everyone save for the butler). But then we have a contradiction: Charles hates no one that Aunt Agatha hates (given), and yet Charles, we've just deduced, hates Agatha, as does she. Ergo, by *reductio ad absurdum*, Charles didn't kill Agatha. Suppose, on the other hand, that the butler killed Agatha. Then the butler hates Agatha, and is not richer than Agatha. Since we are given that the butler hates everyone not richer than Agatha, the butler hates himself (by universal quantifier elimination). The butler also hates Charles, because Agatha hates Charles. But then someone (the butler) hates everyone — a contradiction. Therefore, the butler didn't kill Agatha. Hence (by disjunctive syllogism) Agatha did kill herself; and so all the detectives are correct. This informal proof can easily be specified in the form of a formal proof in first-order logic, given obvious symbolization. E.g., for starters, 'No one hates everyone' could be $\neg\exists x(Hxa \land Hxb \land Hxc)$.

Given this two-part supposition, can one prove that Darla loves Charles? Provide not only your answer, but also a detailed justification.

Problem #2 Suppose that

(1) Charles loves Darla.

(2) No one is such that, if he or she loves Alvin, everyone loves Charles.

Given this two-part supposition, can one prove that Darla loves Charles? Provide not only your answer, but also a detailed justification.

Problem #3 Suppose that

(1) Charles loves Darla.

(2) No one is such that, if he or she loves Darla, everyone loves Charles.

Given this two-part supposition, can one prove that Darla loves Charles? Once again, provide not only your answer, but also a detailed justification.

And now, what would be an example of a *truly* difficult problem, and its solution? We would suggest the problem Gödel solved by way of his famous first incompleteness theorem (= Gödel I). One of us has elsewhere explained in considerable technical detail why the reasoning in Gödel's solution can't even be approximated by today's automated theorem provers (Bringsjord 1998*b*). (Note that this result is quite different than the Penrose-alleged (1994) implication of $\neg\mathcal{C}$ from Gödel I. We treated this implication, of course, in Chapter 2.) Unfortunately, to specify this example here isn't practicable: there is just too much knowledge and expertise presupposed by the full presentation of such a thing, and too little space available here — and hence we refer motivated readers to any or all of (Boolos & Jeffrey 1989, Ebbinghaus et al. 1984, Smullyan 1992). So for present purposes we'll have to make do with a (barbaric!) approximation of the problem Gödel cracked, and the solution he provided. (Actually, we'll leave the solution to you, so that you may gain some appreciation for the cognition involved.[22] Bringsjord regularly gives this problem to students at the start of a course in which they learn Gödel I.) Here, in a form slightly adapted from Raymond Smullyan (1992), is the approximation of the problem Gödel tackled, hereafter referred to simply as $\mathcal{G}$.

[22]A solution can be found in (Smullyan 1992).

$\mathcal{G}$

Suppose there is a machine $\mathcal{M}_0$ which prints out various expressions built from the following five symbols:

$$\sim P\, M\, (\,)$$

By an **expression** we mean any finite non-empty string built from these five symbols. (For example, $PPPPPPMM(($ is an expression, as is $\sim P(P)$.) An expression is called **printable** if the machine can print it. We assume that $\mathcal{M}_0$ is programmed so that any expression it can print will be printed sooner or later.

The **mirror** of an expression ϕ is the expression $\phi(\phi)$ — e.g., the mirror of $P\sim$ is $P\sim(P\sim)$. A **sentence** is an expression having one of the following four forms:

1. $P(\phi)$
2. $PM(\phi)$
3. $\sim P(\phi)$
4. $\sim PM(\phi)$

P stands for "printable;" M stands for "the mirror of" and $\sim$ stands (as it often does in logic) for "not." Hence we define $P(\phi)$ to be **true** iff ϕ is printable. We define $PM(\phi)$ to be true if the **mirror** of ϕ is printable. We call $\sim P(\phi)$ true iff ϕ is not printable, and $\sim PM(\phi)$ is defined to be true iff the mirror of ϕ is not printable.

We are given that the machine $\mathcal{M}_0$ is accurate in that all sentences printed by the machine are true. So, for example, if the machine ever prints $P(\phi)$, then ϕ really is printable (i.e., ϕ will be printed by $\mathcal{M}_0$ sooner or later). Also, if $PM(\phi)$ is printable, so is $M(\phi)$.

Suppose ϕ is printable. Now, as a warm-up to the official problem:

- Do we then know that $P(\phi)$ is printable?

No; here's why. If ϕ is printable then $P(\phi)$ is certainly *true*, but we are not given that the machine is capable of printing *all* true sentences — only that the machine never prints any false ones. And now for your two-part problem:

1. Is it *possible* that the machine *can* print all true sentences? If "No," provide a true sentence S_g that the machine cannot print, and prove that you are correct.

2. What about $PM(\sim PM)$ — can *it* be printed? Prove that your assertion is true.

This is a tricky problem; solving it would seem to require the ability not only to reason *in* first-order logic (and possibly in other logical systems as well), but also the ability to reason *about* first-order logic (or at least elements thereof), and also, specifically, the ability to reason about self-referential formulas. (If you're skeptical, if you don't believe that these abilities are required, we'd love to receive a solution that vindicates your position.)

Now let's return to the second spectrum we need, one for the powers side. Here we similarly invoke a continuum

$$\mathcal{O} = o_1, o_2, o_3, \ldots$$

of powers, running from the simplest of powers (like that had by the table-driven agent), to more impressive ones (as in those possessed by robots able to thrive in the wumpus world, e.g., reasoning in a fragment of the propositional calculus), to powers needed to solve the Dreadsbury Mansion Mystery and the illusions of Chapter 3 and the three-part "Darla" problem we presented above, to powers (e.g.) professional mathematicians and logicians exploit. If you're tackling (or have tackled) $\mathcal{G}$ in earnest, you have a sense for what powers appear at the upper end of the continuum. Specifically, your sense should be that in order to solve even problems as easy as $\mathcal{G}$ (easy relative to the problems solved by professional logicians and mathematicians) you may need to move between, create, reason about (etc.) many different logical systems.

Next we invoke a function e that maps $\mathcal{P}$ to the power set of $\mathcal{O}$ ($2^{\mathcal{O}}$). The idea here is that this function yields, for a given problem p_i, the stretch of contiguous o_i's that are powers that might possibly be "crafted" by evolution to solve p_i.

At this point we can set out Wallace's Paradox in the form of the sort of deductive progression traditionally used in carefully setting out paradoxes:

Wallace's Paradox

(1) If people evolved, then for every power o_i possessed by our foraging ancestors, there exists a problem p_k faced by them or their predecessors such that $o_i \in e(p_k)$.

(2) It's not the case that for every power o_i possessed by our foraging ancestors, there exists a problem p_k faced by them or their predecessors such that $o_i \in e(p_k)$.

(3) ∴ People didn't evolve. (from 1 and 2)

Of course, the idea is that this constitutes a *paradox* because there is supposed to be consensus that people evolved. That is, those who affirm $\mathcal{E}$ (or $\mathcal{C}^{\mathcal{E}}$) are faced with having to swallow a contradiction (while those who are skeptical about $\mathcal{E}$ see in WP a direct argument for $\neg\mathcal{E}$, and therefore for $\neg\mathcal{C}^{\mathcal{E}}$).

7.3.3.2 Classifying Reaction to Wallace's Paradox

Is Wallace's Paradox sound? Well, there's no question that the argument is formally valid: the core inference is *modus tollens*; and it's clear that a formal version of the argument (represented in first-order logic, e.g.) would certify the quantifier reasoning used.[23] But what about the premises? Are they true? There would seem to be four main alternatives in the face of this question; they are shown in Table 7.2.[24] In two of the answers, Exotic Natural Forces and Theism, the premises are accepted; given that the argument is formally valid, this amounts to accepting (3), the conclusion. In each of the other two answers, one of the premises is rejected. Pinker himself rejects premise (2), under the view we call 'Reductionism.' Gould rejects premise (1), under the view known as 'Exaptationism.' According to Reductionism, even the greatest mental powers we have can be reduced to powers used directly by our foraging ancestors in their concrete behaviors. According to Exaptationism, while it's true that the powers possessed by our foraging ancestors far exceeded the problems that needed to be solved to survive in their environment, evolution fortuitously gave them powers that *we* can use to improve *our* lot.

These, then, are the four alternatives. Which one should a rational person affirm? Well, to start, Pinker himself rejects Exaptationism. He writes:

[23] At least potentially, this may place Pinker in a rather ironic situation. As our analysis has just disclosed, in order to grapple with the issues raised by Pinker's book (i.e., WP!), one must use powers high on the spectrum of powers we've invoked. It would seem, specifically, that these powers must make use of parts of a logical systems at and above first-order logic — and yet Pinker seems to limit our foraging ancestors to reasoning in first-order logic (as we soon explain).

[24] An excellent discussion of the Exotic Natural Forces and Theism views, replete with references and quotes (from, in some cases, unpublished conversations) involving Penrose, Chomsky, and Davies can be found in (Dennett 1995*b*). Plantinga's attack on $\mathcal{E}$ can be found in (Plantinga 1993).

Table 7.2: Reaction to Wallace's Paradox

View	Reductionism	Exaptationism	Exotic Nat. Forces	Theism
Resp.	Rejects (2)	Rejects (1)	Accepts (3)	Accepts (3)
Prop.	Pinker	Gould	Penrose, Chomsky	Plantinga, Davies

> Stephen Jay Gould, in an illuminating essay on Darwin and Wallace, sees Wallace as an extreme adaptationist who ignores the possibility of exaptations: adaptive structures that are "fortuitously suited to other roles if elaborated" (such as jaw bones becoming middle-ear bones) and "features that arise without functions ... but remain available for later co-optation" (such as the panda's thumb, which is really a jury-rigged wristbone).
>
> > Objects designed for definite purposes can, as a result of their structural complexity, perform many other tasks as well. A factory may install a computer only to issue the monthly pay checks, but such a machine can also analyze the election returns or whip anyone's ass (or at least perpetually tie them) in tic-tac-toe.
>
> I agree with Gould that the brain has been exapted for novelties like calculus or chess, but this is just an avowal of faith by people like us who believe in natural selection [$= \mathcal{E}$]; it can hardly fail to be true. It raises the issue of who or what is doing the elaborating and co-opting, and why the original structures were suited to being co-opted. The factory analogy is not helpful. A computer that issues paychecks *cannot* also analyze the election returns or play tic-tac-toe, unless someone has reprogrammed it first. (Pinker 1997, p. 301)

Here we're with Pinker to a significant degree, and we think we all should be: he encapsulates some devastating objections. It's easy to see that Exaptationism simply amounts to a leap of self-serving faith. After all, the exaptationist can *always* reply in Gould-like fashion to any candidate for a counter-example to $\mathcal{E}$. Faced with an astonishing power $o^\star$ possessed by a creature C able to thereby solve a problem $p^\star$ now, where $p^\star$ has no precedent in C's ancestors, the exaptationist can *always* claim that $o^\star$ was present but untapped in one or more of these ancestors, and that its presence there was due to good luck. By the lights of Exaptationism, then, $\mathcal{E}$ is unfalsifiable, which would take this view outside the domain of science and philosophy. As Pinker points out, part of the root problem infecting such claims is that they leave aside a meaningful explanation as to "who or what" is responsible for these fortuitous powers.[25]

Pinker is also right that the factory analogy fails. This point against Gould can be broadened and sharpened: Even a simple multicellular organism includes "biological computers" sufficient to allow for the running of any Turing-computable function, given an appropriate program. After all, an abacus, with a suitable set of instructions for how to manipulate beads in it, can compute any Turing-computable function.[26] But again, where would the instructions come from? Who or what would be responsible?

7.3.3.3 The Problems Infecting Reductionism

Now let's turn to Pinker's response to WP, which he gives on pages 358–360. His response, which we've labeled 'Reductionism' (recall Table 7.2), amounts to the claim that in the minds of our foraging ancestors were the very same fundamental powers that (e.g.) professional logicians of today exploit. Whereas our ancestors used these powers to do such things as track and kill animals, today people like Gödel use such powers to solve problems hinted at by our $\mathcal{G}$. Pinker writes:

> The answer to the question [that sums up WP] "Why is the human mind adapted to think about arbitrary abstract entities?" is that it really isn't. Unlike computers and the rules of mathematical logic, we don't think in F's and x's and y's. We have inherited a pad of forms that capture the key features of encounters among objects and forces, and the features of other consequential themes of the human condition such as fighting, food, and health. By erasing the contents and filling in the blanks with new symbols, we can adapt our inherited forms to more abstruse domains. (Pinker 1997, p. 358)

Unfortunately for Pinker, his Reductionism fares no better than Exaptationism. In fact, the former is plagued by some of the same fatal problems as those that explode the latter. For example, take the issue of falsifiability. If Reductionism is is to have any bite at all, it must amount to something that could, at least in principle, be overthrown.

[25]Jim Fahey has pointed out to us that such condemnations of Exaptationism probably presuppose the Principle of Sufficient Reason. Though Bringsjord doesn't believe this is the case, he happens to affirm certain versions of PSR.

[26]The proof can be found in (Boolos & Jeffrey 1989).

Suppose, once again, that we observe, today, an astonishing power $o^\star$ possessed by a creature C able to bring this power to bear in solving a problem $p^\star$, where $p^\star$ has no precedent in C's ancestors. Is it enough, given this observation, for the proponent of Reductionism to say "Oh, no problem, for you see, $o^\star$ is really just an ability to manipulate a pad of forms used in epochs gone by to solve much simpler problems." It can't be enough to simply *say* this kind of thing — because if so, Reductionism is no more falsifiable than Exaptationism: one could say this about any $p^\star$ and $o^\star$. So, let's try to give Pinker a fighting chance by fleshing out Reductionism at least a bit. We can do so by appealing to the machinery we set out earlier.

What kind of powers, then, would be associated with this wondrous "pad of forms" Pinker alludes to? What kind of logical system would the slots be in? For example, the propositional calculus has propositional variables for slots, but these slots and this system aren't sufficient for syllogistic reasoning, which people (to a degree) do *naturally* do. For example, we know that the syllogism (Barbara)

$$\frac{\begin{array}{c}\text{All } A\text{s are } B\text{s.}\\ \text{All } B\text{s are } C\text{s.}\end{array}}{\text{All } A\text{s are } C\text{s.}}$$

is valid, but a "pad of forms" corresponding to the propositional calculus is inadequate to represent this syllogism. For to represent it with such a "pad" would be to plug the statements within it into propositional variables, but the schema that results, viz.,

$$\frac{\begin{array}{c}p\\ q\end{array}}{r,}$$

is about as invalid as deductive reasoning can get.

So what sorts of "blanks" are we talking about? What sorts of blanks and forms and slots does Pinker have in mind? He doesn't tell us; he tells us absolutely nothing about these schemes. And this silence is quite astonishing. After all, given the underlying structure of *HTMW*, the key issue has now boiled down to whether or not Reductionism is a cogent response to Wallace's Paradox. (Of course, it wouldn't help Pinker in the least if either of the remaining two responses is correct.) If it *isn't* a cogent response, not only does the book itself fail, but by Pinker's lights, *psychology* fails. Darwin, faced with WP (delivered to Darwin by Wallace himself), had confidently proclaimed that before long the paradox would be solved by psychology, as this field had now been placed by the theory of evolution on "a new foundation." If Reductionism doesn't solve WP, Darwin looks like little more than a wishful thinker. So, can some set of "blanks" of the type Pinker seems to have in mind do the trick, and save Darwin? We don't think any such set could possibly allow for the kind of cognition at the core of logic and mathematics. In the next section, we explain why.

7.3.3.3.1 Reductionism Can't Reduce Infinitary Reasoning

The basic idea is that the powers of superminds associated with hypercomputation can't be reduced. To pick an example from earlier in our book, consider the infinitary reasoning discussed in Chapter 3. Can this reasoning be captured by filling in the kind of blanks Pinker points toward? No.

Pinker's Reductionism, when charitably understood, can do no better than appeal to first-order logic as the fleshing out of this mysterious "pad of forms." At best, Pinker seems to have this logic in mind when, in the quote above, he says "Unlike computers and the rules of mathematical logic, we don't think in F's and x's and y's" (1997, p. 358). Computers do operate at the level of first-order logic. In fact, mathematically represented, computation can be completely carried out in terms of first-order logic; we explained why this is so, recall, in Chapter 3. But the "pad of forms" that must be used to explain the behavior of logicians and mathematicians who prove things in and about *infinitary* logics is provably beyond first-order logic. One such logic, as we have seen, is $\mathcal{L}_{\omega_1\omega}$.

The basic idea behind $\mathcal{L}_{\omega_1\omega}$, recall, is straightforward. This system allows for infinite disjunctions and conjunctions,[27] where these disjunctions and conjunctions are no longer than the size of the set of natural numbers (let's use ω to denote the size of the set of natural numbers).[28] Here (again) is one simple formula in $\mathcal{L}_{\omega_1\omega}$ which is such

[27] As we pointed out earlier, even finitary logics have underlying alphabets that are infinite in size (the propositional calculus comes with an infinite supply of propositional variables). $\mathcal{L}_{\omega_1\omega}$, however, allows for formulas of infinite length, and for infinitely long derivations.

[28] As we said in Chapter 3, our book isn't the place to baptize some readers into the world of cardinal numbers. Hence we leave the size implications of the subscripts in '$\mathcal{L}_{\omega_1\omega}$,' and other related niceties, such as the precise meaning of 'ω' therein, to the side. For a comprehensive array of the possibilities arising from varying the subscripts, see (Dickmann 1975).

that any world that models it is finite, and any (non-empty) finite model models it:

$$\bigvee_{n<\omega} \exists x_1 \ldots \exists x_n \forall y (y = x_1 \vee \ldots \vee y = x_n).$$

This formula is an infinite disjunction; each disjunct has a different value for n. (One such disjunct is

$$\exists x_1 \exists x_2 \forall y (y = x_1 \vee y = x_2),$$

which says, put informally, there exist at most two things x_1 and x_2 with which everything in the domain is identical, or there are at most two things in the domain.) As discussed in Chapter 3, it is a fact, noted in various textbooks, that the proposition captured by this formula cannot be captured by a formula in a system at or below first-order logic. Since the behavior of some logicians and mathematicians centers around infinitary reasoning that can be accurately described only by formalisms that include such formulas (i.e., formalisms like $\mathcal{L}_{\omega_1 \omega}$), Reductionism is doomed.[29]

Before concluding this section perhaps we should repeat here something we said at the beginning of Chapter 3, viz., that we have focussed on a particular form of infinitary reasoning because it happens to be one we are familiar with. Our case could be made in terms of all sorts of typical mathematical problems. To take an example we happen to be looking at in connection with the busy beaver problem (which was touched upon in Chapters 3 and 6), the problem of trying to determine if the set of twin primes is infinite is not yet solved.[30]

7.3.3.3.2 The Second Problem Plaguing Reductionism But even if we suppose that by some miracle Pinker's Reductionism can

[29]Ebbinghaus et al. (1984) hold that all of mathematical reasoning can be carried out in $\mathcal{L}_I$, because all "background logic" can be carried out in this system, but they offer no argument for this claim.

[30]Two "contiguous" odd numbers that are both primes are **twin** primes. Brun (1919) proved that the sum of the reciprocals of twin primes,

$$B = \left(\frac{1}{3} + \frac{1}{5}\right) + \left(\frac{1}{5} + \frac{1}{7}\right) + \left(\frac{1}{11} + \frac{1}{13}\right) + \left(\frac{1}{17} + \frac{1}{19}\right) + \ldots,$$

is convergent, and some of our colleagues at Rensselaer, Fry, Nesheiwat & Szymanski (2002), have recently made good progress toward determining this number, which is known as Brun's Constant (Brent 1975).

accommodate such powers as those tapped by logicians and mathematicians reasoning in and about infinitary logics, that such reasoning can be reduced to Pinker's "slot system," his response to WP still fails. The reason is that this slot system, whatever it is, didn't need to be such that it *could* be extended to handle such things as infinitary logic. Even if one supposes that our use of (say) infinitary logic is somehow an inscrutable modification of first-order logic, there is a problem facing Pinker (and Darwin): What survival value did this modifiability have at the time our foraging ancestors were running about? Our foraging ancestors could have done just dandy with a slot system frozen at the level of expressiveness required for finding berries and killing animals. Put in terms of the wumpus world discussed earlier, these foragers could have been like the robots in Bringsjord's RAIR Laboratory: they are able to succeed in the wumpus world by way of inflexible powers *specifically engineered* for this world (see Figure 7.6 for a look at such a robot). Pinker's problem here is once again essentially the same as one he showed to be infecting Gould's Exaptationism. Recall Gould's defective factory analogy, intended by him to support the notion that as the computers driving a payroll system could be harnessed to analyze election returns, evolution could have given our foraging ancestors powers for tracking animals that could later be used for the tensor calculus. Just as here Gould fails to understand the profound differences between programs for payroll systems versus those for election analysis, Pinker fails to understand the vast differences between programs dedicated to securing success in the foraging game but not (say) in the tensor calculus game, versus general-purpose programs able to secure success in *both* games.

The point here would seem to be one that AI and robotics makes even tougher for Pinker. We say this because progress in these fields would seem to be fairly interpreted as pressing against Pinker the disturbing possibility that we will soon be able to create a robot able to hunt and gather and communicate just as well as early *Homo sapiens* — but these robots will utterly lack the capacity to do mathematics of a sort suggested by $\mathcal{G}$, and demanded by open questions regarding $\mathcal{L}_{\omega_1\omega}$. Why weren't our ancestors like these robots? Why did they need this strange and wondrous capacity for such things as Gödelian results? These are questions Pinker doesn't, and probably can't, answer.

Along the same lines, what is Pinker to do with the capacity of superminds to enjoy subjective awareness? Pinker leaves discussion of such powers till the final, brief chapter of his book. If you measure the importance of such phenomena to Pinker's case for $\mathcal{C}^{\mathcal{E}}$ by how much space and energy he gives them, the result can be none other than that such phenomena are very nearly irrelevant. Nothing could be further from the truth, for two reasons. The first reason is simply that though Pinker may want to put phenomenal consciousness and the like in the attic while he proceeds to polish up the part of the house that "matters" (the part that deals with perception, emotion, reasoning, and so on), the brute fact of the matter is that no mental power or property is more important to us humans than consciousness. Indeed, the only reason any of us choose to go on living is to try and see to it that we and others experience certain sorts of desirable conscious states! If you knew that starting tomorrow at 6 a.m. "you" were going to forever more have the consciousness of a current computing machine (the new SUN workstation on Bringsjord's desk, e.g.), you would know that at 6:01 a.m. you would be *dead* — despite the fact that "you" *qua* computer would be able to manipulate all sorts of information.[31] The second reason why the phenomena in question are crucial is that such things as qualia are intimately related to that which we've already observed to be at the core of Pinker's enterprise: Wallace's Paradox. This is so because whatever justification there may be for regarding hypercomputational powers such as reasoning to be overkill from the standpoint of evolution would apply, *mutatis mutandis*, to such things as qualia. Why would our foraging ancestors need qualia to thrive? Why couldn't such creatures have had in their behavioral repertoire actions like finding berries and making fires and killing prey, all without qualia like the qualitative inner feelings associated with (say) sitting next to a fire? After all, when we build robots to search out counterparts to fire (power sources, say), do we worry about giving these robots associated inner *feelings*?

7.3.3.4 Devlin's Reductionism

Recall that in Chapter 3 we promised to discuss Devlin's (2000) position on the evolutionary basis of mathematical thinking. To repeat, this view is that mathematical thinking is reducible to language use, and originally emerged from evolutionary reinforcement of what he calls 'offline' thinking — thinking about situations represented in an abstract form removed from their flesh-and-blood instantiations in the

---

[31] For a full treatment of the relationship between such gedanken-experiments and $\mathcal{C}$, see (Bringsjord 1999).

physical world.

We begin the discussion by noting that Devlin does seem to feel, at least to some degree, the challenge of Wallace's Paradox, for he writes:

> Whatever features of our brain enable (some of) us to do mathematics must have been present long before we had any mathematics. Those crucial features, therefore, must have evolved to fulfill some other purpose. What purpose? (Devlin 2000, p. 2)

Part of Devlin's answer to this question is the standard move: point out that some kind of number sense has survival value. For example, we read:

> What evolutionary advantage led to selection for a number sense in rats? One possibility is the need to remember navigational information, such as its hole was the fourth one along after the third tree. It's also useful in keeping track of other animals in the vicinity, be they friends or predators. (Devlin 2000, p. 24) p. 2)

Unfortunately, the standard move is shot down by what we pointed out earlier in the present chapter: viz., a robot operating well below the level of a human mathematician can do just fine. A robotic rat, for example, would no doubt have a capacity to count, but surely it doesn't need a capacity to wrestle with infinitary mathematics! This problem is fatal for Devlin; almost everything he says runs afoul of it. For example, he says

> I will suggest that, in essence, the brain acquired the ability to simulate (in symbolic fashion) the real world to a sufficient extent that the circuits that evolved to enable it to negotiate and survive in the world became able to operate without direct input from the world. (In physical terms, the brain acquired the ability to create activation patterns that resemble those created by direct sensory input from the world.) (Devlin 2000, p. 122–123)

But again, the problem is that a robot able to run all sorts of simulations can be carried out using at most first-order logic and finitary reasoning. This kind of power was all evolution needed to cook up — and we are learning this from the real-life advance of robotics.

Here is Devlin giving a more careful account of his view:

> At the end of the long period of brain growth, somewhere between 75,000 and 200,000 years ago, the human brain acquired

> symbolic thinking. This was caused not by further brain enlargement but by a change in its structure. The brain acquired the ability to simulate the activation patterns normally caused by sensory stimuli, and to isolate from the body's motor centers the outcomes of the resulting thought processes. In short, the brain became able to think "off-line."
>
> Off-line thinking could have developed only after the brain had acquired a sufficient number of types to support an internal "model" of the world that was adequate for reasoning. But a rich collection of types on its own would not be enough. The brain had to be able to represent, internally, a sufficiently rich world *structure*. The "skeletal structure" of the world that the brain simulated in order to think about the world off-line is what we now call syntactic structure, or syntax. (Devlin 2000, p. 250–251)

Here Devlin is giving a story like that given by Pinker: 'skeletal structure' sounds an awful lot like Pinker's 'pad of forms.' And Devlin's view suffers from the same fundamental, fatal defect: Robots operating without qualia, and without a capacity for infinitary reasoning, can easily think off-line in the manner Devlin describes. Robots specifically able to hunt food and gather berries in coordinated groups certainly don't need a capacity for infinitary reasoning. And for the communication between the robots engaged in such feats, there is no reason to think a language at or below the Turing Limit would be inadequate. To put it bluntly, the advance of robotics constitutes an empirical refutation of both Pinker and Devlin, and a vindication of Wallace. This situation is pointed to in Figure 7.9.

7.3.4 What, Then, Should We Conclude?

Let's now take stock of where we've ended up in connection with our "Where'd we come from?" question.

Return, if you will, to the central disjunction expressed in Table 7.2. If Reductionism and Exaptationism are now out of the picture, and if we're unwilling to accept a contradiction, dodging Wallace's Paradox can be accomplished only if we either invoke Theism or Exotic Natural Forces (or some synthesis of the two). We end up in this extraordinary position because personhood, since it entails supermentalism, is beyond the reach of evolution and computation, as shown by our modernized version of Wallace's Paradox.

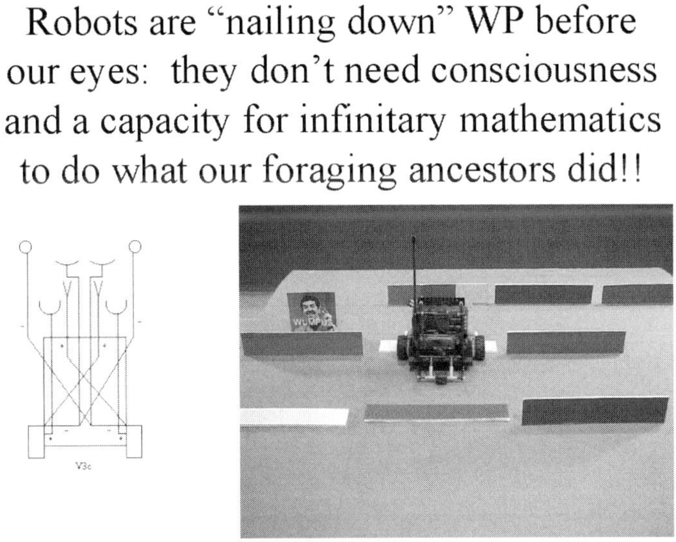

Figure 7.9: The Advance of FOL-level Robots are Vindicating Wallace

7.4 Toward the Second Argument for Doubting that Evolution Produced Us

There is a second argument for doubting that human persons were produced by evolution. As of yet, Bringsjord has not finished specifying the argument; it will appear in a later publication. But we can easily enough convey the gist of the argument here — and given that the argument is so intimately related to hypercomputers and superminds it seems appropriate to do so.

Dennett (1995*b*) has recently pointed out that Darwin postulated: life arose on Earth as a result of an algorithmic process. 'Life' here includes us, of course, as Darwin (1997) explained. As Dennett sums up:

> Darwin introduced an idea of how this evolutionary process took place: via a mindless, mechanical — algorithmic — process he called "natural selection." This idea, that all the fruits of evolution can be explained as the products of an algorithmic process, is Darwin's dangerous idea. (Dennett 1995*b*, p. 60)

Given this, if we are superminds, we can't possibly have evolved; this is easy to see. It is well-known that no algorithm can take as

Genetic Algorithm

begin GA
    g:= 0
    Initialize population P(g)
    Evaluate population P(g) (i.e., compute fitness values)
    while not done **do**
        g:=g+1
        Select P(g) from P(g-1)
        Crossover P(g)
        Mutate P(g)
        Evaluate P(g)
    end while
end GA

Simple Example

Label	**Genome**	**Fitness**	
A	00000110	2	
B	11101110	6	Av Fitness:
C	00100000	1	3
D	00110100	3	

Assume parents {B,D} and {B,C} and crossover at first bit for {B,D} forming E and F, and none for {B,C}. Then:

Label	**Genome**	**Fitness**
E	10110100	4
F	01101110	5
B	11101110	6
C	00100000	1

Proceeding...

Assume offspring E is mutated at the 6th position to form E', Offspring B is mutated at the first bit position for form B', and Offspring F and C are not mutated at all.

Label	**Genome**	**Fitness**	
E'	10110000	3	
F	01101110	5	Av Fitness:
B'	01101110	5	14÷4
C	00100000	1	

.
.
.

Figure 7.10: Example of GA Operating on Strings

input mere algorithms and output a hypercomputer. (Dennett (2001) has elsewhere explained that evolution-as-algorithm works only upon objects that are themselves at most algorithms.) Put a bit more precisely, it's an easily proved theorem that no Turing machine m can take as input a Turing machine m' (encoded) and produce as output a hypercomputer $\mathcal{M}$. (After all, if a TM could do this, then there would be a TM that could solve the halting problem — by simply generating a hypercomputer to do the job.) Given this, and given that we are superminds (and hence include hypercomputers), we cannot be the product of evolution. This argument is quite simple, quite devastating, and, as we've said, will be fully specified after the publication of the present volume. The specification will include discussion of the fact that computer science has actually formalized Dennett's point (e.g., Sipper 1997, from which Figure 7.10 is drawn), and that algorithms (and the schemes that formalize them; e.g., Turing machines) can, despite laic intuitions to the contrary, include true randomness.

Chapter 8

Supermentalism and the Practice of AI

8.1 Toward the Final Stage of the Project

Recall that the present volume is one book within what we've dubbed the 'Supermind Quartet,' which is in turn aimed at securing three objectives:

Objective 1 Overthrow computationalism, essentially the view ($\mathcal{C}$) that minds are ordinary computing machines. (As you may remember, a careful account of computationalism is given in Chapter 1.)

Objective 2 Establish the position ($\mathcal{S}_c^\infty$) that minds are *super*minds, i.e., that though minds aren't computing machines, they *include* such machines, *and* two other main things, viz., information-processing machines capable of hypercomputation (as reflected in the superscript $^\infty$), and a capacity to enjoy and exploit phenomenal consciousness (and other irreducibly mental properties), which can't be expressed in any third-person scheme, computational or otherwise (as reflected in the subscript $_c$).

Objective 3 Specify and defend a practical, concrete, applied approach to AI and CogSci anchored by the theoretical content produced to reach Objectives 1 and 2. This specification and defense is to revolve around implemented systems, not mere speculation; engineering is to flow from philosophizing. The systems will establish (among other things) both that animal behavior can be simulated by ordinary computation ($\mathcal{B}_a$) and that the behavior of persons, to a substantial degree, can also be simulated by such computation ($\mathcal{B}_p$).

As we explained in the Preface, Objective 3 was partially reached through *Articial Intelligence and Literary Creativity: Inside the Mind of* BRUTUS, *A Storytelling Machine* (Bringsjord & Ferrucci 2000). Objective 2 is met by the present volume. Fully reaching Objective 3 will require yet another book, one that will complete the quartet. In this final book, AI engineering undertaken on the basis of the prescription offered in the present chapter will be carried out at a mature level.

8.2 The Eight-fold Prescription for the Practice of AI

The present chapter lays down a bridge to the final book in the supermind quartet. This bridge consists of a eight-fold prescription for the practice of AI in light of supermentalism. Here is a compressed version of this prescription for how AI should henceforth be conducted:

P1 Building harnessable hypercomputers may well be impossible, but strive nonetheless for such devices. To anchor this striving, attack Turing-unsolvable problems head-on. At any rate, formal and conceptual analysis of hypercomputational mentation should continue full-speed ahead — with guidance from technical philosophy. (P1 is intimately related to P7.)

P2 Focus on building artificial animals; they can be engineered, while persons can't. (For reasons explained below, we refer to such artifacts as 'zombanimals.')

P3 Given that "Strong" AI and the attempt to literally build persons is bankrupt, aggressively pursue "*Weak*" AI, in the specific form of the attempt to to build systems able to excel on a wide array of established, validated, empirical tests of all sorts. Following (Bringsjord & Schimanski 2002), we dub this brand of AI **Psychometric AI**. (Turing had the right idea in urging a test. We should generalize, though, and move to all *sorts* of tests.)

P4 When attempting to build systems that behaviorally match the most accomplished human persons, take experimental psychology seriously. For example, when attempting to reach "Simon's Dream" (recall section 3.9), carefully study how the best human reasoners reason (such reasoners clearly engage in reasoning that is rarely even conceptually modeled in AI or CogSci).

P5 In keeping with P4, be brutally honest about the limitations of AI schemes for representation and reasoning. Very advanced hybrid logics are necessary; logic and technical philosophy should be in charge.

Operators	Computation Predicates	Engineeringish Predicates
$\Diamond$: logically possibly	F: FSA computation	$\mathcal{H}^u$: unconsciously harnessable
$\Diamond_p$: physically possibly	T: Turing machine computation	$\mathcal{H}^c$: consciously harnessable
$\Diamond_h$: humanly possibly	O: "Oracle" computation	$\mathcal{A}$: actualizable

Table 8.1: Relevant Operators and Predicates

P6 Investigate, in earnest, language acquisition.

P7 Aggressively pursue the mathematical "ranking" of mentation, independent of physical engineering.

P8 Relegate connectionism to handling but lower aspects of mentation, as only logic-based approaches affirm the essence of the information-processing side of ever more powerful minds.

We now proceed to briefly elaborate on each of P1–P8 in turn.

8.3 P1: Building Consciously Harnessable Hypercomputers is Hard, but Push Ahead Anyway

It seems to us exceedingly improbable that hypercomputational hardware can be built and harnessed. Consider again here the tabular scheme we introduced in the Preface (Tables 8.1 & 8.2). As you'll recall, there are many interesting propositions that can be expressed on the basis of these two tables, but perhaps the most important question is whether it's humanly possible to build a consciously harnessable hypercomputer; that is, whether this proposition is true:

(6) $\Diamond_h \exists c(Oc \wedge \mathcal{A}c \wedge \mathcal{H}^c c)$.

There are many arguments for the falsity of (6); here, in provisional form, are two:

Partitioned Domain
$m_1, m_2, \ldots$: mentations
$c_1, c_2, \ldots$: computations
$p_1, p_2, \ldots$: persons

Table 8.2: Partitioned Domains

8.3.1 The Argument from Infinity

Pessimism about physically realizing and harnessing hypercomputational devices can be grounded in the observation that to build a hypercomputational device would be to somehow harness and compress the power of the infinite in a finitely bounded, physical artifact. Hopefully our little primer on hypercomputation (section 1.8) makes this plain. Given this primer, the problem seems obvious in the case of trial-and-error and Zeus machines. In the former, to build and use them we would need to be either able to see the future (in order to see the in-the-limit verdict; arguably this is exactly what Stannett (forthcoming) has ingeniously formalized) or to compress an infinite period of symbol manipulation into a finite interval. In the latter, once again, to build and use means to achieve such preternatural compression. In the case of analog chaotic neural nets the same problem is there, but camouflaged. An analog chaotic neural net processes information by using irrational numbers (Siegelmann & Sontag 1994). So think of the challenge in these terms: Suppose that you can build a hypercomputational device as long as whenever your device reaches ordinary computational state s, it carries out operation O on s *and* irrational real $r^i \in [0, 1]$. The operation here has in fact been neatly formalized in the form of the analog shift map, which was explained in Chapter 4. How do you capture and manipulate r^i *physically*? There is of course a sense in which every time you operate a physical artifact by in part using readings from an analog dial (the speedometer on a car, e.g.), you are "using irrational numbers." But this is a *very* weak sense of 'use.' Your Volvo may get you to work in the morning, and there may be a sense in which it or parts of it enter into into physical processes that are uncomputable (which is what people like Pour-El and Richards 1981*b*, 1981*a* can be read as having probably established). But to use the Volvo to attack the halting problem seems rather futile. And it's hard to imagine an antidote to this futility from *any engineer*, for *any device*. (We recommend that readers consult (Bringsjord 2001*b*) for a machine of Bringsjord's devising that exploits irrational numbers, but makes clear that it's quite another matter for a human to exploit this machine's exploiting such numbers.)

8.3.2 The Argument from the Absence of Candidates

At best, there are few candidates for physical processes that instantiate hypercomputation. (Stannett (forthcoming)'s "half-life" hypercomputation is one such candidate.) On this issue, Carol Cleland has recently said:

> Some physical processes may only have a few points at which causal intervention is physically possible, thus allowing one to specify an input and output relation but not permitting a *mechanistic account* of what goes on between input and output. An oracle would be such a process. It is important to keep in mind, however, that being *non-mechanistic* is not sufficient for being an oracle. (Cleland forthcoming, p. 18; emphasis ours)

Given what Cleland says here, what would be a likely candidate for physical hypercomputation? It seems that being "non-mechanistic" is a necessary (but not sufficient) condition. Given this, and given the impressionistic description of the type of physical process that could function as an oracle, one would think that quantum mechanical phenomena might work. Indeed, these phenomena are quintessentially those where what goes on between input and output is unavailable (or perhaps it's better to say that causal intervention yields an output). If one has hopes of finding some sort of physical process to function as an oracle, this seems to be a promising place to look. Certainly, given today's science, one can hardly get more exotic.

But attempts to access and artifactually exploit these promising phenomena have not taken us into hypercomputing. Theorists and experimentalists working on quantum computers have long recognized that this type of machine will only give us *faster* Turing computation.[1] Now one could, of course, say that we haven't exploited all the possibilities. This is true — *if* one assumes that other candidates will arrive on the scene in the future. But the issue is plausibility — *today.* Since quantum computing has proved to be a dead end, where else should we look? If no one can present us with any other promising candidates, the most reasonable position would seem to be the denial of $\Diamond_h \exists c O c$ and related propositions.

The situation may in fact be worse than it appears to be. In order to see this, let's grant that physically instantiated oracles exist. We have seen that at present we have no candidates for such things. So, we need to set off in search of candidates. But what should our

[1]Quantum computers are discussed in (Feynman 1986). Brooks (1999) gives a concise discussion of quantum computation and the status of current research. Deutsch (1985) considers a universal quantum computer and argues that it is compatible with the Church-Turing Principle. He shows that this computer wouldn't be able to compute non-recursive functions; however, see (Stannett forthcoming).

search procedure be? It's of course vastly improbable that arbitrary searching will prove successful, so we need help from our best (current and relevant) theories. But our best theories offer no guidance for where to look for "causal openings."

8.3.3 The Busy Beaver Competition

Nonetheless, as reflected in P1, we recommend that Turing-unsolvable problems be tackled directly. If the attempt to do so is sufficiently serious, then "all the stops will be pulled out," and who knows, perhaps hypercomputation will be — to use the key phrase, given what we've said earlier in the book — consciously harnessed. We are attempting to practice what we preach: Bringsjord, Bram van Heuveln, Paul Bello, and Marc Destefano are at present attempting to "crack" $f_{bb}(6)$ in the 4-tuple framework. In Chapter 3 we informed you that Chris Nielsen has demonstrated $f_{bb}(6) \geq 21$; what we didn't tell you is that a quartet of researchers in Portugal, (Pereira, Machado, Costa & Cardoso n.d.) have shown $f_{bb}(6) \geq 25$; their remarkable machine is shown in Figure 8.1. The quartet just mentioned is attempting to find a machine more productive than this one.

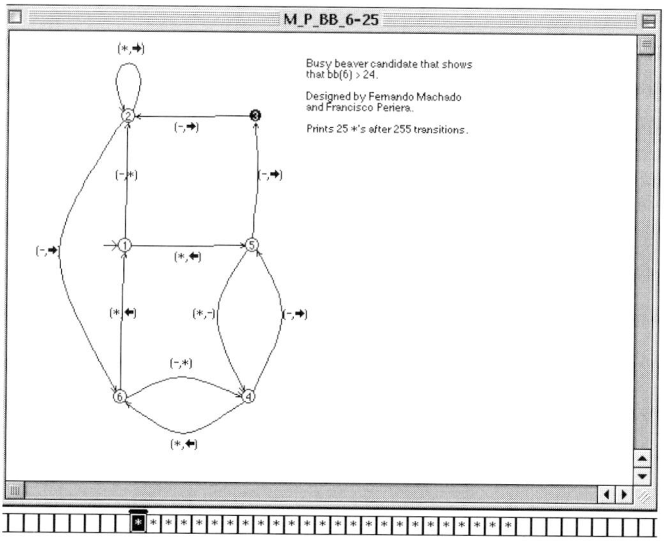

Figure 8.1: The Pereira et al. 6-State Busy Beaver Candidate

8.4 P2: Focus on Building Artificial Animals (Zombanimals)

Back at the beginning of this book, we said that AI, in light of what we have uncovered about the nature of human persons, should strive to engineer "sub-person" artifacts. One convenient way to be more specific about sub-persons is to identify this category with animals. If we make this identification, we can characterize in more detail the brand of "Weak" AI — theriomorphic AI, we might dub it — in question. This characterization now follows; it is a very simplified version of what appears in full form in (Bringsjord 2000*a*).

8.4.1 The Concept of Zombanimalhood Introduced

The animal side of "Weak" AI can be viewed as the attempt to build artificial animals able to pass certain tests. These artificial animals will not have have phenomenal consicousness, so we refer to them as 'zombanimals' (Bringsjord 2000*a*); let us explain.

As was discussed in Chapter 1, Alan Turing (1964) devised his famous test (TT) through a slight modification of the parlor game in which a judge tries to ascertain the gender of two people who are only linguistically accessible. In TT the judge tries to determine — solely by typed queries and corresponding responses: by email, if you will — which of two contestants is a computer and which is a person. As we also discussed in Chapter 1, Stevan Harnad (1991) has encouraged us to leave the pen pal mode: to allow the judge to not only ask the players by email if they can, say, catch a baseball (and if they say they can, what it *feels* like to catch one); so we now permit the judge to *look* at them both, *and* throw each a baseball to catch (or not) before his eyes. Now the test is the so-called *Total* Turing Test (TTT); it challenges the judge to determine which is a robot and which a person. But why stop here? We can flesh out the notion that "Weak" AI, in light of supermentalism, should aim for the engineering of sub-person artifacts. We can do so by confronting the judge with an *animal*, and a robot striving to pass for one, and then challenge him to peg which is which. Now we can index TTT to a particular animal and its synthetic correlate. We might therefore have TTT_{rat}, TTT_{cat}, TTT_{dog}, and so on. So to say that "Weak" AI should attempt to engineer sub-person artifacts is to say that it should strive to engineer artifacts able to pass such tests.

We have started the ball rolling in Bringsjord's RAIR (Rensselaer AI & Reasoning) Lab. In order to understand how we have done so far, you need to first recall that some things that are physically possible are also *humanly* possible. For example, it's physically possible that the laptop on which Bringsjord is currently working fall from eight feet directly to the ground. This is also humanly possible: Bringsjord could stand up, hold the laptop over this head, and let it go. Of course, some physically possible states of affairs are not (at the moment, anyway) humanly possible. For example, it's physically possible that a spaceship travel just short of the speed of light — but NASA can build no such thing.

Now, here's a question we posed to ourselves in connection with the surgery that formed the core of Chapter 5. In this surgery, recall, silicon workalikes were installed in the brain, replacing biological stuff, and a zombie was produced. Here's the question: Is it *humanly* possible for this surgery to be carried out? The answer is clearly "No." You will remember that we relied in Chapter 5 on the talents of Kafka, and on the fact that artificial neural nets are very much like real neural nets. But no actual neurosurgeon, at least today, could pull such a thing off. So we posed another question to ourselves: Could we carry out similar surgery *on animals*? Here our response is different: We believe that zombanimals are humanly (and hence physically) possible. This is our position because someone in Bringsjord's RAIR Laboratory (Clarke Caporale) has successfully approximated the surgery. Why do we say 'approximated'? Well, as in the thought-experiment of Chapter 5, the idea is to study an animal brain, along with associated sensors and effectors, and to create a precise model of how information flows through the sensors, into the brain, and back out to the triggered effectors. But our surgery is a bit different. We don't seek to *supplant* the animal brain; we seek to instantiate the model in new, silicon hardware; in other words, we seek to *duplicate* the biological creature. Let us explain.

8.4.2 Simple Zombanimals

To ease exposition, our discussion will now be based upon hypothetical biological creatures whose information-processing architectures are transparent; that is, the flow of information into, through, and out of these creatures has by hypothesis been **mapped**. The "surgery" (or, as we've explained, duplication) carried out to render these architectures in silicon is real, however. It was carried out in the RAIR Laboratory, the robotics workbench in which is shown in Figure 8.3.

Let's start with a very simple animal. Not a cat or a rat; something simpler. Imagine a simple multi-cellular organism — let's call it a 'bloog.' When you shine a penlight on a bloog, it propels itself energetically forward. If you follow the bloog as it moves, keeping the penlight on it, it continues ahead rather quickly. If you shut the penlight off, the bloog still moves ahead, but very, very slowly: the bloog is — we say — listless. If you keep the penlight on, but shine it a few inches away and in front of the bloog ('front of' being identified with the direction in which it's moving), the bloog gradually accelerates in a straight line in the direction it appears to be facing, but then slows down once it is beyond the light. Given our catalogue of bloog behavior, we can set the TTT_{bloog}, and can attempt to build a zombanimal to pass this test.

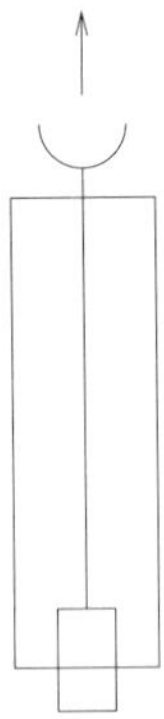

Figure 8.2: V1. *The motor is denoted by the rectangular box at the tail end, the sensor by the half-circle on a stalk.*

Caporale has succeeded in precisely this attempt.

He has played with a bloog for a while with his penlight, and has witnessed the behavior we have just described; he then set to work. He began by scanning the flow of information in a bloog when one is under a microscope. After a bit, he readied his supply of robotics micro-hardware, and initiated duplication. Now that he is done, he presents you with ... a creature he calls 'V1.'[2] V1 is composed of one

[2]Our simple zombanimals are inspired by the vehicular creatures described in Valentino Braitenberg's (1984) *Vehicles: Experiments in Synthetic Psychology*. Our first zombanimal is Braitenberg's Vehicle 1, or just 'V1' for short. Note that turning

tiny sensor and one tiny motor, which are connected, and a structure that supports them. The motor is connected to some device which, when driven by the motor, produces locomotion. V1 is shown in Figure 8.2. The behavior of V1 is straightforward: the more of the source detected by its sensor, the faster its motor runs. Were Caporale to give you a demo,[3] you would see that if V1 is bathed in light from the penlight, it moves forward energetically. If it then enters a darker area it becomes listless. If it detects a light ahead, it accelerates toward the light, passes through it, and then decelerates. Obviously, V1 is not subjectively aware: V1 is a zombanimal.

Figure 8.3: Robotics Workbench in Bringsjord's RAIR Lab.

Caporale has also built zombanimal correlates to these two biological creatures: a 'sneelock' and a 'fleelock.' They are larger than bloogs, a slightly different shade of fleshy brown, and behave differently. A fleelock behaves as follows. If you shine a penlight on the surface on

V*i* into real robots is not new. Other people have constructed such robots, and you can even buy some of such beasts "off the shelf." This isn't a chapter on cutting-edge robotics (though the IQ bot in Bringsjord's RAIR Lab is intended to be cutting-edge); this is a philosophical chapter *informed* by real robotics.

[3]As he did when we presented inchoate elements of the present chapter: "Zombanimals — with Robots from the Minds & Machines Laboratory" (Annual meeting of the Society for Machines and Mentality, at the annual Eastern Division Meeting of the American Philosophical Association, December 1998, Washington, DC).

which the fleelock is located, just ahead and exactly in front of the organism, it moves directly toward the light and passes through it, at which point, like bloogs, it becomes listless. However, if you shine the penlight ahead of the fleelock, but to the left or right, it turns to avoid the light, and then moves forward slowly; fleelocks generally dislike light. Sneelocks are similar. They too dislike light, but there is a difference: sneelocks are aggressive. This can be shown by shining a penlight ahead of a sneelock (and, again, to the left or right). When one does this, the sneelock turns with increasing rapidity toward the light, and moves directly at it, eventually moving frontally into the light, apparently to assault it.

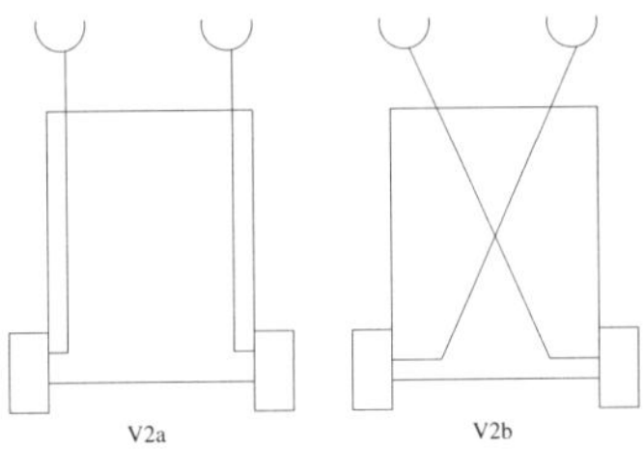

Figure 8.4: V2a and V2b. *V2a orients away from the light; V2b toward it.*

Caporale was once again allowed to perform "surgery." After a bit, he cooked up two new zombanimals, V2a and V2b (see Figure 8.4). Courtesy of micro-sensors and motors, V2a behaves just like a fleelock, V2b just like a sneelock. Both V2a and V2b arise from robotic augmentation of V1 that could not suddenly bestow upon them phenomenal consciousness. Hence V2a and V2b are both, like their predecessor, zombanimals. Were you here in our lab, Caporale could show you V1, V2a, and V2b in action.

8.4.3 From Simple to Complex Zombanimals

You're doubtless thinking that such organisms as bloogs, sneelocks, and fleelocks are excruciatingly simple. Well, you're right. As we've indicated, they're *simple* zombanimals. But Caporale is just warming up.

Consider an animal that can sense and react to not only light, but temperature, oxygen concentration, and amount of organic matter. This biological creature is called a 'multi-moog.' Multi-moogs dislike

Figure 8.5: A Sample Zombanimal — front view.

high temperature, turn away from hot places, dislike light with considerable passion (since it turns toward and apparently attempts to destroy them), and prefers a well-oxygenated environment containing many organic molecules. Caporale has "zombified" a multi-moog: the result is V3c, shown in Figure 8.7. V3c has four pairs of sensors tuned to light, temperature, oxygen concentration, and amount of organic matter. The first pair of sensors is connected to the micro-motors with uncrossed excitatory connections, the second pair with crossed excitatory connections, and the third and fourth pairs with inhibitory connections. It should be obvious that we have no more reason to suppose that V3c is subjectively aware than we have to suppose its predecessors V2a and V2b are: after all, the robotics work that yields V3c from its predecessors consists in a few more wires here and there, and how could such things suddenly bestow upon their bearer phenomenal consciousness?

Before we conclude this section, a reminder regarding our "TTish" theme: Multi-moogs can pass versions of the TTT indexed to animals at this level, that is, they can pass $\mathrm{TTT}_{multi-moog}$.

8.4.4 From Zombanimals to Essence of AI

How powerful is Caporale's toolkit? In the interests of space we will spare you the incremental specification of increasingly complex zombanimals (and we direct you to Braitenberg 1984). Suffice it to say

Figure 8.6: A Sample Zombanimal — side view.

that Caporale's arsenal includes everthing from primitive logic circuits to full-blown artificial neural networks, loads of sensors and effectors, and some pretty darn good electrical engineering. (The Vn zombanimals all have the same core brain, a logic board shown in Figure 8.8. Caporale would soon move to more sophisticated boards.) Caporale is also able to harness evolutionary techniques in order to craft his zombanimals. He will thus have the powers Dennett believes he and others (e.g., Rodney Brooks) have in their attempt to build the humanoid robot COG (Dennett 1994).[4] In sum, it seems to us that Caporale has a toolkit powerful enough to zombify *all* animals. We offer the following inductive argument for this view.

8.4.5 The Inductive Argument

To begin, we need to idealize things a bit; the idealization is essentially the same one that allowed us to express Wallace's Paradox in Chapter 7. Suppose that biological animals fall across a spectrum isomorphic to some initial sequence of the natural numbers.[5] The first animal, b_1,

[4]For a description of these forces in the context of building zombanimals like the Vn creatures, see "Vehicle 6: Selection, the Impersonal Engineer," in (Braitenberg 1984).

[5]The "need to idealize things a bit" may strike some readers as quite a jump, but of course we aren't supposing that this ordering *literally* exists. After all, that's why we say it's an *idealization*. However, there does seem to be an unshakable intuition that animals go from simple to increasingly intelligent at least roughly along some

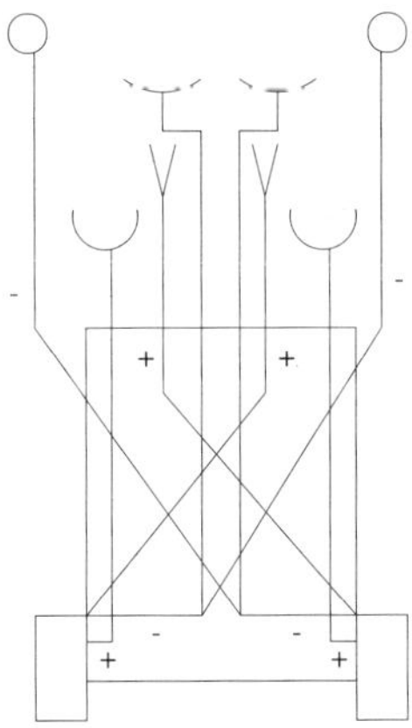

Figure 8.7: V3c. *A multisensorial zombanimal.*

would perhaps be a simple single-cell organism, then we would move on to a more sophisticated single-cell organism b_2, to ..., to a multi-cellular organism b_k, to a more sophisticated multi-cellular organism b_{k+1}, to ..., then perhaps up to an ant b_{k+p}, to ..., to (say) a parrot, to ..., and eventually up to the animals that would seem to be just "beneath" human persons: apes and chimps. The entire spectrum would then be

$$b_1, b_2, \ldots, b_m = \text{ape}.$$

Now, each b_i has been produced by evolutionary forces, mutations reinforced by natural selection, from b_{i-1}; we write $\mathcal{E}(b_i) = b_{i+1}$ to indicate that evolution caused a step. With this primitive scheme, it's easy to express an inductive argument suggested by the progression seen above in passing from V1 to V2a to V2b to V3c. The basis clause is the proposition that b_1 can be zombified, that is, that a zombanimal, an artificial, non-conscious correlate to a real animal, z_1, can be engineered which passes TTT_{b_1}; we denote this by $\mathcal{Z}(b_1)$. It would seem that such a proposition must be true. It would also seem to be the case that the induction hypothesis is true: if b_i can be zombified, then through some rather minor engineering modifications

continuum. Furthermore, it's perhaps not all that implausible to imagine that we could "operationalize" through tests the continuum we invoke. For example, it's well-known that rats can be trained to navigate mazes. It's also well-known that chimps have been subjected to all sorts of challenges; e.g., see (Hauser 2000). Perhaps a battery of tests could be devised to solidify the spectrum we imagine.

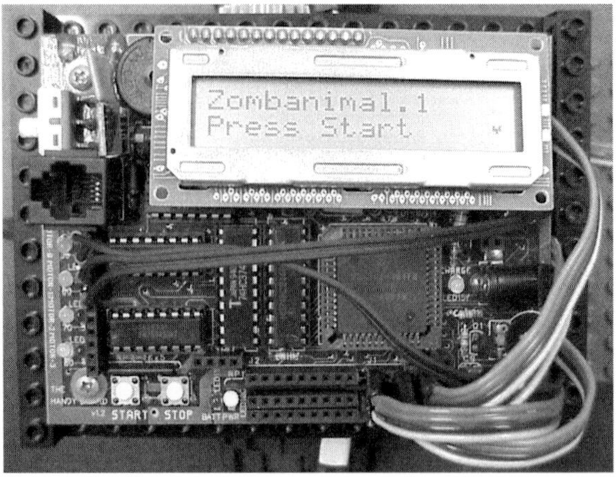

Figure 8.8: The Logic Board the Forms the Brain of the Vn Zombanimals *(This is the so-called "Handy Board.")*

so can b_{i+1}. (Again, how could these tiny engineering tweaks magically generate subjective awareness?) The general principle from which the induction hypothesis follows is this proposition:

$$(*)\ \forall x \forall y(\mathcal{E}(x) = y \rightarrow (\mathcal{Z}(x) \rightarrow \mathcal{Z}(y)))$$

By the principle of mathematical induction it of course follows that *all* animals can be zombified.

Perhaps some readers will object to the basis clause, as in something like: "Even for TTT competence, it is not so clear to me as it apparently is to you two that a simple cell is within the competence of AI as you construe it. In particular, the cell has a continuous dynamics, and there are at least some phenomena that can emerge in such systems that cannot be captured with discrete AI systems. This point would also seem to apply to the induction step, for even if b_1 is accepted at the TTT level, it would not necessarily follow that the induction is valid, since the induction is over tinkerings with AI architectures, not tinkerings in continuous, interactive and metabolic dynamics."

The problem with this objection is that if a set of processes can be diagrammed and measured in the manner of biological science, this set can be digitized and zombified. One need only pick up a biology textbook to see that creatures at the level of b_1 have been diagrammed, measured, manipulated, altered, and so on. Do we really think that aspects crucial to such creatures have been left out of the picture, *and* that such aspects will *never* be rendered in a form amenable to the kind of engineering at the heart of zombification? Now it's of course true that if a process is genuinely and irreducibly continuous, it may be analog and chaotic, and may thus exceed the reach of TM-level computation, and hence may be a process beyond the reach of at least standard AI, whether of the logicist or connectionist sort (see e.g., Siegelmann & Sontag 1994, Siegelmann 1995). But why would anyone think that b_1 harnesses such processes to get by in the world? Surely the burden of proof is on anyone who thinks this.

We imagine that many at this point will also object as follows: "But Caporale is just one lone engineer operating in one isolated laboratory. How do we know that he and his techniques are not idiosyncratic? You yourselves have admitted that he isn't working with real animals, but rather with information processing-based idealizations of real animals. Isn't *that* idiosyncratic? And yet you seek to generalize wildly from what Caporale has done!"

The rebuttal to this objection is given in the next section, where we explain that the essence of AI would seem to be a set of techniques and formalisms for building zombanimals in a manner that coincides remarkably well with Caporale's engineering.

8.4.6 AI as Zombanimal Construction

On page 7 of their 1985 *Introduction to Artificial Intelligence*, Eugene Charniak and Drew McDermott (1985) write: "The ultimate goal of AI (which we are very far from achieving) is to build a person, or, more humbly, an animal." That the more humble goal is all that AI can reach is of course the thesis we're advancing at the moment. When you look in detail at Charniak and McDermott's book, you see there formalisms and techniques sufficient only for creating artificial animals, not persons — that, at least is our position. Unfortunately, even if we're right, the book is over a decade old: it may have captured all of AI in 1985, but may not encompass all of AI *today*.[6]

Fortunately, as we pointed out in the previous chapter, AI has been to an astonishing degree unified around a conception that seems to be coextensive with Caporale's engineering: the conception of an **intelligent agent**. The unification has in large part come courtesy

[6]As a matter of fact, C&M's book doesn't really cover sub-symbolic (e.g., neural net-based) AI, as it was written before the recrudescence of connectionist AI.

of the comprehensive textbook we drew upon so heavily in the previous chapter: one intended to cover literally *all* of AI: Russell and Norvig's (1994) *Artificial Intelligence: A Modern Approach* (*AIMA*), the cover of which, again, also displays the phrase "The Intelligent Agent Book." Again, the overall, informal architecture for an intelligent agent is shown in Figure 8.9; this is taken directly from the *AIMA* text. According to this architecture, agents take percepts from the environment, process them in some way that prescribes actions, perform these actions, take in new percepts, and continue in the cycle.[7]

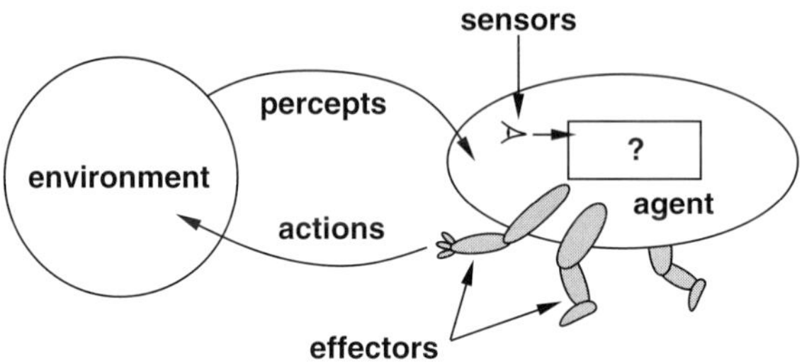

Figure 8.9: The Architecture of an Intelligent Agent

In *AIMA*, just as is the case in the ladder-like scheme we introduced above for animals and zombanimals, intelligent agents fall on a spectrum from least intelligent to more intelligent to most intelligent. The least intelligent artificial agent is a "TABLE-DRIVEN-AGENT," which we described in the previous chapter.

In *AIMA* we reach artificial agents that might strike some as rather smart when we reach the level of a "knowledge-based" agent, a type of agent also described in the previous chapter.

Needless to say, a knowledge-based agent, incarnated in robotic hardware, is a zombie: it has no genuine inner life: there is nothing it is like to be such a thing.

Now someone might respond: "Your wumpus-world robot is no doubt a zombie; no problem there. I agree that it has no genuine inner life. But why do you call it a zomb*animal*? Your zombanimals were non-conscious robotic duplicates of animals. But to what animal does your wumpus-world robot correspond?"

The answer to this question is that zombification could be carried out via the techniques and formalisms that constitute the agent-based

[7] Again, the cycle here is strikingly similar to the overall architecture of cognition described by Pollock (1995).

Table 8.3: Master Table for Incremental Progression Without Consciousness

Animal	Zombanimal	Zomb. via *AIMA*	Relevant TTT
b_1	$\mathcal{Z}(b_1) = z_1$	$\mathcal{A}(b_1)$	z_1 and $\mathcal{A}(b_1)$ pass TTT_{b_1}
$\mathcal{E}(b_1) = b_2$	$\mathcal{Z}(b_2) = z_2$	$\mathcal{A}(b_2)$	z_2 and $\mathcal{A}(b_2)$ pass TTT_{b_2}
$\mathcal{E}(b_2) = b_3$	$\mathcal{Z}(b_3) = z_3$	$\mathcal{A}(b_3)$	z_3 and $\mathcal{A}(b_3)$ pass TTT_{b_3}
$\vdots$	$\vdots$	$\vdots$	$\vdots$
$\mathcal{E}(b_n) = b_{n+1}$	$\mathcal{Z}(b_{n+1}) = z_{n+1}$	$\mathcal{A}(b_{n+1})$	z_{n+1} and $\mathcal{A}(b_{n+1})$ pass $\text{TTT}_{b_{n+1}}$

approach preached in *AIMA*. Where b_i is some animal in the continuum invoked earlier, let $\mathcal{A}(b_i)$ denote the process of zombification — except now the process uses the programs and formalisms in the agent-based approach, along with requisite robotics. Put schematically, the situation so far can be summed up in the progression shown in Table 8.3, where the items in each row are equivalent with respect to both the information processing that drives them, and the phenomenal consciousness that they wholly lack.

8.4.7 Yes, but What about People?

Of course, AIniks will be drawn irresistibly toward the goal of building not just zombanimals, but persons. Courtesy of the arguments we've presented in this book (and in other places: see e.g. Table 1), we know that this goal is a pipe dream. We are persons; so are you. On the other hand, we are also biological creatures; in a real and undeniable sense we are animals: *homo sapiens sapiens*. So how is it that Table 8.3, in reaching to animal b_{n+1}, fails to reach *us*? When we take biological creatures of greater and greater sophistication, and present them to Caporale and an *AIMA*-based engineer for zombification, why will they eventually succeed when we bring them a rat or parrot or a chimp, but fail when we bring them a person? They will fail because they will be impotent in the face of the properties that distinguish persons. What are these properties? They are listed in L_p, enumerated in Chapter 1. For a discussion of these properties in connection with zombanimal construction see (Bringsjord 2000*a*).

8.5 P3: Pursue What We have Dubbed "Psychometric AI"

In this section we give an extremely brief description of a specific form of "Weak" AI that we suggest should be pursued on the basis of supermentalism. This brand of AI we call **Psychometric AI**. The driving idea behind Psychometric AI is that we should construe "Weak" AI as the field devoted not to striving only for TT-passing machines, but machines able to excel on all *sorts* of tests — preferably tests that are more precisely described than the one test all seem to agree is already at the heart of AI: the Turing Test.

In order to introduce Psychometric AI, consider the question: What is AI? We seem to have a fairly good handle on what it means to say that something is an artifact. (We can ignore here conundrums arising from self-reproducing systems, systems that evolve without human oversight, etc.) It's the 'I' part of the query that seems to throw us for a bit of a loop. What's intelligence? *This* is the big, and hard, question. Innumerable answers have been given, but most seem to forget that there is a particularly clear and straightforward answer available, courtesy of the field that has operationalized the concept in question; that field is psychometrics. The answer is this:

> Some agent is intelligent if and only if it excels at all established, validated tests of intelligence.

This answer is somewhat problematic.[8] For example, consider:

> If one were offered a machine purported to be intelligent, what would be an appropriate method of evaluating this claim? The most obvious approach might be to give the machine an IQ test. ... However, [good performance on tasks seen in IQ tests would not] be completely satisfactory because the machine would have to be specially prepared for any specific task that it was asked to perform. The task could not be described to the machine in a normal conversation (verbal or written) if the specific nature of the task was not already programmed into the machine. Such considerations led many people to believe that the ability to

[8]Nonetheless, there seems to have been a time, in the very early days of AI, when this answer was taken seriously. All those who would understand (let alone contribute to) Psychometric AI, should begin by reading the seminal (Evans 1968). When it debuted, this was perhaps the largest and most sophisticated computer program on the planet.

> communicate freely using some form of natural language is an essential attribute of an intelligent entity. (Fischler & Firschein 1987, p. 12)

One way to interpret this is to say that Fischler & Firschein are simply noting the standard impetus for the Turing Test.[9] But perhaps moving from their point directly to TT is premature. Is there a response that doesn't give up an emphasis on real, particular, validated tests? (The TT is said to be a test, and it is, but not in the sense of 'test' used in psychology when referring to such things as the Graduate Record Exam.) Indeed, it would seem there is; our own response qualifies as such. This response is to move to Psychometric AI, defined as follows:

> Psychometric AI is the field devoted to building information-processing entities (some of which will be robots) capable of at least solid performance on all established, validated tests of intelligence and mental ability, a class of tests that includes IQ tests, tests of reasoning, of creativity, mechanical ability, and so on.

Psychometric AI may strike you as preposterously narrow, but this isn't the place to mount a full defense. However, we will say a bit more.

A full account and defense of Psychometric AI can be found in (Bringsjord & Schimanski 2002). It is explained therein that Bringsjord and Schimanski start their R&D by replicating, using an AI methodology, the work of (Carpenter, Just & Shell 1990), which was devoted to discovering how humans solve problems on the Raven's (1962) Progressive Matrices intelligence test (and hence was more cognitive modeling than AI). (Figure 8.13 shows a sample proof from the program written

[9] They also echo here Descartes' tests — briefly mentioned in Chapter 1 — for separating people and machines:

> ... although machines can perform certain things as well as or perhaps better than any of us can do, they infallibly fall short in others, by the which means we may discover that they did not act from knowledge, but only for the disposition of their organs. For while reason is a universal instrument which can serve for all contingencies, these organs have need of some special adaptation for every particular action. From this it follows that it is morally impossible that there should be sufficient diversity in any machine to allow it to act in all the events of life in the same way as our reason causes us to act. (Descartes 1911, p. 116)

by Bringsjord; it is a Common Lisp program that calls OTTER.) Their objective, however, is to "crack" *all* intelligence tests. (The full space of such tests is shown in Figure 8.10; this is taken from (Carpenter et al. 1990), wherein this rather rich graphic is explained (it shows correlations between tests, e.g.) Bringsjord and Schimanski are currently focusing on tests that involve physical manipulation of objects, and are seeking to have a robot actually carry out the manipulation, after begin presented with such a test. One of these tests is shown in Figure 8.11; Bettina Schimanski has written a program that cracks this test. In terms of intelligence tests, one of the most interesting manipulation tasks is so-called "picture completion." In this task, the subject must arrange pictures in such a way as to tell a story. An example, created by Schimanski, is shown in Figure 8.12. Both cube assembly and picture completion are sub-tests in the Wechsler Adult Intelligence Scale.

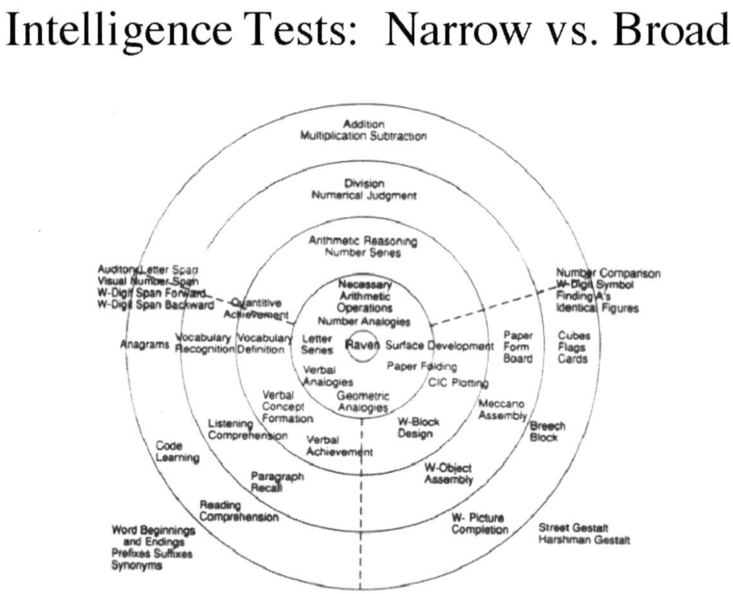

Figure 8.10: The Space of IQ Tests

For present purposes it suffices to point out an artificial agent able to perform well on all established tests would no doubt be able to do a lot. After all, if we believe that psychology has, through tests, isolated, in gem-like fashion, much of what's most important in cognition, then

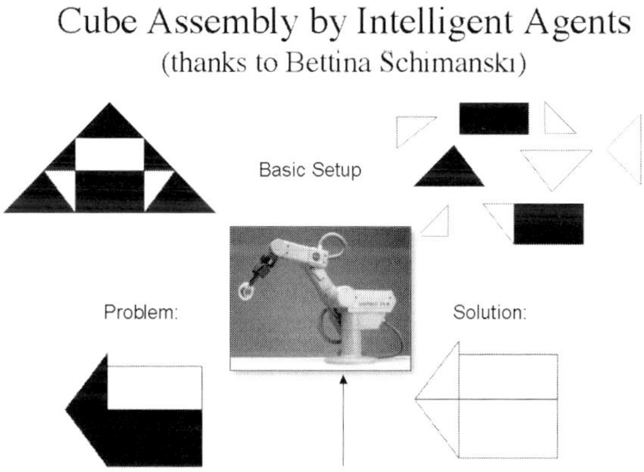

Figure 8.11: Bettina Schimanski's Cube Assembly Test

powerful agents in Psychometric AI will be powerful agents, period. We should probably also mention that Psychometric AI is committed to building artificial agents capable of meeting new challenges in the form of new tests. Bringsjord, with others, has proposed a new test for creativity to be given to machines: the **Lovelace Test**; see (Bringsjord, Bello & Ferrucci 2001.)

8.6 P4: Take Experimental Psychology of Reasoning Seriously

The fourth prescription was anticipated by what we said at conclusion of Chapter 3; it prescription runs as follows.

There is consensus among the relevant luminaries in AI, theorem proving, psychology of reasoning, and cognitive modeling that machine reasoning stands to the best of human reasoning as a rodent stands to the likes of Kurt Gödel. In the summer before Herb Simon died, in a presentation at the annual Computing and Philosophy Conference at Carnegie Mellon University, he essentially acknowledged this fact — and set out to change the situation by building a machine reasoner with the power of first-rate human reasoners (e.g., professional logicians). Unfortunately, Simon passed away. Now, the only way to fight toward his dream (which of course some others before him expressed) is to affirm the primacy of psychology of reasoning. Otherwise we will end up building systems that are anemic. The fact is that first-rate human

Figure 8.12: Bettina Schimanski's Picture Completion Example

reasoners use techniques that haven't found their way into machine systems (e.g., humans use extremely complicated, temporally extended mental images and associated emotions to reason). No machine, no theorem prover, no cognitive architecture, uses such a thing. The situation is different than chess — radically so. In chess, we knew that brute force could eventually beat humans (Bringsjord 1998*a*). In reasoning, brute force shows no signs of exceeding human reasoning. Therefore, unlike the case of chess, in reasoning we are going to have to stay with the attempt to understand and replicate in machine terms what the best human reasoners do.

Of course, some readers are likely to point out that reasoning is far from the only way to pursue AI. Well, there is indeed another paradigm: one based not on declarative information and reasoning over it, but rather on sub-declarative information (e.g., numerical information). However, for reasons we explain below in our eighth prescription, the connectionist route (pursued by itself) is one we find bankrupt. Assuming that we're right about this, and right about the need to get serious about expert human reasoners, it follows that psychology of reasoning and "Weak" AI (if the latter is to be successful) must in the future walk hand in hand.

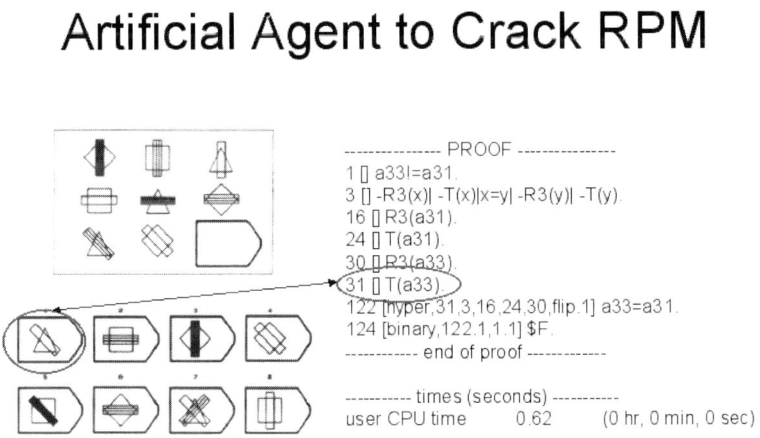

Figure 8.13: A Sample Problem and Proof. *This figure shows a problem of the sort seen on Raven's Progressive Matrices. The challenge is to pick, from the option, the figure that completes the matrix, in the sense of abiding by the underlying horizontal and vertical patterns. In this case, the problem is accompanied by a proof establishing that the circled option isn't correct. The proof is in the theorem prover* OTTER, *and specifically shows that an assumption that option 1 is correct leads to a contradiction, and hence is inconsistent with the rules in this case. The contradiction is indicated by* $F.

8.7 P5: Be Brutally Honest about the Limitations of Standard Schemes for Knowledge Representation and Reasoning

We declare that "Weak" AI carried out in light of supermentalism ought to be brutally honest with itself about the high degree of expressiveness needed by AI's knowledge representation schemes. Bringsjord, with Ferrucci, has made this point, in softer form, elsewhere: (Bringsjord & Ferrucci 1998, Bringsjord & Ferrucci 2000). The bottom line is that the declarative information, and the reasoning over that information required by full-blown natural language, pushes us well beyond extensional first-order logic (which after all was developed out of an attempt to formally capture not human reasoning expressed in natural language, but rather reasoning in mathematics). What is required are systems based on the *confluence* of various kinds of logics — systems that combine, say, modal, deontic, epistemic, temporal, and other logics. For a look at such logics, and related issues, we recommend that those in AI (and related fields) wanting to design and build (what we have referred to as) zombanimals and zombies go to (Goble 2001) as a nice starting place.

8.8 P6: Investigate Language Acquisition

As of now, AI hasn't managed to produce a computer system able to converse in sustained fashion at the level of an average three-year-old. We are now well after the start of the new millennium, and hence Turing (1964)'s starry-eyed prediction that by that time we'd have TT-passing computers has turned out to ring decidedly hollow. AI's failure to produce a conversational computer, combined with the fact that the "in the limit" characterization of hypercomputation is a natural model of human language acquisition (e.g., see Osherson, Stob & Weinstein 1986, Gold 1967, Gold 1965), suggests to us that our brand of "Weak" AI should devote time trying to simulate language acquisition.

8.9 P7: Pursue the Mathematical Modeling of Mentation, Independent of Even Future Implementation

Our seventh prescription should come as no surprise. After all, to a considerable degree the book you're holding reflects our following of this prescription. What does the prescription amount to? The idea is simple. If what we have said is right, then though we cannot *build* superminds (we may not even be able to build the least powerful supermachines), we can certainly explore, *formally*, the *nature* of such minds. That is to say, mentation can be studied from the standpoint of technical philosophy, logic, and mathematics, without a (fruitless) desire to replicate (or even, for that matter, simulate) mentation in information processing artifacts. There are others following this prescription; for example, we applaud the efforts of Steinhart (forthcoming) and Doyle (1991).

8.10 P8: Put Connectionism in its Place

Connectionism (save perhaps for exotic brands based on analog chaotic neural nets, referred to earlier in this chapter) is of no help in dodging the arguments we specify in this book. We have longstanding skepticism about connectionism. We have argued (Bringsjord & Zenzen 1991) that a thoroughly logicist approach to the mind, even one that eschews any mind/environment symbiosis, is far from obviously wrongheaded. Yet another part of our skepticism about connectionism derives from certain *gedanken*-experiments devised with Dave Ferrucci which seem to show that connectionism will have a hard time explaining, scientifically, how it is we do what we do. We conclude the book with a brief version of such a thought-experiment; take it for what it's worth, since it is not an integral part of our case for superminds. Rather, it simply supports our final prescription, offered *in light* of our case for superminds. Though our presentation of the thought-experiment here is a miniature version of that found in (Bringsjord & Ferrucci 2000), we do include here something new: a rebuttal to published objections to our arguments in (Bringsjord & Ferrucci 2000) based upon this thought-experiment.

Suppose the year is 2019, and that connectionism has produced remarkable offspring — in the form of a robot (or android), SHER-COG, capable of the sort of behavior associated with Sherlock Holmes.[10] Consider perhaps Holmes' greatest triumph, namely solving the mystery surrounding the disappearance of the racehorse known as "Silver Blaze" (Doyle 1984); and suppose that SHER-COG is asked (by an analogue for Dr. Watson), after cracking this case, how it accomplished the feat. What options does our robotic sleuth have for communicating an answer?

One thing that would surely fail to enlighten would be to allow humans to examine the neural nets of SHER-COG. After all, how would information about the states of nodes and the weights on connections between them help you divine how SHER-COG deduced that the culprit in this mystery could not be a stranger to dogs on the farm that was Silver Blaze's home?

Of course, SHER-COG could resort to natural language. It could proceed to explain its solution in (e.g.) English, in much the same way that Sherlock Holmes often explains things to the slower Dr. Wat-

[10] SHER-COG is intended to trigger thought about the robot COG, who Dennett and others plan on evolving into a humanoid robot (Dennett 1994).

son. But this route seems to invoke the traditional logicist machinery that the connectionist approach was supposed to replace. This is so because in order to really understand what SHER-COG is telling us in English, it will be necessary to analyze this English formally; the formal analysis will bring to bear the logical systems at the heart of computational linguistics and other relevant fields (including logicist AI itself, of course).

For example, to truly understand Holmes' explanation (conveyed to the nonplussed Watson) concerning the mystery of Silver Blaze, one must grasp the following chain of reasoning (which involves the famous clue about the "dog doing nothing in the night-time").

Arg$_1$

	(1)	If the dog didn't bark, then the person responsible for lacing the meal with opium couldn't be a stranger.	
	(2)	The dog didn't bark.	
∴	(3)	The person responsible for lacing the meal with opium couldn't be a stranger.	(1), (2)
	(4)	Simpson was a stranger.	
∴	(5)	Simpson was not responsible.	(3), (4)

At work here, of course, are the rules *modus ponens* and *modus tollens* (and standard quantifier rules), cornerstones of logicist AI.[11] Absent these cornerstones, and the enlightening analysis they allow when brought to bear on what cognizers think and say, SHER-COG's success will be impenetrable, and will thus fail to advance our understanding of how detectives do what they do.

The philosopher Ronald de Sousa disagrees; he writes:

> Bringsjord and Ferrucci stress the fact that the nature of [Holmes'] reasoning couldn't be "gleaned from neural nets" any more than

[11] For a more detailed analysis of the Silver Blaze story, see (Bringsjord 1992). Lest it be thought that the ratiocination of Sherlock Holmes is a phenomenon confined to the world of fiction, we direct readers to the remarkable reasoning used by Robert N. Anderson (1996) to solve the 80-year-old mystery of what caused the fire that destroyed Jack London's "Wolf House" in 1913. Wolf House was to be London's "manly" residence, a 15,000 square foot structure composed of quarried volcanic rock and raw beams from ancient redwoods. The conflagration occurred just days before London was to move in, and though London vowed to rebuild, he died three years later with the house still in ruins. The sort of painstaking reasoning Anderson carried out is of a type routinely reported in newspapers. This sort of reasoning is needed nearly every time there is a plane crash, and investigators attempt to determine the cause of the tragedy.

> our own could be read off a brain scan (p. 30). Quite, but so what? That doesn't show that the logical reasoning didn't supervene on the lower-level activity of both nets and neurons. ... [Their position] is reminiscent of the old joke about looking for your keys under the lamp, because that is where the light is. If that just is the way the brain works, and logical inferences, like conscious states and creative innovations, are just supervenient on those unintelligible processes, then consciousness, creativity, and logic will all be emergent properties of a system the inner workings of which are simply too complex to be understood in detail. That may well be regrettable, but this hardly constitutes a good reason for thinking it's not true. (de Sousa 1955, pp. 644–645)

This may be a fatal objection, but only to a straw man. The claim isn't that the mental phenomena de Sousa cites don't arise from lower-level processing. Rather, the claim is a two-part one, viz., that (i) the lower-level processing is impenetrable, and (ii) given (i), anyone wanting to understand higher-level cognition *at least to some degree* should model it in logicist terms. De Sousa himself grants (i). And nothing he says indicates that he rejects (ii).

However, de Sousa offers another argument of some interest in the present context. This argument is designed to capitalize on certain mathematical facts discussed in (Bringsjord & Ferrucci 2000), such as that standard (artificial) neural networks are equivalent to Turing machines. The argument runs as follows:

1. Humans are creative.
2. Human brains are neural nets.
3. ∴ Neural nets can be creative.
4. Neural nets are logically equivalent to Turing machines.
5. ∴ Turing machines can be creative.

(de Sousa 1955, pp. 645)

This argument can of course be generalized: creativity can be supplanted with any of the capacities listed in L_p (recall, again, this list from Chapter 1). For example, the argument can be modified so that its final conclusion becomes 'Turing machines can have the kind of autonomy that people enjoy.' But is the argument a good one? Unfortunately, it's not. The first problem is that the first inference is patently invalid. For it to go through, either 'Humans' in premise 1 would need to be replaced with 'Human brains,' or 'Human brains' in premise 2 would need to be replaced with 'Humans.' Presumably de Sousa seeks to gain most of his leverage from the intuitive idea that the brain — by definition a neural structure, and moreover a structure that seems, courtesy of all sorts of direct evidence gathered in neuroscience, to be composed of networks of interconnected neurons — is in some sense composed of neural networks. So let's go with:

1′. Human brains are creative.

2. Human brains are neural nets.

3. ∴ Neural nets can be creative.

4. Neural nets are logically equivalent to Turing machines.

5. ∴ Turing machines can be creative.

What about *this* argument? It's still exceptionally poor, for many reasons. For example, premise 4, which is supposed to be a mathematical fact, is a fact only if 'neural nets' therein is taken to refer to *non-analog artificial* neural networks (ANNs). ANNs are idealized mathematical creatures set out in textbooks. You do not find them in real, pulsing brains. We have no proofs about the computational power of the wet, fleshy stuff inside our heads. In light of this, for de Sousa to have a chance, we need to further modify the argument: we need 'neural nets' to be replaced by 'non-analog ANNs.' You may ask: "Why non-analog?" The answer should be clear in light of our explanation of hypercomputation earlier in this book, for in that explanation we explained that there are ANNs that are *more powerful* than Turing machines: viz., **analog chaotic** ANNs. If we neaten up the argument accordingly, it will have this premise:

2′. Human brains are non-analog ANNs.

Now, why should we assume that this premise is true? Perhaps the brain is constituted by analog chaotic neural networks, a view that would seem to be in line with Penrose (1994)'s views about the brain. As a matter of fact, at least one of us is happy to say, with Gödel, that the brain is in fact merely an ordinary digital computer (or an ordinary ANN). But there is something *else* wrong with the argument: it presupposes that human persons are human brains, and at the very least that the accomplishments of persons immediately imply that their brains are to be credited in parallel fashion. For example, both of us regard Mark Helprin (1995)'s novel *Memoir From Antproof Case* to be extraordinarily good, and we are therefore quite willing to hold that Helprin is creative. But would we agree immediately that therefore Helprin's *brain* is creative? No. Anyone reserving judgment as to whether Helprin is his brain will not ascribe F to a person p's brain because they have ascribed F to p himself. de Sousa's argument simply begs the question. Moreover, anyone having an open mind as to the truth or falsity of supermentalism will resist identifying persons with their brains, if those brains are in fact known to be standard ANNs.

So, to return to the issue of whether a non-logicist approach is viable, if in the future we desire not only to *build* person-matching robots, but to *understand* them to a significant degree, it seems to us that AI must sustain the logicist approach. Of course, by our lights, for reasons explained in this book (and elsewhere; recall Table 1), *any* information-processing approach designed to replicate (rather than merely simulate) human persons, given that they are superminds, is doomed.

Bibliography

Aleven, V., Koedinger, K. & Cross, K. (1999), Tutoring answer explanation fosters learning with understanding, *in* S. Lajoie & M. Vivet, eds, 'Artificial Intelligence in Education', IOS, Amsterdam, The Netherlands, pp. 199–206.

Anderson, J. R. & Lebiere, C. (1998), *The Atomic Components of Thought*, Lawrence Erlbaum, Mahwah, NJ.

Ashcraft, M. (1994), *Human Memory and Cognition*, HarperCollins, New York, NY.

Averbach, E. & Coriell, A. S. (1961), 'Short-term memory in vision', *Bell System Technical Journal* **40**, 309–328.

Barr, A. (1983), Artificial intelligence: Cognition as computation, *in* F. Machlup, ed., 'The Study of Information: Interdisciplinary Messages', Wiley-Interscience, New York, NY, pp. 237–262.

Barwise, J. & Etchemendy, J. (1993), *Turing's World 3.0*, CSLI, Stanford, CA.

Barwise, J. & Etchemendy, J. (1994), *Hyperproof*, CSLI, Stanford, CA.

Barwise, J. & Etchemendy, J. (1999), *Language, Proof, and Logic*, Seven Bridges, New York, NY.

Beall, J. C. (2001), 'Is Yablo's paradox non-circular?', *Analysis* **60**, 176–187.

Bennett, C. H. (1973), 'Logical reversibility of computation', *IBM Journal of Research Development* **November**, 525–532.

Bennett, C. H. (1982), 'The thermodynamics of computation — a review', *International Journal of Theoretical Physics* **21**, 905–940.

Bennett, C. H. (1984), 'Thermodynamically reversible computation', *Phys. Rev. Lett.* **53**, 1202.

Berkeley, G. (1949*a*), Three dialogues between Hylas and Philonous, *in* A. A. Luce & T. E. Jessop, eds, 'The Works of George Berkeley Bishop of Cloyne, vol. 2', Thomas Nelson and Sons, London, UK.

Berkeley, G. (1949*b*), A treatise concerning the principles of human knowledge, *in* A. A. Luce & T. E. Jessop, eds, 'The Works of George Berkeley Bishop of Cloyne, vol. 2', Thomas Nelson and Sons, London, UK.

Berlekamp, E., Conway, J. & Guy, R. (1982), *Winning Ways Vol. 2*, Academic Press, New York, NY. See chapter 25 for Conway's description of Life.

Bishop, M. & Preston, J., eds (2002), *Views into the Chinese Room: New Essays on Searle and Artificial Intelligence*, Oxford University Press, Oxford, UK.

Block, N. (1980), Troubles with functionalism, *in* N. Block, ed., 'Readings in Philosophy of Psychology Volume 1', Harvard University Press, Cambridge, MA.

Block, N. (1995), 'On a confusion about a function of consciousness', *Behavioral and Brain Sciences* **18**, 227–247.

Boolos, G. S. & Jeffrey, R. C. (1989), *Computability and Logic*, Cambridge University Press, Cambridge, UK.

Braine, M. (1998*a*), How to investigate mental logic and the syntax of thought, *in* M. Braine & P. O'Brien, eds, 'Mental Logic', Lawrence Erlbaum, Mahwah, NJ, pp. 45–61.

Braine, M. (1998*b*), Steps toward a mental predicate-logic, *in* M. Braine & D. O'Brien, eds, 'Mental Logic', Lawrence Erlbaum Associates, Mahwah, NJ, pp. 273–331.

Braitenberg, V. (1984), *Vehicles: Experiments in Synthetic Psychology*, Bradford Books, Cambridge, MA.

Brent, R. (1975), 'Irregularities in the distribution of primes and twin primes', *Math. Comp.* **29**(129), 43–56.

Bringsjord, S. (1991), 'Is the connectionist-logicist clash one of AI's wonderful red herrings?', *Journal of Experimental & Theoretical AI* **3.4**, 319–349.

Bringsjord, S. (1992), *What Robots Can and Can't Be*, Kluwer, Dordrecht, The Netherlands.

Bringsjord, S. (1994), 'Computation, among other things, is beneath us', *Minds and Machines* **4.4**, 469–488.

Bringsjord, S. (1995*a*), Could, how could we tell if, and why should–androids have inner lives?, *in* K. Ford, C. Glymour & P. Hayes, eds, 'Android Epistemology', MIT Press, Cambridge, MA, pp. 93–122.

Bringsjord, S. (1995*b*), 'In defense of impenetrable zombies', *Journal of Consciousness Studies* **2**(4), 348–351.

Bringsjord, S. (1995*c*), Pourquoi hendrik ibsen est-il une menace pour la littérature générée par ordinateur?, *in* A. Vuillemin, ed., 'Littérature et Informatique la Littérature Générée Par Orinateur', Artois Presses Universite, Arras, France, pp. 135–144.

Bringsjord, S. (1997*a*), *Abortion: A Dialogue*, Hackett, Indianapolis, IN.

Bringsjord, S. (1997*b*), 'Consciousness by the lights of logic and common sense', *Behavioral and Brain Sciences* **20.1**, 227–247.

Bringsjord, S. (1998*a*), 'Chess is too easy', *Technology Review* **101**(2), 23–28.

Bringsjord, S. (1998*b*), 'Is Gödelian model-based deductive reasoning computational?', *Philosophica* **61**, 51–76.

Bringsjord, S. (1998*c*), Philosophy and 'super' computation, *in* J. Moor & T. Bynam, eds, 'The Digital Phoenix: How Computers are Changing Philosophy', Blackwell, Oxford, UK, pp. 231–252.

Bringsjord, S. (1999), 'The zombie attack on the computational conception of mind', *Philosophy and Phenomenological Research* **59.1**, 41–69.

Bringsjord, S. (2000*a*), 'Animals, zombanimals, and the total Turing test: The essence of artificial intelligence', *Journal of Logic, Language, and Information* **9**, 397–418.

Bringsjord, S. (2000*b*), 'Clarifying the logic of anti-computationalism: Reply to Hauser', *Minds and Machines* **10**, 111–113.

Bringsjord, S. (2000*c*), 'Review of John Searle's *The Mystery of Consciousness*', *Minds and Machines* **10**(3), 457–459.

Bringsjord, S. (2001*a*), 'Are we evolved computers? A critical review of Steven Pinker's *How the Mind Works*', *Philosophical Psychology* **14**(2), 227–243. A more detailed version of this paper is available from the author, and is currently available online at `http://www.rpi.edu/~faheyj2/SB/SELPAP/PINKER/pinker.rev2.pdf`.

Bringsjord, S. (2001*b*), 'In computation, parallel is nothing, physical everything', *Minds and Machines* **11**, 95–99.

Bringsjord, S. (n.d.), The mental eye defense of an infinitized version of Yablo's paradox. Available from the author.

Bringsjord, S. & Bringsjord, E. (1993), 'Animal communication of private states doesn't illuminate the human case', *Behavioral and Brain Sciences* **16**(4), 645–646.

Bringsjord, S. & Bringsjord, E. (1996), 'The case against AI from imagistic expertise', *Journal of Experimental and Theoretical Artificial Intelligence* **8**, 383–397.

Bringsjord, S. & Ferrucci, D. (1998), 'Logic and artificial intelligence: Divorced, still married, separated...?', *Minds and Machines* **8**, 273–308.

Bringsjord, S. & Ferrucci, D. (2000), *Artificial Intelligence and Literary Creativity: Inside the Mind of Brutus, a Storytelling Machine*, Lawrence Erlbaum, Mahwah, NJ.

Bringsjord, S., Ferrucci, D. & Bello, P. (2001), 'Creativity, the Turing test, and the (better) Lovelace test', *Minds and Machines* **11**, 3–27.

Bringsjord, S. & Noel, R. (2002), Real robots and the missing thought experiment in the chinese room dialectic, *in* J. Preston

& M. Bishop, eds, 'Views into the Chinese Room: New Essays on Searle and Artificial Intelligence', Oxford University Press, Oxford, UK, pp. 144–166.

Bringsjord, S., Noel, R. & Ferrucci, D. (2002), Why did evolution engineer consciousness?, *in* J. Fetzer & G. Mulhauser, eds, 'Evolving Consciousness', Benjamin Cummings, San Francisco, CA, pp. 111–138.

Bringsjord, S. & Schimanski, B. (2002), Introducing psychometric AI. The current version is available from the authors upon request.

Bringsjord, S. & Xiao, H. (2000), 'A refutation of Penrose's gödelian case against artificial intelligence', *Journal of Experimental and Theoretical Artificial Intelligence* **12**, 307–329.

Bringsjord, S. & Yingrui, Y. (2002), Multi-agent reasoning: A new frontier old theories cannot tame. The current version is available from the authors upon request.

Bringsjord, S. & Zenzen, M. (1991), In defense of hyper-logicist AI, *in* 'IJCAI 91', Morgan Kaufman, Moutain View, CA, pp. 1066–1072.

Bringsjord, S. & Zenzen, M. (forthcoming), 'Toward a formal philosophy of hypercomputation', *Minds and Machines* .

Brngsjord, S. (1982), 'Strong ai is simply silly', *AI Magazine* **18**(1), 9–10.

Brooks, M. (1999), *Quantum Computing and Communications*, Springer-Verlag, Berlin, Germany.

Brun, V. (1919), 'La série 1/5 + 1/7 + 1/11 + 1/13 + 1/17 + 1/19 + 1/29 + 1/31 + 1/41 + 1/43 + 1/59 + 1/61 ... où les dénominateurs sont 'nombres premieres jumeaux' est convergente ou finie', *Bull. Sci. Math.* **43**, 124–128.

Carpenter, P., Just, M. & Shell, P. (1990), 'What one intelligence test measures: A theoretical account of the processing in the Raven progressive matrices test', *Psychological Review* **97**, 404–431.

Chalmers, D. (1995), 'Minds, machines, and mathematics', *Psyche* **2.1**. This is an electronic publication. It is available at http://psyche.cs.monash.edu.au/psyche/volume2-1/psyche-95-2-7-shadows-7-chalmers.html.

Chalmers, D. (1996), *The Conscious Mind: In Search of a Fundmental Theory*, Oxford, Oxford, UK.

Charniak, E. & McDermott, D. (1985), *Introduction to Artificial Intelligence*, Addison-Wesley, Reading, MA.

Chellas, B. F. (1980), *Modal Logic: An Introduction*, Cambridge University Press, Cambridge, UK.

Chisholm, R. (1989), Bolzano on the simplicity of the soul, *in* W. L. Gombocz, H. Rutte & W. Sauer, eds, 'Traditionen und Perspektiven der Analytischen Philosophie', Holder-Pichler-Tempsky, Vienna, Austria.

Chisholm, R. (1991), On the simplicity of the soul, *in* J. Tomberlin, ed., 'Philosophical Perspectives 5: Philosophy of Religion', Ridgeview, Atascadero, CA.

Chisholm, R. (1997), Reply to philip quinn, *in* L. E. Hahn, ed., 'The Philosophy of Roderick Chisholm, The Library of Living Philosophers Volume XXV', Open Court, Chicago, IL, pp. 68–71.

Church, A. (1936), An unsolvable problem of elementary number theory, *in* M. Dave, ed., 'The Undecidable', Raven Press, New York, NY, pp. 89–100.

Clark, A. & Chalmers, D. (1998), 'The extended mind', *Analysis* **58**(1), 7–19.

Cleland, C. (1993), 'Is the Church-thesis true?', *Minds and Machines* **3**, 283–312.

Cleland, C. (1995), 'Effective procedures and computable functions', *Minds and Machines* **5**, 9–23.

Cleland, C. (forthcoming), 'Effective procedures and causal processes', *Minds and Machines* .

Cole, D. & Foelber, R. (1984), 'Contingent materialism', *Pacific Philosophical Quarterly* **65**(1), 74–85.

Darwin, C. (1859), *On the Origin of the Species by Means of Natural Selection*, Murray, London, UK.

Darwin, C. (1997), *The Descent of Man*, Prometheus, Amherst, NY. A reprint edition.

Davidson, D. (1987), 'Knowing one's mind', *Proceedings and Addresses of the American Philosophical Association* **60**, 441–458.

Davis, M. (1980), 'How subtle is Gödel's theorem?', *Behavioral and Brain Sciences* **16**, 611–612.

Davis, M. & Hersh, R. (1972), 'Nonstandard analysis', *Scientific American* **June**, 79–86.

Davis, M., Sigal, R. & Weyuker, E. (1994), *Computability, Complexity, and Languages: Fundamentals of Theoretical Computer Science*, Academic Press, New York, NY.

Davis, W. (1985), *The Serpent and the Rainbow*, Simon and Shuster, New York, NY.

Davis, W. (1988), *Passage of Darkness: The Ethnobiology of the Haitian Zombie*, University of North Caronlina Press, Chapel Hill, NC.

de Sousa, R. (1955), 'Review of *Artificial Intelligence and Literary Creativity: Inside the Mind of Brutus, a Storytelling Machine*', *Computational Linguistics* **26**(4), 642–647.

Dennett, D. (1978), *Brainstorms*, MIT Press, Cambridge, MA.

Dennett, D. (1991), *Consciousness Explained*, Little, Brown, Boston, MA.

Dennett, D. (1993), 'Review of Searle's *The Rediscovery of the Mind*', *Journal of Philosophy* **90**(4), 193–205.

Dennett, D. (1994), 'The practical requirements for making a conscious robot', *Philosophical Transactions of the Royal Society of London* **349**, 133–146.

Dennett, D. (1995*a*), 'The unimagined preposterousness of zombies', *Journal of Consciousness Studies* **2**(4), 322–326.

Dennett, D. (1996), 'Cow-sharks, magnets, and swampman', *Mind and Language* **11**(1), 76–77.

Dennett, D. (2001), 'In Darwin's wake, where am I?', *Proceedings and Addresses of the American Philosophical Association* **75**(2), 13–30.

Dennett, D. C. (1995*b*), *Darwin's Dangerous Idea*, Simon and Shuster, New York, NY.

Descartes, R. (1911), *The Philosophical Works of Descartes, Volume 1. Translated by Elizabeth S. Haldane and G.R.T. Ross*, Cambridge University Press, Cambridge, UK.

Desmond, A. & Moore, J. (1991), *Darwin*, Michael Joseph, London, UK.

Deutsch, D. (1985), 'Quantum theory, the Church-Turing principle, and the universal quantum computer', *Proceedings of the Royal Society of London, Series A* **400**, 87–117.

Devlin, K. (2000), *The Math Gene*, Basic Books, New York, NY.

Dickmann, M. A. (1975), *Large Infinitary Languages*, North-Holland, Amsterdam, The Netherlands.

Dietrich, E. (1990), 'Computationalism', *Social Epistemology* **4**(2), 135–154.

Doyle, A. C. (1984), The adventure of Silver Blaze, *in* 'The Celebrated Cases of Sherlock Holmes', Amarenth Press, Minneapolis, MN, pp. 172–187.

Doyle, J. (1991), The foundations of psychology: A logico-computational inquiry into the concept of mind, *in* 'Philosophy and AI: Essays at the Interface', MIT Press, Cambridge, MA, pp. 39–78.

Dretske, F. (1996), 'Absent qualia', *Mind & Language* **11**(1), 78–85.

Earman, J. (1986), *A Primer on Determinism*, D. Reidel, Dordrecht, The Netherlands.

Ebbinghaus, H. D., Flum, J. & Thomas, W. (1984), *Mathematical Logic*, Springer-Verlag, New York, NY.

Eco, U. (1979), *The Role of the Reader: Explorations in the Semiotics of Texts*, Indiana University Press, Bloomington, IN.

Eicholz, R. E., O'Daffer, P. G., Charles, R. I., Young, S. I., Barnett, C. S., Clemens, S. R., Gilmer, G. F., Reeves, A., Renfro, F. L., Thompson, M. M. & Thornton, C. A. (1995), *Grade 7 Addison-Wesley Mathematics*, Addison-Wesley, Reading, MA.

Evans, G. (1968), A program for the solution of a class of geometric-analogy intelligence-test questions, *in* M. Minsky, ed., 'Semantic Information Processing', MIT Press, Cambridge, MA, pp. 271–353.

Fahey, J. & Zenzen, M. (n.d.), Emergentism: Dead again? An ancestor of this paper was presented at the xxxx Eastern APA Meeting. The current draft is available from the authors.

Feferman, S. (1995), 'Penrose's Göodelian argument', *Psyche* **2.1**. This is an electronic publication. It is available at http://psyche.cs.monash.edu.au/psyche/volume2-1/psyche-95-2-7-shadows-5-feferman.html.

Feltovich, P., Ford, K. & Hayes, P., eds (1997), *Expertise in Context*, AAAI Press, Menlo Park, CA.

Fetzer, J. (1994), 'Mental algorithms: Are minds computational systems?', *Pragmatics and Cognition* **2.1**, 1–29.

Fetzer, J. (2001*a*), *Computers and Cognition: Why Minds are not Machines*, Kluwer Academic Publishers, Dordrecht, The Netherlands.

Fetzer, J. (2001*b*), *Computers and Cognition: Why Minds are not Machines*, Kluwer, Dordrecht, The Netherlands, chapter People Are Not Computers: (Most) Thought Processes are Not Computational Procedures, pp. 153–180.

Feynman, R. (1986), 'Quantum mechanical computers', *Foundations of Physics* **16**, 507–531.

Fischler, M. & Firschein, O. (1987), *Intelligence: The Eye, the Brain, and the Computer*, Addison-Wesley, Reading, MA.

Fjelde, R. (1965), Foreward, *in* 'Four Major Plays — Ibsen', New American LIbrary, New York, NY, pp. ix–xxxv.

Flanagan, O. & Polger, T. (1995), 'Zombies and the function of consciousness', *Journal of Consciousness Studies* **2**(4), 313–321.

Folina, J. (1993), Commentary on Selmer Bringsjord's 'Church's thesis, contra Mendelson, is unprovable ... and worse: It may be false'. Presented at the Annual Eastern Division APA Meeting, Atlanta, GA, December 27, 1993.

Forster, T. (n.d.), The significance of yablo's paradox without self-reference. Available from `http://www.dpmms.cam.ac.uk`.

Fry, P., Nesheiwat, J. & Szymanski, B. (2002), Computing twin primes and brun's constant: A distributed approach.

Giunti, M. (1997), *Computation, Dynamics, and Computation*, Oxford University Press, Oxford, UK.

Glasgow, J., Narayanan, N. & Chandrasekaran, B., eds (1995), *Diagrammatic Reasoning: Cognitive and Computational Perspectives*, MIT Press, Cambridge, UK.

Glymour, C. (1992), *Thinking Things Through*, MIT Press, Cambridge, MA.

Goble, L., ed. (2001), *The Blackwell Guide to Philosophical Logic*, Blackwell Publishers, Oxford, UK.

Gold, M. (1965), 'Limiting recursion', *Journal of Symbolic Logic* **30**(1), 28–47.

Gold, M. (1967), 'Language identification in the limit', *Information and Control* **10**, 447–474.

Graesser, A., VanLehn, K., Rose, C., Jordan, P. & Harter, D. (2001), 'Intelligent tutoring systems with conversational dialogue', *AI Magazine* **22**(4), 39–51.

Graphic Art Materials Reference Manual (1981), Letraset, New York, NY.

Grzegorczyk, R. (1955), 'Computable functionals', *Fundamentals of Mathematics* **42**, 168–202.

Grzegorczyk, R. (1957), 'On the definitions of computable real continuous functions', *Fundamentals of Mathematics* **44**, 61–71.

Hardy, J. (1995), 'Is yablo's paradox liar-like?', *Analysis* **55**, 197–198.

Harnad, S. (1991), 'Other bodies, other minds: A machine incarnation of an old philosophical problem', *Minds and Machines* **1**(1), 43–54.

Harnad, S. (1995), 'Why and how we are not zombies', *Journal of Consciousness Studies* **1**, 164–167.

Haugeland, J. (1981), *Artificial Intelligence: The Very Idea*, MIT Press, Cambridge, MA.

Haugeland, S. (1998), *Having Thought: Essays in the Metaphysics of Mind*, Harvard University Press, Cambridge, MA.

Hauser, M. (2000), *Wild Minds*, Henry Holt, New York, NY.

Helprin, M. (1995), *Memoir from Antproof Case*, Harcourt Brace & Company, New York, NY.

Hoffman, R.R. & Klein, G. (1993), Seeing the invisible: Perceptual-cognitive aspects of expertise, *in* 'Cognitive Science Foundations of Instruction', Lawrence Erlbaum Associates, Mahwah, NJ, pp. 203–226.

Hofstadter, D. (1982), 'Metafont, metamathematics, and metaphysics', *Visible Language* **14**(4), 309–338.

Hofstadter, D. (1985), Waking up from the Boolean dream, *in* 'Metamagical Themas: Questing for the Essence of Mind and Pattern', Bantam, New York, NY, pp. 631–665.

Holt, J. (2001), 'The Ada perplex: How Byron's daughter came to be celebrated as a cyber-visionary', *The New Yorker* **March 5**, 88–93.

Hopcroft, J. & Ullman, J. (1979), *Introduction to Automata Theory, Languages and Computation*, Addison-Wesley, Reading, MA.

Jacquette, D. (1987), 'Metamathematical criteria for minds and machines', *Erkenntnis* **27**, 1–16.

Jacquette, D. (1994), *Philosophy of Mind*, Prentice Hall, Englewood Cliffs, NJ.

Jacquette, D. (1997), 'Psychology and the philosophical shibboleth', *Philosophy and Rhetoric* **30**(3), 312–331.

Johnson-Laird, P. (1988), *The Computer and the Mind*, Harvard University Press, Cambridge, MA.

Johnson-Laird, P. (1997*a*), 'Rules and illusions: A criticial study of Rips's *The Psychology of Proof*', *Minds and Machines* **7**(3), 387–407.

Johnson-Laird, P. N. (1983), *Mental Models*, Harvard University Press, Cambridge, MA.

Johnson-Laird, P. N. (1997*b*), 'And end to the controversy? A reply to Rips', *Minds and Machines* **7**, 425–432.

Johnson-Laird, P. & Savary, F. (1995), How to make the impossible seem probable, *in* 'Proceedings of the 17th Annual Conference of the Cognitive Science Society', Lawrence Erlbaum Associates, Hillsdale, NJ, pp. 381–384.

Kalmár, L. (1959), An argument against the plausibility of Church's thesis, *in* A. Heyting, ed., 'Constructivity in Mathematics', North-Holland, Amsterdam, The Netherlands, pp. 72–80.

Karp, C. (1964), *Languages with Expressions of Infinite Length*, North-Holland, Amsterdam, The Netherlands.

Keisler, H. (1971), *Model Theory for Infinitary Logic*, North-Holland, Amsterdam, The Netherlands.

Kim, J. (1984), 'Concepts of supervenience', *Philosophy and Phenomenological Research* **45**.

Kim, J. (1992), 'downward Causation' in emergentism and nonreductived physicalism, *in* A. B. et al., ed., 'Essays on the Prospects of Nonreductive Physicalism', Walter de Gruyter, New York, NY, pp. 119–138.

Kleene, S. C. (1983), 'General recursive functions of natural numbers', *Math. Annalen* **112**, 727–742.

Koetsier, T. & Allis, V. (forthcoming), 'Assaying supertasks', *Logique et Analyse* **159**, 291–313.

Kreisel, G. (1965), Mathematical logic, *in* T. Saaty, ed., 'Lectures in Modern Mathematics', John Wiley, New York, NY, pp. 111–122.

Kreisel, G. (1968), Church's thesis: A kind of reducibility thesis for constructive mathematics, *in* A. Kino, J. Myhill & R. Vesley, eds, 'Intuitionism and Proof Theory: Proceedings of a Summer Conference at Buffalo, N.Y.', North-Holland, Amsterdam, The Netherlands, pp. 219–230.

Kugel, P. (1986), 'Thinking may be more than computing', *Cognition* **18**, 128–149.

Kugel, P. (1990), Is it time to replace turing's test? Paper presented at *Artificial Intelligence: Emerging Science or Dying Art Form?*, sponsored by AAAI and the State University of New York's program in Philosophy and Computer and Systems Sciences, the Univeristy at Binghamton, NY, June 27.

LaForte, G., Hayes, P. & Ford, K. (1998), 'Why Gödel's theorem cannot refute computationslism', *Artificial Intelligence* **104**, 265–286.

Lambek, J. (1961), 'How to program an infinite abacus', *Canadian Mathematical Bulletin* **4**, 295–302.

Lewis, H. & Papadimitriou, C. (1981), *Elements of the Theory of Computation*, Prentice Hall, Englewood Cliffs, NJ.

Lucas, J. R. (1964), Minds, machines, and Gödel, *in* A. R. Anderson, ed., 'Minds and Machines', Prentice-Hall, Englewood Cliffs, NJ, pp. 43–59. Lucas' paper is available online at http://users.ox.ac.uk/~jrlucas/mmg.html.

Marxen, H. & Buntrock, J. (1990), 'Attacking the busy beaver 5', *Bulletin of the European Association for Theoretical Computer Science* **40**, 247–251.

Maudlin, T. (1989), 'Computation and consciousness', *Journal of Philosophy* **84**, 407–432.

McCulloch, W. & Pitts, W. (1943), 'A logical calculus of the ideas immanent in nervous activity', *Bulletin of Mathematical Biophysics* **5**, 115–137.

McLaughlin, W. (1994), 'Resolving Zeno's paradoxes', *Scientific American* **November**, 84–89.

McMenamin, M. (1992), Deciding uncountable sets and Church's thesis.

Meehan, J. (1981), Tale-spin, *in* R. Schank & C. Reisbeck, eds, 'Inside Computer Understanding: Five Programs Plus Miniatures', Lawrence Erlbaum, Englewood Cliffs, NJ, pp. 197–226.

Melzak, Z. A. (1961), 'An informal arithmetical approach to computability and computation', *Canadian Mathematical Bulletin* **4**, 279–293.

Mendelson, E. (1963), 'On some recent criticism of Church's thesis', *Notre Dame Journal of Formal Logic* **4**(3), 201–205.

Mendelson, E. (1986), 'Second thoughts about Church's thesis and mathematical proofs', *Journal of Philosophy* **87.5**, 225–233.

Millikan, R. G. (1996), 'On swampkinds', *Mind and Language* **11**(1), 103–117.

Moor, J. (1976), 'An analysis of the Turing test', *Philosophical Studies* **30**, 249–258.

Moore, A. W. (1990), *The Infinite*, Routledge, New York, NY.

Moravec, H. (1999), *Robot: Mere Machine to Transcendant Mind*, Oxford University Press, Oxford, UK.

Moschovakis, Y. (1968), 'Review of four recent papers on Church's thesis', *Journal of Symbolic Logic* **33**, 471–472. One of the four papers is Kalmár (1959) "An Argument Against the Plausibility of Church's Thesis," in Heyting, A., ed., *Constructivity in Mathematics* (Amsterdam, The Netherlands: North-Holland), pp. 72–80.

Nelson, E. (1977), 'Nonstandard analysis', *Bulletin of the American Mathematical Society* **83**(6), 1165–1198.

Nelson, R. J. (1987), 'Church's thesis and cognitive science', *Notre Dame Journal of Formal Logic* **28**(4), 581–614.

Newell, A. (1973), Production systems: Models of control structures, *in* W. Chase, ed., 'Visual Information Processing', Academic Press, New York, NY, pp. 463–526.

Newell, A. (1980), 'Physical symbol systems', *Cognitive Science* **4**, 135–183.

Osherson, D., Stob, M. & Weinstein, S. (1986), *Systems That Learn*, MIT Press, Cambridge, MA.

Partee, B., Meulen, A. & Wall, R. (1990), *Mathematical Methods in Linguistics*, Kluwer, Dordrecht, The Netherlands.

Pelletier, F. J. (1986), 'Seventy five problems for testing automatic theorem provers', *Bell System Technical Journal* **2**, 191–216.

Penrose, R. (1989), *The Emperor's New Mind*, Oxford, Oxford, UK.

Penrose, R. (1994), *Shadows of the Mind*, Oxford, Oxford, UK.

Penrose, R. (1996), 'Beyond the doubting of a shadow: A reply to commentaries on *Shadows of the Mind*', *Psyche* **2.3**. This is an electronic publication. It is available at `http://psyche.cs.monash.edu.au/v2/psyche-2-23-penrose.html`.

Pereira, F., Machado, P., Costa, E. & Cardoso, A. (n.d.), Busy beaver: An evolutionary approach. citeseer.nj.nec.com/pereira99busy.html.

Peters, R. S., ed. (1962), *Body, Man, and Citizen: Selections from Hobbes' Writing*, Collier, New York, NY.

Pinker, S. (1997), *How the Mind Works*, Norton, New York, NY.

Plantinga, A. (1993), *Warrant and Proper Function*, Oxford University Press, Oxford, UK.

Plum, F. & Posner, J. B. (1972), *The Diagnosis of Stupor and Coma*, F. A. Davis, Philadelphia, PA.

Pollock, J. (1989), *How to Build a Person: A Prolegomenon*, MIT Press, Cambridge, MA.

Pollock, J. (1995), *Cognitive Carpentry: A Blueprint for How to Build a Person*, MIT Press, Cambridge, MA.

Post, E. (1944), 'Recursively enumerable sets of positive integers and their decision problems', *Bulletin of the American Mathematical Society* **50**, 284–316.

Poundstone, W. (1985), *The Recursive Universe*, William Morrow, New York, NY.

Pour-El, M. & Richards, I. (1981*a*), 'A computable ordinary differential equation which possesses no computable solution', *Annals of Mathematical Logic* **17**, 61–90.

Pour-El, M. & Richards, I. (1981*b*), 'The wave equation with computable initial data such that its unique solution is not computable', *Advances in Mathematics* **39**, 215–239.

Priest, G. (1997), 'Yablo's paradox', *Analysis* **57**(4), 236–242.

Putnam, H. (1965), 'Trial and error predicates and a solution to a problem of mostowski', *Journal of Symbolic Logic* **30**(1), 49–57.

Putnam, H. (1975), Minds and machines, *in* 'Mind, Language, and Reality: Philosophical Papers Vol. 2', Cambridge University Press, Cambridge, UK.

Quinn, P. (1997), Tiny selves: Chisholm on the simplicity of the soul, *in* L. E. Hahn, ed., 'The Philosophy of Roderick Chisholm, The Library of Living Philosophers Volume XXV', Open Court, Chicago, IL, pp. 55–67.

Rado, T. (1963), 'On non-computable functions', *Bell System Technical Journal* **41**, 877–884.

Rapaport, W. (1998), 'How minds can be computational systems', *Journal of Experimental and Theoretical Artificial Intelligence* **10**, 403–419.

Raven, J. C. (1962), *Advanced Progressive Matrices Set II*, H. K. Lewis, London, UK. Distributed in the United States by The Psychological Corporation Inc., San Antonio, Texas.

Richards, R. J. (1987), *Darwin and the Emergence of Evolutionary Theories of Mind and Behavior*, University of Chicago Press, Chicago, IL.

Rips, L. (1994), *The Psychology of Proof*, MIT Press, Cambridge, MA.

Rogers, H. (1967), *Theory of Recursive Functions and Effective Computability*, McGraw-Hill, New York, NY.

Rosenthal, D. M. (1986), 'Two concepts of consciousness', *Philosohical Studies* **49**, 329–359.

Rosenthal, D. M. (1989), Thinking that one thinks, Technical Report 11, ZIF Report Zentrum für Interdisziplinäre Forschung, Bielefeld, Germany.

Rosenthal, D. M. (1990*a*), A theory of consciousness?, Technical Report 40, ZIF Report Zentrum für Interdisziplinäre Forschung, Bielefeld, Germany.

Rosenthal, D. M. (1990*b*), Why are verbally expressed thoughts conscious?, Technical Report 32, ZIF Report Zentrum für Interdisziplinäre Forschung, Bielefeld, Germany.

Rosenthal, D. M. (forthcoming), State consciousness and what it's like, *in* 'Title TBA', Clarendon Press, Oxford, UK.

Russell, B. (1936), 'The limits of empiricism', *Proceedings of the Aristotelian Society* **36**, 131–150.

Russell, S. & Norvig, P. (1994), *Artificial Intelligence: A Modern Approach*, Prentice Hall, Saddle River, NJ.

Salmon, W. C. (1975), *Space, Time and Motion: A Philosophical Introduction*, Dickenson, Encino, CA.

Schacter, D. L. (1989), On the relation between memory and consciousness: Dissociable interactions and conscious experience, *in* H. Roediger & F. Craik, eds, 'Varieties of Memory and Consciousness: Essays in Honour of Endel Tulving', Lawrence Erlbaum, Mahwah, NJ.

Schank, R. (1995), *Tell Me a Story*, Northwestern University Press, Evanston, IL.

Searle, J. (1980), 'Minds, brains and programs', *Behavioral and Brain Sciences* **3**, 417–424. This paper is available online at http://members.aol.com/NeoNoetics/MindsBrainsPrograms.html.

Searle, J. (1983), *Intentionality*, Cambridge University Press, Cambridge, UK.

Searle, J. (1992), *The Rediscovery of the Mind*, MIT Press, Cambridge, MA.

Searle, J. (1997), Roger Penrose, Kurt Gödel, and the cytoskeletons, *in* J. Searle, ed., 'The Mystery of Consciousness', New York Review of Books, New York, NY, pp. 53–93.

Shoemaker, S. (1975), 'Functionalism and qualia', *Philosophical Studies* **27**, 291–315.

Sieg, W. & Byrnes, J. (1996), K-graph machines: Generalizing Turing's machines and arguments, *in* 'Gödel 96, Lecture Notes in Logic', Springer-Verlag, New York, NY, pp. 98–119.

Siegelmann, H. (1995), 'Computation beyond the Turing limit', *Science* **268**, 545–548.

Siegelmann, H. & Sontag, E. (1994), 'Analog computation via neural nets', *Theoretical Computer Science* **131**, 331–360.

Simon, H. (1980), 'Cognitive science: The newest science of the artificial', *Cognitive Science* **4**, 33–56.

Simon, H. (1981), 'Study of human intelligence by creating artificial intelligence', *American Scientist* **69**(3), 300–309.

Sipper, M. (1997), *Evolution of Parallel Cellular Machines*, Springer, New York, NY.

Slezak, P. (1982), 'Gödel's theorem and the mind', *British Journal for the Philosophy of Science* **33**, 41–52.

Smolensky, P. (1988*a*), 'On the proper treatment of connectionism', *Behavioral & Brain Sciences* **11**, 1–22.

Smolensky, P. (1988*b*), 'Putting together connectionism — again', *Behavioral & Brain Sciences* **11**, 59–70.

Smullyan, R. (1992), *Gödel's Incompleteness Theorems*, Oxford University Press, Oxford, UK.

Soare, R. (1980), *Recursively Enumerable Sets and Degrees*, Springer-Verlag, New York, NY.

Sperling, G. (1960), 'The information available in brief visual presentations', *Psychological Monographs* **74**, 48.

Stannett, M. (forthcoming), 'Computation and hypercomputation', *Minds and Machines* .

Stanovich, K. E. & West, R. F. (2000), 'Individual differences in reasoning: Implications for the rationality debate', *Behavioral and Brain Sciences* **23**(5), 645–665.

Steinhart, E. (forthcoming), 'Logically possible machines', *Minds and Machines* **xx**, xx–xx.

Stillings, N., Weisler, S., Chase, C., Feinstein, M., Garfield, J. & Rissland, E. (1995), *Cognitive Science*, MIT Press, Cambridge, MA.

Thomas, W. (1973), Doubts about some standard arguments for Church's thesis, *in* 'Papers of the Fourth International Congress for Logic, Methodology, and Philosophy of Science, Bucharest', D. Reidel, Amsterdam, The Netherlands, pp. 13–22.

Trabasso, T. (1996), 'Review of *Knowledge and Memory: The Real Story*', *Minds and Machines* **6**, 399–403.

Turing, A. (1938), *Dissertation for the PhD: "Systems of Logic Based on Ordinals"*, Princeton University, Princeton, NJ.

Turing, A. (1939), 'Systems of logic based on ordinals', *Proceedings of the London Mathematical Society (series 2)* **45**, 161–228.

Turing, A. (1964), Computing machinery and intelligence, *in* A. R. Anderson, ed., 'Minds and Machines', Prentice-Hall, Englewood Cliffs, NJ, pp. 4–30.

Turing, A. M. (1936), 'On computable numbers with applications to the entscheidung-problem', *Proceedings of the London Mathematical Society* **42**, 230–265.

Wang, H. (1974), *From Mathematics to Philosophy*, Keagan Paul, London, UK.

Webb, J. (1980), *Mechanism, Mentalism and Metamathematics*, D. Reidel, Dordrecht, The Netherlands.

Weyl, H. (1949), *Philosophy of Mathematics and Natural Science*, Princeton University Press, Princeton, NJ.

Wyer, R. S. (1995), *Knowledge and Memory: The Real Story*, Lawrence Erlbaum, Hillsdale, NJ.

Yablo, S. (1993), 'Paradox without self-reference', *Analysis* **53**, 251–252.

Yang, Y., Braine, M. & O'Brien, D. (1998), Some empirical justification of one predicate-logic model, *in* M. Braine & D. O'Brien, eds, 'Mental Logic', Lawrence Erlbaum Associates, Mahwah, NJ, pp. 333–365.

Yang, Y. & Bringsjord, S. (2001), Mental metalogic: A new paradigm for psychology of reasoning, *in* 'Proceedings of the Third International Conference on Cognitive Science (ICCS 2001)', Press of the University of Science and Technology of China, Hefei, China, pp. 199–204.

Ybarra, M. J. (1996), 'Discovering an answer in the flames', *New York Times* **Sunday February 4 Section A**, 13.

Zenzen, M. & Hollinger, H. (1985), *The Nature of Irreversibility*, D. Reidel, Dordrecht, The Netherlands.

Index

STUDIES IN COGNITIVE SYSTEMS

1. J.H. Fetzer (ed.): *Aspects of Artificial Intelligence.* 1988
ISBN 1-55608-037-9; Pb 1-55608-038-7
2. J. Kulas, J.H. Fetzer and T.L. Rankin (eds.): *Philosophy, Language, and Artificial Intelligence.* Resources for Processing Natural Language. 1988 ISBN 1-55608-073-5
3. D.J. Cole, J.H. Fetzer and T.L. Rankin (eds.): *Philosophy, Mind and Cognitive Inquiry.* Resources for Understanding Mental Processes. 1990 ISBN 0-7923-0427-6
4. J.H. Fetzer: *Artificial Intelligence: Its Scope and Limits.* 1990
ISBN 0-7923-0505-1; Pb 0-7923-0548-5
5. H.E. Kyburg, Jr., R.P. Loui and G.N. Carlson (eds.): *Knowledge Representation and Defeasible Reasoning.* 1990 ISBN 0-7923-0677-5
6. J.H. Fetzer (ed.): *pistemology and Cognition.* 1991 ISBN 0-7923-0892-1
7. E.C. Way: *Knowledge Representation and Metaphor.* 1991 ISBN 0-7923-1005-5
8. J. Dinsmore: *Partitioned Representations.* A Study in Mental Representation, Language Understanding and Linguistic Structure. 1991 ISBN 0-7923-1348-8
9. T. Horgan and J. Tienson (eds.): *Connectionism and the Philosophy of Mind.* 1991
ISBN 0-7923-1482-4
10. J.A. Michon and A. Akyürek (eds.): *Soar: A Cognitive Architecture in Perspective.* 1992
ISBN 0-7923-1660-6
11. S.C. Coval and P.G. Campbell: *Agency in Action.* The Practical Rational Agency Machine. 1992 ISBN 0-7923-1661-4
12. S. Bringsjord: *What Robots Can and Can't Be.* 1992 ISBN 0-7923-1662-2
13. B. Indurkhya: *Metaphor and Cognition.* An Interactionist Approach. 1992
ISBN 0-7923-1687-8
14. T.R. Colburn, J.H. Fetzer and T.L. Rankin (eds.): *Program Verification.* Fundamental Issues in Computer Science. 1993 ISBN 0-7923-1965-6
15. M. Kamppinen (ed.): *Consciousness, Cognitive Schemata, and Relativism.* Multidisciplinary Explorations in Cognitive Science. 1993 ISBN 0-7923-2275-4
16. T.L. Smith: *Behavior and its Causes.* Philosophical Foundations of Operant Psychology. 1994
ISBN 0-7923-2815-9
17. T. Dartnall (ed.): *Artificial Intelligence and Creativity.* An Interdisciplinary Approach. 1994
ISBN 0-7923-3061-7
18. P. Naur: *Knowing and the Mystique of Logic and Rules.* 1995 ISBN 0-7923-3680-1
19. P. Novak: *Mental Symbols.* A Defence of the Classical Theory of Mind. 1997
ISBN 0-7923-4370-0
20. G.R. Mulhauser: *Mind Out of Matter.* Topics in the Physical Foundations of Consciousness and Cognition. 1998 ISBN 0-7923-5103-7
21. K.L. Butler: *Internal Affairs.* Making Room for Psychosemantic Internalism. 1998
ISBN 0-7923-5261-0
22. B.A. Thyer (ed.): *The Philosophical Legacy of Behaviorism.* 1999 ISBN 0-7923-5736-1
23. D. Livingstone Smith. *Freud's Philosophy of the Unconscious.* 1999 ISBN 0-7923-5882-1
24. M. Perlman: *Conceptual Flux.* Mental Representation, Misrepresentation, and Concept Change. 2000 ISBN 0-7923-6215-2
25. J.H. Fetzer: *Computers and Cognition: Why Minds are Not Machines.* 2001
ISBN 0-7923-6615-8

STUDIES IN COGNITIVE SYSTEMS

26. H. Cruse, J. Dean and H. Ritter (eds.): *Prerational Intelligence: Adaptive Behavior and Intelligent Systems Without Symbols and Logic, Volume 1.* 2000 ISBN 0-7923-6665-4
    H. Ritter, H. Cruse and J. Dean (eds.): *Prerational Intelligence: Adaptive Behavior and Intelligent Systems Without Symbols and Logic, Volume 2.* 2000 ISBN 0-7923-6670-0
    J. Dean, H. Ritter and H. Cruse (eds.): *Prerational Intelligence: Interdisciplinary Perspectives on the Behavior of Natural and Artificial Systems.* 2000 ISBN 0-7923-6669-7
    ISBN Indivisible Set 0-7923-6666-2
27. H.R. Holcomb III (ed.): *Conceptual Challenges in Evolutionary Psychology.* Innovative Research Strategies. 2001 ISBN 1-4020-0133-9
28. R. Cordeschi: *The Discovery of the Artificial.* Behavior, Mind and Machines Before and Beyond Cybernetics. 2002 ISBN 1-4020-0606-3
29. S. Bringsjord and M. Zenzen: *Superminds.* People Harness Hypercomputation, and More. 2003 ISBN 1-4020-1094-X; Pb 1-4020-1095-8

KLUWER ACADEMIC PUBLISHERS – DORDRECHT / BOSTON / LONDON